New Empirical Industrial Organization & the Food System

PETER LANG
New York • Washington, D.C./Baltimore • Bern
Frankfurt am Main • Berlin • Brussels • Vienna • Oxford

New Empirical Industrial Organization & the Food System

Edited by
Harry M. Kaiser
& Nobuhiro Suzuki

PETER LANG
New York • Washington, D.C./Baltimore • Bern
Frankfurt am Main • Berlin • Brussels • Vienna • Oxford

Library of Congress Cataloging-in-Publication Data

New empirical industrial organization and the food system /
edited by Harry M. Kaiser, Nobuhiro Suzuki.
p. cm.
Includes bibliographical references and index.
1. Food industry and trade—Econometric models. 2. Industrial organization (Economic theory)—Case studies. 3. Industrial concentration—Case studies. 4. Competition, Imperfect—Case studies.
I. Kaiser, Harry Mason. II. Suzuki, Nobuhiro. III. Title.
HD9000.5.N454 338.1—dc22 2006008959
ISBN 0-8204-8144-0

Bibliographic information published by **Die Deutsche Bibliothek**.
Die Deutsche Bibliothek lists this publication in the "Deutsche Nationalbibliografie"; detailed bibliographic data is available on the Internet at http://dnb.ddb.de/.

Cover photograph by Jack Dykinga provided courtesy
of USDA Agricultural Research Service

The paper in this book meets the guidelines for permanence and durability of the Committee on Production Guidelines for Book Longevity of the Council of Library Resources.

Printed in Germany

Table of Contents

Illustrations

Tables

Preface

An important economic trend that has been occurring throughout the past and present century is increased concentration of firms in various industries. This phenomenon has occurred in virtually every sector of the economy, including the food sector. Over time, industry consolidation through mergers and absorptions has occurred, resulting in fewer but larger-sized firms. Evidence of the increasing concentration of industries can be gleaned by examining any time-series data on largest-firm concentration ratios, which have risen steadily. For instance, the concentration ratio of the largest one hundred food manufacturers in the United States has more than doubled since the mid-1950s.

Traditionally, high firm-concentration ratios have been frowned upon by industrial organization economists. For instance, the traditional Harvard School theory of industrial organization argued that high firm-concentration ratios in an industry produce welfare losses and therefore should be reduced by government intervention such as strict merger restrictions and division of a monopoly firm into two or more firms. However, competing views rejected the notion that concentration is always bad for the economy and have advocated for *laissez-faire* in the market place. For instance, proponents of the Chicago School's efficiency doctrine and Baumol's contestability theory argue that even a monopoly can achieve a competitive equilibrium under certain conditions. The policy prescription advocated by Chicago economists is to remove all government intervention policies, including procompetition policies.

In addition, as the economy has become more globalized and geographic borders less important, the issue of domestic monopolies/monopsonies has become less relevant as firms face competition from foreign and multinational enterprises.

Therefore, the focus of industrial organization (IO) studies has shifted from analyzing concentration ratios to empirically measuring how far a market diverges from perfect competition (price = marginal cost = average cost in the long term). Measurement of the degree to which an actual market strays from perfect competition is therefore important in determining whether competition is precluded.

New empirical industrial organization (NEIO) is a pioneering framework developed by economists to measure the degree of competitiveness of economic markets. The primary contribution of NEIO is generalization of perfect-competition and monopoly models to intermediate imperfect-competition models that can be empirically estimated. This framework has been applied to many sectors to provide policy-makers dealing with antitrust issues with empirical evidence of market power throughout the marketing channel. Yet, to our knowledge, the underlying theoretical framework of NEIO has not been systematically presented in textbooks of industrial organization. This is the first book to provide a detailed, systematic overview and explanation of NEIO.

Broadly, the book is divided into two sections. Chapter 1 presents a thorough and exhaustive review of the NEIO framework and its development within the context of IO. This overview provides the foundation for the remaining chapters, which present a sampling of important empirical applications of the framework to the food system in greater detail. These applications provide valuable information regarding how the technique can be used in real-world applications.

This book is intended for economists with graduate training, primarily economists interested in industrial organization. The book offers the reader both quantitative lessons in NEIO and empirical evidence of market power in various industries. While the empirical focus is on the domestic and international food system, economists interested in other industries should find the book useful as well since the techniques are generalizable to any economic sector.

A debt of thanks is owed to a number of individuals and organizations who helped to bring this book to fruition. First, the authors of the individual chapters have done an excellent job. It has been our pleasure to work with them. We thank Anita Vogel for her assistance in managing the volumes of correspondence and other managerial tasks involved in producing this book. Anita performed all these functions in her usual efficient and friendly fashion. Thanks go to Natalie Karst who was the technical and copy editor of the book. Natalie did a wonderful job of editing all chapters and putting them into camera ready form for the publisher. Thanks are also extended to the academic journals that granted us permission to republish several previous articles as chapters in the book.

In particular, the following journals allowed us republish articles as book chapters free of charge: *Agribusiness: An International Journal*, *Agricultural and Resource Economics Review*, *Journal of Agricultural Economics*, *Journal of Agricultural and Resource Economics*, and *Review of Industrial Organization*. Finally, this book would not have been possible without the financial support of the National Institute of Commodity Promotion and Research Evaluation. We very much appreciate that support. The institute has sponsored several research projects involving NEIO models to evaluate the impact of generic advertising on food markets.

Harry M. Kaiser – Cornell University

Nobuhiro Suzuki – Kyushu University

Part I
Theory and tools

1

New Empirical Industrial Organization Theories and Their Application to Food System Analyses

Nobuhiro Suzuki and Harry M. Kaiser

An important economic trend that has been occurring throughout the past and present century is increased concentration of firms in various industries. This phenomenon has occurred in virtually every sector of the economy, including the food sector. Over time, industry consolidation through mergers and absorptions has resulted in fewer but larger sized firms. Evidence of the increasing concentration of industries can be gleaned by examining any time-series data on largest-firm concentration ratios, which have risen steadily. For instance, Rogers (1997) estimated that the concentration ratio of the largest one hundred food manufacturers in the United States doubled between 1954 and 1995, rising from 40 percent to 80 percent (Sexton and Zhang 2001; see Chapter 7).

Traditionally, high firm-concentration ratios have been frowned upon by industrial organization economists. For instance, the traditional Harvard School *theory of industrial organization* argued that high firm-concentration ratios in an industry produce welfare losses and therefore should be reduced by government intervention such as strict merger restrictions and dividing a monopoly firm into two or more firms. However, competing views rejected the notion that concentration is always bad for the economy and advocated for *laissez-faire* in the marketplace. For instance, proponents of the Chicago School's *efficiency doctrine* and Baumol's *contestability theory* (Baumol 1982; Baumol, Panzar, and Willig 1982) argue that even a monopoly can achieve a competitive equilibrium under certain conditions (outlined later in this chapter). The policy prescription advocated by Chicago School economists is to remove any government intervention policies, including procompetition policies. In addition, as the economy has become more globalized and geographic borders less important, the issue of domestic monopolies/monopsonies has become less relevant as firms face competition from foreign and multinational enterprises.

Therefore, the focus of industrial organization (IO) studies has shifted from analyzing concentration ratios to empirically measuring how far a market diverges from perfect competition (price = marginal cost = average cost in the long term). Measurement of the degree to which an actual market strays from perfect competition is therefore important in determining whether competition is precluded.

A useful approach in measuring the degree of market imperfection has been the *new empirical industrial organization* (NEIO) framework (Iwata 1974; Bresnahan 1981; Appelbaum 1982). The primary contribution of the NEIO framework is generalizing and extending conventional perfect-competition (monopoly) models to intermediate imperfect-competition models that can be empirically estimated. Many researchers have used this approach to analyze the degree of market imperfection in various industries with the food industry being the most heavily studied. Yet, to our knowledge, the underlying theoretical framework of NEIO has not been systematically presented in textbooks of industrial organization (with the exception of Sexton and Lavoie (2001), who cover NEIO as part of a broader survey of IO and the food industry). Accordingly, the overall aim of this chapter is to provide a detailed and systematic explanation of the NEIO framework.

We begin by presenting the NEIO framework and its development within the context of IO. We divide the NEIO models into two categories: quantity-setting homogeneous oligopoly, which is a generalization of the Cournot model, and price-setting product-differentiated oligopoly, which is a generalization of the Bertrand model.[1] We present a number of generalization methods for extending conventional perfect-competition

---

1 The original Cournot model assumes that firms set output quantity, as opposed to price, and each firm believes that its rivals will hold their outputs constant if it changes its output. On the other hand, the original Bertrand model assumes that firms set price, as opposed to output quantity, and each firm believes that its rivals will hold their prices constant if it changes its price. Although both models are applicable to competition among both homogeneous and differentiated goods, the Cournot model is often used for homogeneous oligopoly and the Bertrand model for product-differentiated oligopoly.

(monopoly) models to intermediate competition. We then review some important studies that have applied NEIO to the food sector. The overview of the NEIO framework in this chapter serves as the foundation for the remaining chapters, which present a sampling of important empirical applications to the food system in greater detail.

A Brief History of the Industrial Organization Approach

Industrial organization is a practical, applied economic approach intended to guide policies that control competition in markets. Competition policies endeavor to eliminate welfare losses caused by imperfect competition or a situation in which players are not price-takers in the market.

An Era of Concentration Ratios

The Harvard School, also known as Bain's group, established the industrial organization or IO framework based on a paradigm known as *Structure – Conduct – Performance* (SCP) in the early 1950s (Bain 1956). The Harvard School framework is sometimes called traditional IO or TIO. TIO calls for severe restrictions on mergers to avoid excessive concentration in a market. This position is based on the theory that the structure of a market (S) determines market conduct (C), which then determines market performance (P), and that higher concentration ratios generate welfare losses by competition-restricting activities such as cartels. The Harvard School considers the market function of automatic adjustment towards competitive equilibrium to be very weak, making governmental-intervention competition policies necessary and welfare improving.

Based on the principles of Harvard's TIO, mergers in the U.S. were largely restricted by policies introduced in the 1950s. Some have pointed out that one effect of those merger restrictions was that U.S. industries lost international competitive power because they prevented efficiency through scale economies.

The Harvard School is often contrasted with another framework for market mechanisms known as the Chicago School. Each school's benchmark or ideal is perfect competition based on neoclassical price theory. However, the Chicago School trusts the power of the market mechanism to allow for proper adjustments and therefore opposes government intervention and restrictions on mergers.

Can Competitive Equilibrium be Realized by a Single Firm?

Beginning in the 1980s with the Reagan Administration, the Chicago School's *efficiency doctrine* and Baumol's *contestability theory* gained influence in shaping competition policy in the United States. During this period, previous stringent policies that had been restricting mergers were dismantled. The logic behind these doctrines was that competitive equilibrium could be realized even when there is only one firm (or few firms) in a market.

Although the Chicago School and the Harvard School share the same benchmark of perfect competition or competitive equilibrium, the Chicago School views disequilibrium as a temporal phenomenon and asserts that market mechanisms bring about competitive equilibrium if all interventions, including procompetition policies, are eliminated. Under the Chicago School's approach, there are no entry barriers other than governments' entry-approval regulations and a monopoly firm is assumed to be under competitive pressure from potential entrants according to Stigler's definition (Stigler 1971, p. 67) of an entry barrier: "A cost of producing (at some or every rate of output) which must be borne by a firm which seeks to enter an industry but is not borne by firms already in the industry." Thus, a high concentration ratio is not problematic according to the Chicago School and monopolization is an acceptable way of seeking superior efficiency.

The contestability theory, which was proposed by Baumol (Baumol 1982; Baumol, Panzar, and Willig 1982), also defends monopolization because the theory has been used to show that a monopoly firm is under competitive pressure from potential entrants if there are no sunk costs (costs that cannot be recovered by transferring assets to other uses or selling them when exiting from the market). In other words, an incumbent in the market cannot raise prices higher than a competitive level because potential entrants can, in the absence of sunk costs, quickly enter the market, garner profits, and exit before the incumbent can respond (a so-called hit-and-run entry) in a perfectly contestable market. A perfectly contestable market means a market with no barriers to entry and exit.

In fact, we can easily illustrate a case of competitive equilibrium under high concentration ratios using the following equation:

(1) $(P - MC) / P = kH / e$

where P is price, MC is average industry marginal costs weighted by market share, k is dQ / dq_i (i.e., conjectural variation + 1) where dQ / dq_i is assumed to be the same for all firms ($dQ / dq_i \equiv k$), $H = \Sigma(q_i / Q)^2$ (i.e., a Herfindahl Index), and e is the price elasticity of demand (in absolute value). The derivation of Equation (1) is shown later in this chapter.

Equation (1) tells us that a high concentration ratio is indeed a factor that raises the price-cost margin (PCM) ratio, but the PCM ratio can be zero (competitive equilibrium can be achieved) if $k = 0$ (conjectural variation $= -1$) even if $H = 1$ (monopoly). So mathematically, it is possible for a monopoly industry to behave like a perfectly competitive industry.

An excellent real-world example of this is a domestic monopolist operating in a market that has no import restrictions. In this case, suppose that the monopoly firm desires to raise the price of its product above its marginal costs by controlling supply. Since there are no import restrictions, output reductions by the domestic monopolist will be perfectly offset by an increase in imports into the market. This is equivalent to the conjectural variation being equal to –1 in Equation (1). This example illustrates the need to expand the definition of a market from a single country to the world and to recognize that a "monopoly" in one country is no longer a monopoly in a global market.

Implications for Treatment of Farmer Cooperatives under Competition Law

Efforts toward market concentration and collective marketing by farmer cooperatives are exempt from laws that are intended to foster greater competition in the market. The Capper-Volstead Act (7 USC § 291), for example, exempts domestic agricultural producers from U.S. antitrust laws. However, agricultural cooperatives are not permitted to engage in competition-restricting activities (such as undue price enhancement) that generate welfare losses (Baumer, Masson, and Masson 1986; Madhavan, Masson, and Lesser 1994). Therefore, differences in the way cooperatives and propriety firms are treated in regards to competition have been disappearing because the measures of competition policy have shifted from the degree of concentration to case-by-case reviews of welfare losses.

Other Related Theories

Another important theory that is related to contestability is asset specificity in Williamson's *transaction cost theory* (Williamson 1981). Asset specificity describes the degree to which investments are specialized to a particular producing/transaction activity. High asset specificity means that an investment is valuable only for a specific producing/transaction activity and not valuable for other purposes, resulting in large sunk costs. A nuclear power plant is a good example of very high asset specificity. In contrast, the contestability theory treats sunk costs as minor.

Another line of thinking that is similar to the Chicago School is the *Neo-Austrian School*. Like the Chicago School, the Neo-Austrian School defends monopolization and advocates *laissez-faire* in the market. However, this theory differs from the Chicago School in its attack on the neoclassical price theory of competitive equilibrium. Specifically, the assumption of perfect information is rejected in favor of a theory that considers all activities to be part of a dynamic disequilibrium process of continuous competition among firms under imperfect information. Under this theory, monopolies are assumed to have no ill side effects because of the monitoring function provided by consumers and potential competitive pressures. According to this framework, monopoly is a temporal ideal goal of competition and free markets should be allowed to operate without intervention.

According to Schumpeter (1942), innovation can be created not by a perfectly competitive market with a large number of atomistic firms but by competition among oligopolistic large firms investing large amounts of capital in research and development (R&D) activities. Schumpeter suggests that imperfect competition is indispensable for innovation (Odagiri 2001).

Strategic Trade Policy and Competition Policies

According to *strategic trade policy*, imposing export subsidies and tariffs can increase economic welfare in the imperfectly competitive world market (for example, see Ito et al. (1988) and Yanagawa (1998)). Similarly, it has also been argued that the dumping of exports (price discriminating) can increase world economic welfare in some imperfectly competitive cases (Yano 2001). These arguments are made using the actual, imperfectly competitive market situation as a benchmark for welfare comparison rather than a free-trade market under perfect competition. However, if one uses the free-trade market under perfect competition as a benchmark, export subsidies, tariffs, and dumping policies unambiguously reduce economic welfare.[2] Therefore, contrary to strategic trade policy, if the current situation is already imperfectly competitive, it seems reasonable for us to endeavor to improve the market by way of competition-enhancing policies rather than accepting the market as is and using it as a benchmark for welfare comparison. This is the role of competition policies in bringing markets closer to perfect competition.

Increasing numbers of multinational firms in the food system are generating an international oligopoly/oligopsony problem (Suzuki 2002), and international harmonization of competition policies is required to solve the problem. Accepting the current, imperfectly competitive situation in the world means giving up harmonizing competition policies and makes any discussion of World Trade Organization (WTO) negotiations based on free-trade benefits under perfect competition meaningless.

A Paradoxical Relationship between External Diseconomy and Competition Policy

There are cases in which maintaining imperfect competition (not exercising procompetition laws) is better for economic welfare when there are externalities (e.g., pollution) in the market. In a case where there is a gap between aggregate private marginal costs of firms (*MC*) and social marginal costs (*SMC*), as shown in Figure 1.1, the gap (marginal polluting costs) typically should be imposed on firms as Pigovian taxes[3] and the welfare losses (the triangle *ACD* in Figure 1.1) can be recovered. In the case of an imperfectly competitive market, suppose that the aggregate perceived marginal revenue curve is *MR* in Figure 1.1. In this case,

2 In theory, there is a counterexample to the general proposition that a free-trade market under perfect competition is always welfare improving. Consider, for example, two countries that trade with one another initially with no government intervention and under free market conditions. Let country A be the exporter and country B be the importer and, assuming no transfer costs, the initial equilibrium price is the same in both countries. Now suppose that country A implements an export subsidy to reduce the export price. However, in response, country B imposes a "perfect" countervailing import tariff against country A. In this case, the two trade restrictions cancel each other out; there is no impact on the equilibrium price in the two countries and there is no change in producer and consumer welfare. The only impact from the change from a free market to trade restrictions in this case is a transfer of revenue from country A's government (the cost of the export subsidy) to country B's government (revenue from the import tariff). Therefore, world welfare is unchanged in this example of switching from a free-trade to a trade-barrier world.

3 The imposed taxes can be transferred to consumers as shown in Figure 1.1.

Figure 1.1. External Diseconomy and Competition Policy

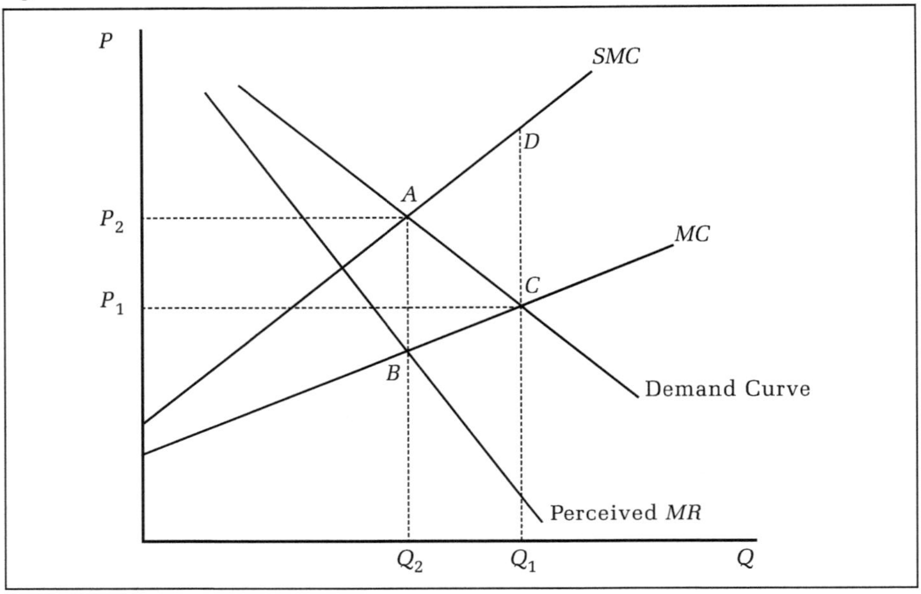

the welfare losses (the triangle *ACD*) can be recovered without imposing Pigovian taxes if we accept this imperfectly competitive situation without exercising procompetition policies. This is a paradoxical case because accepting imperfect competition improves economic welfare.[4] An empirical example of this is discussed by Gopinath and Wu (1999) using data from the U.S. agricultural chemical industry.

Summary of Developments in Frameworks and Tools for Economic Analyses of Degree of Market Power

Estimation of Statistical Relationships between Concentration and Profits

As previously noted, TIO (traditional industrial organization) attaches importance to the relationship expressed by $S \rightarrow P$: S (market structure) determines P (market performance) based on the SCP ($S \rightarrow C \rightarrow P$) paradigm. The Harvard School tried to prove the relationship empirically by statistical regression analyses. The dependent variable in these regression analyses was often a profit ratio as an indicator of P. Explanatory

4 However, the firms never pay their polluting costs in this case.

variables as indicators of *S* often included the top-four-firm concentration ratio (CR4) and several indicators of entry barriers, such as capital intensity, advertising/sales ratio, minimum efficiency scale, and market growth rate. In general, TIO regression analyses have shown a positive relationship between concentration ratios and profit ratios. These results have been used to advocate policies designed to reduce concentration ratios such as strict limits on mergers and division of a monopoly into two or more firms. However, the TIO empirical analyses were criticized for their loose theoretical background in that they did not explicitly incorporate market conduct (*C*) in the model.

Modeling Interdependency and Strategic Behavior between Firms

Unlike TIO, which focuses on the $S \rightarrow P$ relationship, new industrial organization (NIO) focuses on the $C \rightarrow P$ and $C \rightarrow S$ relationships. Proponents of NIO have built detailed models of strategic behavior between firms based on repeated-game theory (Schmalensee 1982). For example, while the Chicago School regards predatory pricing[5] as innocent price competition, strategic behavior analyses based on repeated-game theory can prove that predatory pricing generates welfare losses under some conditions but not under others (see, e.g., Milgrom and Roberts (1982) and Brennan (2000) for a more detailed discussion). As such, NIO contributes to recognition of the necessity of case-by-case reviews rather than reliance on a no-exceptions conclusion that predatory pricing is always detrimental or always innocent. However, NIO does not provide an empirical model capable of estimating the actual magnitude of welfare gains or losses.

NEIO, on the other hand (unlike TIO and NIO), directly estimates the degree of market power empirically. This is accomplished by introducing conduct parameters called *conjectural variations*—conjectures about rivals' responses to changes in a firm's own price or quantity supplied.

NEIO is based on the philosophy that a theory should be represented by a model capable of being empirically tested. Some NIO researchers criticize NEIO for modeling a dynamic process such as a repeated game in a static framework (Tirole 1988). Actually, either Cournot or Bertrand

---

5 The predatory firm first lowers its price below the average cost of its competitors to a point at which they will be driven out of business. After competitors are forced out of the market, the predatory firm raises its price, thereby compensating itself for the money lost while it was engaged in predatory pricing.

behavior is often assumed in NIO game-theoretic models. However, empirical economists understand well that Cournot and Bertrand behaviors are only special cases of many possible conjectures about rivals' behavior and that the actual degree of market competition cannot be approximated by Cournot or Bertrand behavior in most cases. Even if a model is theoretically perfect and highly sophisticated, it is unacceptable to empirical economists if it veers too far from reality. So in general, NIO models are useful to show that a certain market situation can be generated under certain assumptions. Thus, NIO models contribute to understanding mechanisms of market conduct and discussions of possibilities in market performance.

If, however, one needs to know the actual degree of market competition, market power, and welfare losses, NEIO approaches should be employed based on estimation of actual conjectural variations or conduct parameters. Martin (1993, p. 30) stated that "conjectural oligopoly models, in any event, have been more useful than game-theoretic oligopoly models in guiding the specification of empirical research in industrial economics."[6]

We note that there are some recent movements toward harmonization of the two contrasting approaches—conjectural-variation models and game-theoretic models (Cabral 1995; Pfaffermayer 1999; Itaya and Shimomura 2001). For example, Cabral (1995) showed that an exact correspondence exists between the conjectural-variations solution and the solution of a quantity-setting repeated game under some conditions. This justifies use of the conjectural-variations solution as the reduced form of the equilibrium of a dynamic game in some cases.

A dynamic modeling of conjectural variations is also important to respond to the criticism that NEIO models a dynamic process in a static framework. Karp and Perloff (1989, 1993) did pioneering work in this area (see Chapter 10). Their dynamic model of oligopoly was extended to a dynamic model of oligopoly and oligopsony by Katchova, Sheldon, and Miranda (2005). Another pioneering approach was developed by Amos,

---

6 Martin (1993, 2002) is one of several good texts on modern industrial organization. Another is Carlton and Perloff (2004), which includes numerous references to articles on the food system, such as Azzam and Pagoulatos (1990), Buschena and Perloff (1991), Deodhar and Sheldon (1995), Karp and Perloff (1989, 1993), Lopez (1984), Schroeter and Azzam (1991), Suzuki et al. (1994b), and Wann and Sexton (1992).

Karp, and Perloff (2000), who estimated games directly instead of using conjectural-variations models.

The NEIO Framework and Analytical Tools for the Food System

NEIO approaches are divided broadly into two categories: (1) a homogeneous oligopoly with quantity as a strategic variable, known as a generalized Cournot model; and (2) a product-differentiated oligopoly model with price as the strategic variable, known as a generalized Bertrand model.

Homogeneous Oligopoly with Quantity as a Strategic Variable: A Generalized Cournot Model

In the following sections, we explain basic NEIO theories; analytical tools for detailed problems such as processors' oligopoly/oligopsony, cooperatives' oligopoly, and bilateral oligopoly between cooperatives and processors and between processors and retailers; expansion of the NEIO framework to spatial-equilibrium models; and an example of a practical application of the NEIO approach to the WTO debates.

1. Basic Theory. Firm i is assumed to be in an industry where all firms supply a homogeneous commodity and use quantity as a strategic variable. Firm i's profit-maximization problem is expressed as:

(2) Max: $R_i = Pq_i - c_i$,

(3) subject to $P = P(Q)$,

(4) $Q = q_i + \Sigma_{j \neq i} q_j$,

(5) $\Sigma_{j \neq i} q_j = f(q_i)$, and

(6) $c_i = c_i(q_i)$.

Here, R_i is i's profit, P is price, Q is industry total supply, q_i is i's supply, $\Sigma_{j \neq i} q_j$ is $Q - q_i$, and c_i is i's total cost. Equation (3) is the (inverse) demand function, Equation (5) is i's conjecture about how the other firms change supply in response to a change in i's supply (called a reaction function), and Equation (6) is the cost function. Considering conditions (3) through (6), the objective function of Equation (2) can be rewritten as:

(7) $$\text{Max: } R_i(q_i) = P[q_i + \Sigma_{j\neq i} q_j\,(q_i)]q_i - c_i(q_i).$$

The first-order condition for maximizing Equation (7) for q_i is

(8) $$dR_i / dq_i = P + q_i(dP / dQ)(1 + d\Sigma_{j\neq i} q_i/dq_i) - dc_i/dq_i = 0$$

where $d\Sigma_{j\neq i} q_j \,/\, dq_i$ is conjectural variation or the conduct parameter (Bowley 1924; Frisch 1933; Hicks 1935). Because $1 + d\Sigma_{j\neq i} q_j / dq_i = dQ / dq_i$, we can express it in an elasticity term: $(1 + d\Sigma_{j\neq i} q_j \,/\, dq_i\,)(q_i/Q) = (dQ \,/\, dq_i)\;(q_i \,/\, Q) \equiv \theta_i$ (conjectural elasticity by Appelbaum (1982) or market-response elasticity by Tsujimura (1981); $0 \leq \theta \leq 1$).[7] Using the expressions $dc_i / dq_i = mc_i$ (marginal cost) and $-(dQ / dP)(P / Q) = e$ (price elasticity of demand in absolute terms) as well, the equality condition between marginal revenue and marginal cost is:

(9) $$P + \theta_i\,(dP / dQ)\,Q = mc_i, \text{ or}$$

(10) $$P(1 - \theta_i/e) = mc_i.$$

When $q_i / Q \approx 0$ (perfect competition), $\theta_i = 0$, or

(11) $$P = mc_i.$$

When $q_i = Q$ (monopoly), $\theta_i = 1$, or

(12) $$P(1 - 1/e) = mc_i.$$

Even if $0 < q_i < Q$, in cases where $d\Sigma_{j\neq i} q_j \,/\, dq_i = -1$ or $dQ \,/\, dq_i = 0$,

$$\theta_i = 0 \text{ or } P = mc_i \text{ (price-taking behavior).}$$

Also, even if $0 < q_i < Q$, in cases where $dQ \,/\, dq_i = Q \,/\, q_i$,

---

7 Dixit (1988) used the term $(dP / dQ)\,(1 + d\Sigma_{j\neq i} q_j / dq_i) = dP / dq_i$ as the conjectural variation in his product-differentiated oligopoly model.

$\theta_i = 1$ or $P(1 - 1/e) = mc_i$ (perfect collusion).

Because the rivals' supply is given when Cournot behavior is assumed, $d\Sigma_{j \neq i} q_j / dq_i = 0$ or $\theta_i = q_i / Q$. Therefore, Equation (9) is replaced by:

(13) $P + (dP / dQ)\, q_i = mc_i$.

Equation (10) can be rewritten to show the PCM ratio:

(14) $(P - mc_i) / P = \theta_i / \mathrm{e}$.

Multiplying q_i / Q and summing up for all n firms,

(15) $[P - \Sigma mc_i(q_i / Q)] / P = \Sigma\theta_i(q_i / Q) / e$.

Using the expressions $\Sigma mc_i(q_i / Q) \equiv MC$ and $\Sigma\theta_i(q_i / Q) \equiv \theta$, we get the following equations as market-average relations weighted by market shares:

(16) $(P - MC) / P = \theta / e$, or

(17) $P + (dP / dQ)\theta Q = MC$.

In addition, when we assume that dQ / dq_i is the same for all firms ($dQ / dq_i \equiv k$), the right-hand side of Equation (15) can be rewritten as:

(15') $\Sigma\theta_i\,(q_i / Q) / e = \Sigma(dQ / dq_i)(q_i / Q)^2 / e = kH / e$

where $H = \Sigma(q_i / Q)^2$ (Herfindahl Index). Thus,

(18) $(P - MC) / P = kH / e$.

Because $k = 1$when Cournot behavior is assumed,

(19) $(P - \mathrm{MC}) / P = H / e$.

In other words, because $d\Sigma_{j \neq i} q_j / dq_i = 0$ or $dQ / dq_i = 1$ when Cournot behavior is assumed, $\theta_i = q_i / Q$ (market share) and $\theta = \Sigma\theta_i\,(q_i / Q) = \Sigma(q_i / Q)^2 = H$ at the market-average level. In addition, assuming that the marginal cost (mc_i) is the same for all firms, $q_i / Q = 1 / n$ in the Cournot-Nash equilibrium. Therefore,

(20) $\theta = H = 1 / n$.[8]

When the cost function is specified by $c_i = vq_i + K$ (where K is fixed cost), v = unit variable cost (same for all firms) = marginal cost (mc_i). Therefore,

(21) $(PQ - vQ) / PQ = (P - v) / P = kH / e$.

This means that

(22) (sales – total variable costs) / sales = kH / e.

---

8 From Equation (20),

(a1) $n = 1 / \theta$.

The relationship (a1) can be used as an indicator to show how many identical Cournot firms are in the market if we assume that the degree of imperfection (θ) in the market is generated only by Cournot firms with the same cost structure (Suzuki and Kaiser 1998; Hyde and Perloff 1995). Suzuki and Kaiser (1998) called this indicator the Cournot firm number equivalent. This is the same concept as that of Dixit (1988, p. 146), who called it the Cournot-equivalent number of firms.

In addition, according to Clarke and Davies (1982), we can assume that firm i's conjecture is proportional to rival j's supply ratio to i's own supply or $dq_j / dq_i = A(q_j / q_i)$:

(a2) $d\Sigma_{j \neq i} q_j / dq_i = A\Sigma_{j \neq i} q_j / q_i = A(Q/q_i - 1)$

where A is Frisch's (1933) conjectural elasticity and A is assumed to be constant in (a2). Using (a2), Equation (8) can be rewritten as

(a3) $P - q_i(1 / e)(P / Q)[1 + A (Q/q_i - 1)] - mc_i = 0$.

Multiplying q_i,

(a4) $Pq_i - q_i^2(1 / e) (P / Q) [1 + A (Q/q_i - 1)] - mc_i q_i = 0$.

Summing up for all n firms using $H = \Sigma q_i^2 / Q^2$ (Herfindahl Index),

(a5) $PQ - (1 / e)(H \cdot PQ + A \cdot PQ - H \cdot A \cdot PQ) - \Sigma mc_i q_i = 0$, and

(a6) $(PQ - \Sigma mc_i q_i) / PQ = (P - MC) / P = (1 / e)[A + (1 - A) H]$.

The relationship (a6) is sometimes used instead of Equation (18).

Because $k = 1$ and Equation (20) holds when Cournot behavior is assumed,

(23) $(PQ - vQ) / PQ = (P - v) / P = H / e = 1 / (ne)$.

In a case of perfect collusion,

(24) $(PQ - vQ) / PQ = (P - v) / P = 1 / e$.

Using the expression $dQ / dq_i \equiv k_i$, changes in P in response to a change in $mc_i = v$ are shown by:

(25) $dP / dv = \Sigma(1 / k_i) / [\Sigma (1 / k_i) + 1]$.[9]

Therefore, $dP / dv = n / (n + 1)$ when Cournot behavior is assumed, $dP / dv = 1/2$ when $n = 1$ (monopoly), and $dP / dv = 1$ when $n \to \infty$ (perfect competition).

2. Consistency of Conjectures and Rationality of Cournot Behavior. A consistent conjecture means that firm i's conjecture corresponds to its rival j's actual reaction (Bresnahan 1981). Duopoly with a linear demand function is assumed in the following discussion. As shown in the previous section, firm 1's necessary condition for its profit maximization is:

(26) $P + q_1(dP / dQ)(1 + r_{12}) - mc_1 = 0$

where $r_{12} = dq_2 / dq_1$ (conjectural variation) and $mc_1 = dc_1 / dq_1$ (assumed to be constant).

Specifying the inverse demand function as

(27) $P = a - bQ$,

Equation (26) can be reorganized as firm 1's reaction function:

(28) $q_1 = -1/(2 + r_{12})q_2 + (a - mc_1)/[b\ (2 + r_{12})]$.

9 For example, when $n = 2$, Equation (25) can be expressed as $dP / dv = (k_1 + k_2) / (k_1 + k_1k_2 + k_2)$. When $n = 3$, $dP / dv = (k_1k_2 + k_2k_3 + k_3k_1) / (k_1k_2 + k_2k_3 + k_3k_1 + k_1k_2k_3)$.

Similarly, firm 2's reaction function is

(29) $$q_2 = -1/(2 + r_{21})q_1 + (a - mc_2)/[b(2 + r_{21})].$$

Therefore, firm 1's actual reaction to a change in firm 2's supply is expressed as

(30) $$dq_1 / dq_2 = -1 / (2 + r_{12}),$$

which includes firm 1's conjectural variation about firm 2's reaction to a change in firm 1's supply. Similarly,

(31) $$dq_2 / dq_1 = -1 / (2 + r_{21}).$$

Consequently, the condition for consistent conjectures is that

(32) $$r_{21} = -1 / (2 + r_{12}) \text{ and}$$

(33) $$r_{12} = -1 / (2 + r_{21})$$

hold simultaneously. The condition is calculated as

(34) $$r_{12} = r_{21} = -1.$$

This implies that consistent conjectures can be realized when both firms are price-takers. Note that this conclusion assumes a linear demand function and constant marginal costs. The results can differ by specification of demand and cost functions (Hosoe 1990).

Is the Cournot conjecture consistent? The Cournot conjecture can be expressed as

(35) $$r_{12} = r_{21} = 0$$

while the actual reaction (the slope of the reaction function) is –1/2. Thus, the Cournot conjecture is inconsistent. In other words, Cournot firm *i* continues to believe that its rival *j* will not change its supply in response to a change in *i*'s supply although firm *i* can actually observe that firm *j* reduces its supply by a half unit for each one-unit increase in *i*'s supply. Cournot behavior is not reasonable in this sense.

3. A Generalization of the Dorfman-Steiner Theorem. The optimal advertising strategy of a monopoly firm can be specified as the following maximization problem:

(36) Max: $R = PQ - C - A$,

(37) subject to $Q = Q(P, A)$, and

(38) $C = C(Q)$

where R is profit, P is price, Q is supply, C is costs other than advertising, and A is advertising expenditure and where Equation (37) is the demand function and (38) is the cost function. When the unit advertising price is one dollar, A can be regarded as the quantity of advertising as well. Solving this problem, the necessary condition for P, Q, C, and A to maximize R is

(39) $A / (PQ) = \eta / e = \eta\, (P - MC) / P$

using the expressions $dC / dQ = MC$ (marginal cost), $-(\partial Q / \partial P)(P / Q) = e$ (price elasticity of demand in absolute terms), and $(\partial Q / \partial A)(A / Q) = \eta$ (advertising elasticity of demand).

Dividing (39) into two parts, we get:

(40) Dorfman-Steiner Theorem: $A / (PQ) = \eta / e$, and

(41) Lerner Index: $(P - MC) / P = 1 / e$.

Thus, the Dorfman-Steiner Theorem, $A / (PQ) = \eta / e$, holds in monopoly. How about oligopoly? Can we extend the theorem to the general case of imperfect competition?

As an example of imperfectly competitive cases, we consider the case where homogeneous oligopoly firms implement generic advertising using per-unit assessments imposed on each firm, such as dairy check-off programs in the U.S and Japan. Firm i is assumed to be in an industry where all firms supply a homogeneous product and use quantity as a strategic variable. Firm i's profit-maximization problem is expressed as:

(42) Max: $R_i = Pq_i - c_i - A_i$,

(43) subject to $Q = Q(P, A)$,

(44) $Q = q_i + \Sigma_{j \neq i} q_j$,

(45) $\Sigma_{j \neq i} q_j = f(q_i)$,

(46) $c_i = c_i(q_i)$,

(47) $A = A_i + \Sigma_{j \neq i} A_j$,

(48) $A_i = tq_i$, and

(49) $\Sigma_{j \neq i} A_j = t\Sigma_{j \neq i} q_j$

where R_i is i's profit, P is price, Q is industry total supply, q_i is i's supply, $\Sigma_{j \neq i} q_j$ is $Q - q_i$, c_i is i's costs other than advertising, A_i is i's advertising expenditure, and t is the per-unit assessment rate. Equation (43) is the demand function, Equation (45) is i's conjecture about how the other firms change supply in response to a change in i's supply (called "reaction function"), and Equation (46) is the cost function. Using the expressions $dc_i / dq_i = mc_i$ (marginal cost), $-(\partial Q / \partial P)(P / Q) = e$ (price elasticity of demand in absolute terms), $(\partial Q / \partial A)(A / Q) = \eta$ (advertising elasticity of demand), and $(1 + d\Sigma_{j \neq i} q_j / dq_i)(q_i / Q) = (dQ / dq_i)(q_i / Q) \equiv \theta_i$ (conjectural elasticity), the necessary conditions for P, Q, q_i, $\Sigma_{j \neq i} q_j$, A, A_i, $\Sigma_{j \neq i} A_j$, and t to maximize R_i are

(50) Dorfman-Steiner Theorem: $A / (PQ) = t / P = \eta / e$, and

(51) Generalized Lerner Index: $(P - mc_i) / P = [1 - (\theta_i - 1)(\eta - 1)] / e$.

Using (50) and (51), the equality condition between marginal revenue and marginal cost can be expressed as:

(52) $P[1 - \theta_i(1 - \eta) / e] = mc_i + t,$

which is a transformation of the ordinal relationship $P(1 - \theta_i/e) = mc_i$. Equation (52) indicates that more market power creates lower optimal check-off rates (see Zhang and Sexton (2002) for a different way to reach the same result). Note that the Dorfman-Steiner Theorem holds even in imperfectly competitive markets in general while the Lerner Index is transformed.

Among the earliest studies to evaluate the effectiveness of generic advertising using an imperfect-competition model are Suzuki et al. (1994b; see Chapter 3) and Suzuki and Kaiser (1997), which found that conventional perfect-competition models underestimate the effectiveness of milk check-off programs in Japan and the U.S.

Although we need not consider conjectural variations about advertising here because we use generic advertising as an example, we should incorporate conjectures about rivals' reactions to advertising activity when we discuss branded advertising by a firm. In such cases, the Dorfman-Steiner Theorem is transformed (See Clarke (1985) for a more thorough discussion).

4. Two Interpretations of Conjectural Variations. One interpretation of conjectural variations is its literal and original definition—the firm's conjecture about its rivals' reaction to a change in its supply. The second interpretation is that conjectural variations is a measure of the degree of market power and, more specifically, how far price differs from marginal cost (Hyde and Perloff 1995).

Interpreting conjectural variations literally is subject to criticism from game theorists who point out that estimating dynamically varying conjectures through the repeated game as a constant static parameter is questionable in some time periods (see Corts' 1999 discussion on Iwata (1974) and Suzuki, Lenz, and Forker (1993)). Interpreting conjectural variations as a summarized parameter of market competitiveness is free from such criticism.

In addition, as previously mentioned, we note that an exact correspondence exists between the conjectural-variations solution and the solution of a quantity-setting repeated game in some conditions, as shown by Cabral (1995). This justifies use of the conjectural-variations solution as the reduced form of the equilibrium of a dynamic game in some cases.

5. Two Approaches for Estimating Conjectural Variations. *Calibration* is a method of identifying parameter values as theoretically required "residuals" by assuming that necessary equilibrium conditions hold in the system. Recall Equation (10). When we assume that Equation (10) holds in each period, values for θ_i in each period can be calculated as values that should satisfy Equation (10) using observed values for P and values for e and mc_i obtained from estimation of demand and cost functions. This type of estimation is used in Iwata (1974), which is known as a pioneering NEIO study, and in Suzuki (1994); Suzuki, Lenz, and Forker (1993); Suzuki et al. (1994a, 1994b; see Chapter 2); Liu, Sun, and Kaiser (1995; see Chapter 4); Matsuda and Kurokawa (1996); Suzuki and Kaiser (1997, 1998); Kano (1999); Dixit (1988); Baldwin and Krugman (1988); McCorriston and Sheldon (1996); Thursby and Thursby (1990); and Tiffin and Dawson (1997; see Chapter 13). Studies that also calculated standard errors for conjectural-variation estimates are Iwata (1974); Liu, Sun, and Kaiser (1995); Suzuki (1994); Suzuki, Lenz, and Forker (1993); Suzuki et al. (1994b); and Thursby and Thursby (1990).

An advantage of the calibration method is that one can obtain time-varying values for θ_i with this approach. For example, Suzuki (1994) showed that conjecture by the Japanese province of Hokkaido about the reaction of another region, Tofuken, was gradually adjusted by observing Tofuken's actual reaction over time.[10] According to Suzuki (1994), Hokkaido believed in 1979 that Tofuken would reduce its supply by 0.62 units when Hokkaido decreased one unit of its supply. However, Tofuken actually increased its supply by 0.8 units when Hokkaido decreased its supply one unit. Therefore, Hokkaido's conjecture decreased drastically over time—from 0.62 in 1979 to –0.61 in 1988—after observing Tofuken's noncooperative actual reactions.

10 Japan's milk-producing region is divided into two regions, Hokkaido and Tofuken, that compete with each other.

Thus, this method is somewhat useful in overcoming criticism that the conjectural-variations model treats a dynamic game as a constant parameter in some time periods. One drawback is that this method does not use any simultaneous estimation method, although the demand and cost functions and Equation (10), which represents the equality condition between marginal revenue and cost, should be simultaneously determined.

Direct Estimation as Regression Parameters. The demand and cost functions and Equation (10) (the equality condition between marginal revenue and cost) are simultaneously estimated using a system estimation method. Values for θ_i can be obtained as regression parameters in the direct estimation method, which is used in Appelbaum (1982), one of the most frequently cited articles in the NEIO literature, as well as in many other articles not previously mentioned (see Bresnahan (1989) for reviews prior to 1990). A drawback is that the estimated value for θ_i is an average value over the estimation periods. Additional means are necessary to obtain time-varying values for θ_i. Also, identification problems sometimes occur for θ_i (see Bresnahan (1982) and Suzuki et al. (1994b)).

6. Estimation of Welfare Losses Caused by Processor Oligopoly. There are four key players in the food system: farmers, processors, retailers, and consumers. The relationships and channel among these players can be expressed as:

$$\begin{array}{ccccccc}[\text{Farmers}] & \Rightarrow\ \Leftarrow & [\text{Processors}] & \Rightarrow\ \Leftarrow & [\text{Retailers}] & \Rightarrow\ \Leftarrow & [\text{Consumers}]. \\ & \text{(a) (b)} & & \text{(c) (d)} & & \text{(e) (f)} & \end{array}$$

The symbol $\Rightarrow$ indicates the degree of oligopoly power and the symbol $\Leftarrow$ indicates the degree of oligopsony power in the system. If a group has no market power, it is a price-taker.

Since Appelbaum's work (1982), many researchers have tried to estimate the degree of oligopoly power in U.S. food-processing sectors. Most recently, Bhuyan and Lopez (1997, 1998a, 1998b; see Chapter 5) did comprehensive studies estimating welfare losses generated by oligopoly for each sector of the U.S. food-processing industry. They assumed that (c) = price-setter and (d) = price-taker.

Using the relationship

$$(53) \qquad (P - MC) / P = \theta / e,$$

which was shown previously as Equation (16), numerous studies in recent years estimated welfare losses by each sector of the U.S. food-processing industry (see, for example, Bhuyan and Lopez (1995, 1997, 1998a, 1998b)). Bhuyan and Lopez standardized the data for all food industries to compare welfare losses by sector. To do this, the authors indexed the perfectly competitive price and output to equal one and specified the industry demand function as $Q = P^{-e}$ and the industry inverse marginal cost function as $Q = MC^{\eta}$. Therefore, under this setup, $P = Q = 1$ if $P = MC$ in perfect competition. Using this normalized data, the deadweight loss (*DWL*) can be estimated by

(54) $$DWL = \int_{Q_0}^{1} [Q^{-1/e} - Q^{1/\eta}]\, dQ$$

because the oligopoly price and quantity are given respectively as

(55) $$P_0 = [e / (e - \theta)]\, Q_0^{1/\eta} \text{ and}$$

(56) $$Q_0 = P_0^{-e} = [(e - \theta) / e]^{e\eta / (e+\eta)}.$$

Thus, if we estimate e (price elasticity of demand in absolute terms), η (inverse of elasticity of marginal costs with respect to supply), and θ (conjectural elasticity or market-power parameter), welfare losses can be calculated by sector.

Instead of estimating the actual degree of market competitiveness, Peterson and Connor (1995) estimated welfare losses using several different assumptions for market conduct (e.g., the Cournot type and the Stackelberg type), and they found that estimated welfare losses vary widely depending on the assumptions used. This strongly suggests that we should try to estimate the actual degree of competition (θ) in order to accurately measure welfare losses from oligopoly instead of assuming an *a priori* value for θ.

7. Processor Oligopsony Power over Raw Materials. The meat-packing industry in the United States is perhaps the foremost agricultural industry that has undergone drastic concentration. For instance, the CR4 for the beef-packing industry grew from 30 percent in 1978 to 83.5 percent in 2003.[11]

11 The data source is *Cattle Buyer's Weekly*: Steer and Heifer Slaughter reported in *Feedstuffs* 16 June 2003.

Because of this tremendous increase in concentration, this industry has been the focus of many NEIO studies. More attention has been paid to oligopsony than to oligopoly in these studies, with the key issue being whether meat packers are exercising their oligopsony power to lower the raw-material input prices they pay to farmers. Studies addressing this topic include Azzam (1992); Azzam and Pagoulatos (1990, 1999; see Chapter 6); Azzam and Schroeter (1991, 1995); Gohin and Guyomard (2000); Koontz and Garcia (1997); Koontz, Garcia, and Hudson (1993); Muth and Wohlgenant (1999a, 1999b); Paul (1999a, 1999b, 2001); Purcell (1990); Schroeter (1988, 1990); Schroeter and Azzam (1990, 1991); Schroeter, Azzam, and Zhang (2000); Stiegert, Azzam, and Brorsen (1993); and Weliwita and Azzam (1996). Azzam and Schroeter (1995) modeled a trade-off between efficiency and market power caused by mergers of meat packers, and Koontz, Garcia, and Hudson (1993) presented a model in which price-taking behavior and price-setting behavior switched depending on market conditions. Gohin and Guyomard (2000) reviewed dairy processors' oligopsony power as well.

The earliest study in this area was done by Schroeter (1988), who assumed that meat packers' oligopoly power in the output market is equal to their oligopsony power in the input market. Azzam and Pagoulatos (1990; see Chapter 6) subsequently introduced a method for estimating different values for oligopoly and oligopsony parameters. They focused on

<= [Processors] =>
(b) (c)

and assumed that (a) = price-taker, (b) = price-setter, (c) = price-setter, and (d) = price-taker. Since then, many studies of the meat-packing industry have been done, but Schroeter (1988) and Azzam and Pagoulatos (1990) developed the basic model, which is a generalization of the monopoly/monopsony model to oligopoly/oligopsony. The model for monopsony power over farmers and monopoly power over retailers/consumers can be expressed as

(57) $$P\,(1 - 1/e_d)(dQ\,/\,dL) = w\,(1 + 1/e_s).$$

Then, a generalized model can be expressed as

$$(58) \quad P\,(1 - \theta/e_d)(dQ\,/\,dL) = w\,(1 + \lambda/e_s)$$

where P is product price, w is input price, Q is product quantity, L is input quantity, e_d is price elasticity of demand in absolute terms, e_s is the inverse of elasticity of marginal cost with respect to supply, θ is conjectural elasticity (a market-power parameter) in the input market, and λ is conjectural elasticity (a market-power parameter) in the product market. In the left-hand side of (58), $P\,(1 - \theta/e_d)$ is the perceived marginal revenue from the processed product and $(dQ\,/\,dL)$ is the marginal product generated from the input, and therefore $P\,(1 - \theta/e_d)\,(dQ\,/\,dL)$ is the marginal product value generated from the input. The term $w\,(1 + \lambda/e_s)$ in the right-hand side is the marginal expense to buy the input. Values for θ and λ can be obtained from (simultaneous) estimation of the demand and cost functions and Equation (58).

Many of the studies have used the following specification:

$$(59) \quad P\,(1 - \theta/e_d) = w\,(1 + \lambda/e_s) + c.$$

The marginal input costs are divided into two categories—raw material from farmers, $w\,(1 + \lambda/e_s)$, and other variable inputs, c, where w is the raw-material price and c is the per-unit cost for other variable inputs that are assumed to be constant. In addition, the term $(dQ\,/\,dL)$ is eliminated in (59) because the product quantity is converted into the raw-material base. Equation (59) also can be used as a model of an oligopolistic/oligopsonistic nonprocessing distributor that pushes a low price (w) on sellers and resells the product to buyers at a high price (P). The per-unit reselling cost is c in that case.

Equation (59) can be transformed to:

$$(60) \quad \theta MR(Q) + (1 - \theta)D(Q) = \lambda MC(Q) + (1 - \lambda)S(Q) + c$$

where $MR(Q) = P + (dP\,/\,dQ)\,Q$ (the marginal revenue in monopoly or perfect collusion), $MC(Q) = w + (dw\,/\,dQ)\,Q$ (the marginal expense to inputs in monopsony or perfect collusion), $P = D(Q)$ (the inverse demand function showing marginal revenue in perfect competition), and $w = S(Q)$ (the inverse supply function showing the marginal expense to inputs in perfect competition) (Huang and Sexton 1996). The marginal revenue and marginal expense to inputs in intermediate imperfect competition (between perfect competition and monopoly/monopsony (or perfect collusion)) can be expressed as an average weighted by the market-power parameters ($0 \leq \theta, \lambda \leq 1$).

The trade-off between efficiency and market power addressed by Azzam and Schroeter (1995) is also an important topic. If welfare gains that are generated by cost efficiency through scale economies offset welfare losses caused by oligopsony and/or oligopoly, consolidation is desirable for society. The models proffered by Azzam and Schroeter (1995) and Sexton (2000) can be used to calculate the cost reductions that are necessary to offset the anticompetitive effects of market power.

8. Cooperatives' Market Power: Controlling Production and Marketing. Many researchers suggest that farmers are typical price-takers, but this is doubtful when we consider the existence of farmer cooperatives. One of the primary purposes of organizing a farmer cooperative is to exercise countervailing power to the oligopsonic power of buyers of the product (Galbraith 1952). Therefore, farmer cooperatives naturally create an imperfectly competitive market by controlling production and/or selling quantities.[12] If farmers have two or more separate markets, such as fresh and processing or private and government, they may exercise price-discriminating marketing by reducing sales to the price-inelastic market.

When both production and selling controls are implemented, we can use a generalization of the discriminated monopoly model. The discriminated monopoly model can be expressed as

$$(61) \qquad P_1\,(1 - 1/e_1) = P_2\,(1 - 1/e_2) = MC.$$

The generalized model for discriminated oligopoly can then be expressed as:

$$(62) \qquad P_1\,(1 - \theta_1/e_1) = P_2\,(1 - \theta_2/e_2) = MC$$

12 Compared to production control, marketing control that creates discriminated prices is more frequently criticized by competition-policy authorities, but production control is more often detrimental in that it can have a greater negative effect on production efficiency.

where P is price, MC is marginal cost, e is price elasticity of demand, θ is conjectural elasticity, and the subscripts 1 and 2 stand for the two separate markets. Values for θ_1 and θ_2 can be obtained from (simultaneous) estimation of the demand and cost functions and Equation (62) (see, for example, Kano (1999)).

There are many cases in which only marketing control (without production control) is implemented by a cooperative. The cooperative can be thought of as a consignment seller of the farmer members' commodity. Under marketing control, the cooperative charges different prices for the commodity based on market segmentation and pays a weighted-average price (or a pooled return) to farmers who produced the commodity with the price as given. Hence, the farmer is a price-taker in producing but a price-setter through the cooperative in selling. In this case, the model can be revised as:

(63) $P_1 (1 - \theta_1/e_1) = P_2 (1 - \theta_2/e_2)$ and

(64) $(P_1Q_1 + P_2Q_2) / (Q_1 + Q_2) = MC.$

When incorporating this "dual structure" process (i.e., farmer as a price-taker in producing and a price-setter in selling) into the model, the largest production is generated in the dual-structure monopoly while monopoly creates the lowest volume of production in the ordinal market. The dual-structure model developed by Suzuki, Lenz, and Forker (1993) and Suzuki et al. (1994a, 1994b)[13] was one of the first to treat the farmer side of the food system as a price-setter to estimate the degree of oligopoly power on the farmer side. These authors assumed that (a) = price-setter and (b) = price-taker.

9. Bilateral Oligopoly between Cooperatives and Processors and between Processors and Retailers. When oligopsonistic processors face oligopolistic farmer cooperatives instead of price-taking individual farmers, we must consider a bilateral oligopoly model. This implies that models of processor oligopoly that assume that farmers are price-takers should be reconsidered. Although Suzuki, Lenz, and Forker (1993) and Suzuki et al. (1994a, 1994b) made a notable step regarding treatment of the farmer side, they

13 The term "dual structure" was coined by Kawaguchi, Suzuki, and Kaiser (1997).

avoided incorporating the bilateral oligopoly problem by assuming that processors are price-takers.

The case of competition between processors and retailers is similar. Most previous models have assumed that one of the two is a price-taker. Retailers are often assumed to be price-takers when processors' oligopoly power is estimated. However, that assumption is questionable because the market share of large supermarkets has been rapidly increasing.

Sexton and Zhang (2001) discussed welfare losses caused by successive oligopoly and oligopsony models covering the whole food system from farmer to consumer. They assumed that farmers and consumers are always price-takers. Then, they presented two alternative models for the entire food system:

[Farmers] $\Rightarrow\Leftarrow$ [Processors] $\Rightarrow\Leftarrow$ [Retailers] $\Rightarrow\Leftarrow$ [Consumers].
(a) (b) (c) (d) (e) (f)

In the first model, they assumed that (a) = price-taker, (b) = price-setter, (c) = price-setter, (d) = price-taker, (e) = price-setter, and (f) = price-taker. In the second model, they assumed that (a) = price-taker, (b) = price-setter, (c) = price-taker, (d) = price-setter, (e) = price-setter, and (f) = price-taker.

The successive oligopoly model used by Sexton and Zhang (2001) is as follows:

Farmers' (price-takers') inverse supply function: $P^f = S(Q)$.

Processors' marginal expense to raw materials + unit processing costs = marginal revenue from sales to retailers: $P^f(1 + \theta^f/e^f) + c^w = P^w(1 - \lambda^w/\eta^w)$.

Supermarkets' purchase price from processors + unit retailing costs = marginal revenue from retail sales: $P^w + c^r = P^r(1 + \lambda^r/\eta^r)$.

Consumers' (price-takers') inverse demand function: $P^r = D(Q)$.

Similarly, the successive oligopsony model used by Sexton and Zhang (2001) is as follows:

Farmers' (price-takers') inverse supply function: $P^f = S(Q)$.

Processors' marginal expense to raw material from farmers + unit processing costs = wholesale price to retailers: $P^f (1 + \theta^f/e^f) + c^w = P^w$.

Supermarkets' marginal expense to the product from processors + unit retailing costs = marginal revenue from retail sales: $P^w (1 + \theta^w/\eta^w) + c^r = P^r (1 + \lambda^r/\eta^r)$.

Consumers' (price-takers') inverse demand function: $P^r = D(Q)$.

These are not empirical models, but, by assuming various degrees of conjectural elasticities, they illustrate how the distribution of economic welfare moves in response to degrees of successive oligopoly and oligopsony power at multiple stages of the market channel. Results from this simulation show that even modest market power can enable the marketing sector to capture large shares of market surplus from farmers and consumers. Although Sexton and Zhang's (2001) models are innovative because the entire food system is incorporated into them, one drawback is that the models avoided the bilateral oligopoly problems that exist between processors and retailers and between processors and farmers.

If retailers have market power and processors are price-takers, then the product price determined between processors and retailers should be at the lowest level that retailers can pay to processors based on the current degree of horizontal competition among retailers. Alternatively, if processors have market power and retailers are price-takers, then the market price should be at the highest level that processors can obtain from retailers based on the degree of horizontal competition among processors. In reality, since both retailers and processors likely have some degree of market power (i.e., bilateral oligopoly), the actual price lies somewhere between the highest and lowest level. A similar argument is true for market competition between farmer cooperatives and processors.

In these situations, estimates of the degree of balance in vertical power between sellers and buyers, the degree of horizontal competition among sellers, and the degree of competition among buyers should be made simultaneously. It becomes difficult to produce a unique solution in modeling such markets. Consequently, as previously mentioned, most studies have assumed that one side of the market is a price-taker while the other has completely dominant vertical power. However, because the actual observable price is at neither the highest nor the lowest possible level, such approaches are inherently biased. Azzam (1996) presented an innovative modeling approach for solving this problem between packers and retailers in the U.S. beef market. He examined

[Processors] $\Rightarrow\Leftarrow$ [Retailers]
(c) (d)

and assumed that (c) and (d) are both price-setters.

There are several hurdles such as identification problems to overcome in simultaneously estimating vertical and horizontal competition parameters for the entire food system.[14] We explain them using a model that includes farmer, processor, and retailer levels. We first introduce a dual structure in the farmer side. When farmer cooperatives have two markets (the private market and the government market) and the government's purchase price is given, then the necessary condition for optimal allocation between the two markets that will maximize sales is:

$$(65) \qquad PF\,(1 - \theta^f/\eta^f) = PG$$

where *PF* is the price paid by processors to cooperatives, *PG* is the exogenously given government purchase price, η^f is the price elasticity of the processor demand that cooperatives face (in absolute value), and θ^f is a parameter representing cooperatives' degree of horizontal competition and ranges from zero to one.

It is important to note that Equation (65) assumes that processors have no vertical market power; in other words, the power balance between cooperatives and processors is 1:0. Therefore, the price determined by Equation (65) means that the price cooperatives or farmers receive (PF^U)

14 Azzam (1996) also had an identification problem because there were six equations for seven variables in his model.

based on their degree of horizontal competition (θ^f) is at the maximum level. PF^U can be rewritten as:

(66) $PF^U = PG / (1 - \theta^f/\eta^f)$.

Next, consider the processor level of the industry. The processors' profit-maximizing condition can be expressed as:

(67) $PW^U (1 - \theta^w/\eta^w) = PF^L (1 + \lambda^w/\varepsilon^w) + MC^w$

where PW^U is the maximum or upper-limit price processors can receive from retailers when retailers have no vertical market power relative to processors, PF^L is the minimum or lower-limit price processors will pay to cooperatives when cooperatives have no vertical market power relative to processors, MC^w is the marginal processing and selling costs other than the raw material, η^w is the price elasticity of the retailer demand that processors face (in absolute value), θ^w is a parameter representing the processors' degree of horizontal competition in the processor-versus-retailer market (ranging from zero to one), ε^w is the price elasticity of the cooperative supply that processors face, and λ^w is a parameter representing the processors' degree of horizontal competition in the processor-versus-retailer market (ranging from zero to one). The minimum (lower-limit) price processors can pay to cooperatives (PF^L) is:

(68) $PF^L = [PW^U(1 - \theta^w/\eta^w) - MC^w] / (1 + \lambda^w/\varepsilon^w)$.

The actual price (PF) determined by transaction between cooperatives and processors is somewhere between PF^U (the upper limit) and PF^L (the lower limit):

(69) $PF = W^f PF^U + (1 - W^f)\, PF^L$ or

(70) $PF = W^f PG / (1 - \theta^f/\eta^f) + (1 - W^f)\, [PW^U (1 - \theta^w/\eta^w) - MC^w] / (1 + \lambda^w/\varepsilon^w)$

where W^f is the parameter (ranging from zero to one) for the degree of vertical market power of cooperatives relative to processors and $(1 - W^f)$ is the parameter for the degree of vertical market power of processors relative to cooperatives.

Next, consider the retailer stage of the market. The retailers' profit-maximizing condition can be expressed as

(71) $$PR\,(1 - \theta^r/\eta^r) = PW^D\,(1 + \lambda^r/\varepsilon^r) + MC^r$$

where *PR* is the actual retail price or the maximum (upper-limit) price retailers can get from consumers since consumers likely have no market power, PF^L is the lower-limit price retailers pay to processors assuming processors have no vertical market power relative to retailers, MC^r is marginal retailing costs, η^r is the price elasticity of consumers' demand (in absolute value), θ^r is the retailers' degree-of-horizontal-competition parameter in the retailer-versus-consumer market (ranging from zero to one), ε^r is the price elasticity of the processors' supply, and λ^r is the retailers' degree-of-horizontal-competition parameter in the retailer-versus-processor market (ranging from zero to one). The lower-limit price retailers can pay to processors (PW^L) is:

(72) $$PW^L = [PR\,(1 - \theta^r/\eta^r) - MC^r] \,/\, (1 + \lambda^r/\varepsilon^r).$$

On the other hand, from Equation (67), the maximum or upper-limit price processors can obtain from retailers (PW^U) is:

(73) $$PW^U = [PF^L\,(1 + \lambda^w/\varepsilon^w) + MC^w] \,/\, (1 - \theta^w/\eta^w).$$

The actual price (*PW*), determined by transactions between processors and retailers, is somewhere between PW^U and PW^L and is described as:

(74) $$PW = W^w PW^U + (1 - W^w)\,PW^L \text{ or}$$

(75) $$PW = W^w\,[PF^L\,(1 + \lambda^w/\varepsilon^w) + MC^w] \,/\, (1 - \theta^w/\eta^w) + (1 - W^w)\,[\mathrm{PR}\,(1 - \theta^r/\eta^r) - MC^r] \,/\, (1 + \lambda^r/\varepsilon^r)$$

where W^w is the degree of vertical market power of processors against retailers (ranging from zero to one) and $(1 - W^w)$ is the degree of vertical market power of retailers against processors.

Equations (70) and (75) have some common parameters. Therefore, simultaneous estimation of Equations (70) and (75) with parameter constraints provides for all values the degree of vertical and horizontal competition among cooperatives, processors, and retailers (W^f, W^w, θ^f, θ^w, θ^r, λ^w, and λ^r). However, the variables PF^L in Equation (70) and PW^U in Equation (75) are unobservable. Consequently, while this approach is good in theory, it is not a practical way to estimate these vertical and horizontal competition parameters.

Using some assumptions, we can treat Equations (70) and (75) separately. Consider the case of bilateral oligopoly between processors and retailers by adjusting Equation (75). Assuming that processors' vertical market power against cooperatives is 1:0, PF^L can be replaced by the actual *PF* in Equation (75). When it can be assumed that processors' vertical market power against retailers is 0:1, then the term PW^U $(1 - \theta^w/\eta^w)$ can be replaced by the actual *PW* in Equation (70). In addition, we can obtain price elasticities of supply and demand from extraneous sources by following Azzam (1996). Using these methods, we can find a practical way to identify the competition parameters.

A successful example of this approach appears in Kinoshita, Suzuki, and Kaiser (2004b). The authors developed a practical model that estimated the degree of balance of vertical power between fluid-milk processors and retailers and between dairy cooperatives and processors with simultaneous estimation of the degree of horizontal competition in each stage of the market. The authors' results were tentative due mainly to data constraints, but their study constituted the first econometric evidence supporting the general perception that retailers, though facing nearly perfect horizontal competition with each other, have extremely dominant vertical market power over fluid-milk processors. Kinoshita, Suzuki, and Kaiser (2004b) also offered the first econometric evidence that processors may have some vertical market power over dairy cooperatives. Once more complete data sets are available, the method proposed by Kinoshita, Suzuki, and Kaiser would be a practical and useful way to analyze bilateral oligopoly in many situations. Their results are summarized in Figure 1.2.

10. Price Transmission with Imperfect Competition. Imperfect farm-to-retail price transmission generated by imperfect competition is also an important topic in NEIO studies, especially in analyzing the effect of reducing government price supports for farmers on consumer welfare. If there is oligopoly and/or oligopsony at any stage of the food-market channels, gains from deregulation may not be transmitted fully to consumers. On the other hand, regulations that provide higher prices to farmers, such as the Northeast Dairy Compact in the U.S., may not significantly raise consumer prices due to the existing market power of processors and retailers. Various types of NEIO studies that include conjectural variation have been conducted to analyze imperfect price transmission, including Holloway (1991); Schroeter and Azzam (1991); Suzuki, Lenz, and Forker (1993); Suzuki et al. (1994b); McCorriston, Morgan, and Rayner (1998, 2001); Bettendorf and Verboven (2000); Dhar and Cotterill (2002); Weldegebriel (2004); and Kojima (2005). Most of these studies considered only oligopoly. Schroeter and Azzam (1991) and Weldegebriel (2004) considered oligopoly in output markets and oligopsony in input markets; Dhar and Cotterill (2002) analyzed differentiated product markets. None of the studies introduced bilateral oligopoly where sellers' oligopoly power could cancel out buyers' oligopsony power.

Figure 1.2. The Degree of Vertical and Horizontal Competition in Japan's Milk Markets

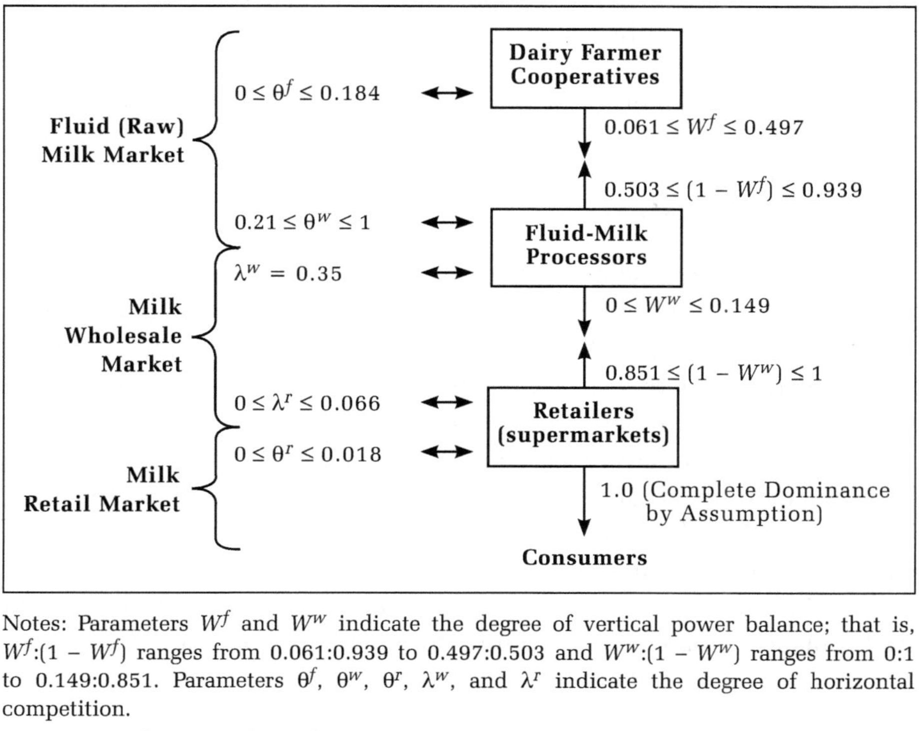

Notes: Parameters W^f and W^w indicate the degree of vertical power balance; that is, W^f:$(1 - W^f)$ ranges from 0.061:0.939 to 0.497:0.503 and W^w:$(1 - W^w)$ ranges from 0:1 to 0.149:0.851. Parameters θ^f, θ^w, θ^r, λ^w, and λ^r indicate the degree of horizontal competition.

Source: Kinoshita, Suzuki, and Kaiser (2004b).

Suzuki, Lenz, and Forker (1993) and Suzuki et al. (1994b) provided a simple expression for the degree of price transmission between regulated and unregulated markets when oligopolistic firms (cooperatives) allocate a homogeneous commodity to a private market and a market with a guaranteed price. When the government reduces the guaranteed price (GP), the degree of reduction in the unregulated market price (P) can be expressed as follows. Using the expression $dQ / dq_i \equiv k_i$ (1 + conjectural variation), the change in P in response to a change in GP is shown by

(76) $$dP / dGP = \Sigma (1 / k_i) / [\Sigma (1 / k_i) + 1].$$

In Suzuki, Lenz, and Forker (1993), Equation (76) was expressed as

(77) $$dP / dGP = (k_1 + k_2) / (k_1 + k_1k_2 + k_2)$$

because they used a duopoly model.

When all firms are identical, (76) can be expressed more simply as

(78) $$dP / dGP = 1 / (\theta + 1)$$

where θ is the common conjectural elasticity for all firms (Suzuki et al. 1994b). Therefore, $dP / dGP = n / (n + 1)$ when Cournot behavior is assumed, $dP / dGP = 1/2$ when $n = 1$ (monopoly), and $dP / dGP = 1$ when $n \rightarrow \infty$ (perfect competition).

11. Expansion to Spatial-Equilibrium Models. NEIO models have also been applied to spatial-equilibrium analyses. For instance, we can generalize Takayama and Judge's (1971) spatial-equilibrium model to allow for incorporation of any degree of market structure from perfect competition to monopoly using the following notations: D_j as quantity demanded in market j ($j = 1, 2, \ldots, n$), S_i as the quantity supplied in producing region i ($i = 1, 2, \ldots, n$), X_{ij} as the quantity shipped from region i to market j, and PD_j as the price in market j. MC_i is region i's marginal cost and FC_i is region i's fixed cost (constant); $D_j = \alpha_j - \beta_j PD_j$ is the demand function in market j; $S_i = -v_i + \eta_i \mathrm{MC}_i$ is region i's (inverse) marginal cost function; and T_{ij} is the unit transportation cost of shipping from region i to market j (constant). Region i's sales revenue in market j ($PD_j X_{ij}$) can be written as:

(79) $PD_jX_{ij} = [\alpha_j/\beta_j - (1/\beta_j)\,D_j]\,X_{ij} = [\alpha_j/\beta_j - (1/\beta_j)(\Sigma_{i=1}{}^{n}X_{ij}\,)]X_{ij}$
$= [\alpha_j/\beta_j - (1/\beta_j)(\Sigma_{m\neq i}X_{mj} + X_{ij})]X_{ij}$

where m ($m \neq i$) indicates all regions other than i. When region i believes that a change in its supply to market j will cause changes in all other regions' supplies to market j, region i's "perceived" marginal revenue in market j is:

(80) $\partial(PD_jX_{ij})/\partial X_{ij} = [\alpha_j/\beta_j - (1/\beta_j)\,D_j] - (1/\beta_j)\,(\partial\Sigma_{m\neq i}X_{mj}/\partial X_{ij} + 1)X_{ij}$
$= PD_j - (1/\beta_j)\,(r_{ij} + 1)X_{ij}$

where r_{ij} is region i's conjectural variation regarding changes in all other regions' supplies to market j caused by a change in region i's supply.

Using this relationship, the total-revenue maximization problem for all n regions can be respecified as the following net social payoff adjusted for imperfectly competitive markets (*ANSP*):

(81) Max: $ANSP = \Sigma_{j=1}{}^{n}\int[\alpha_j/\beta_j - (1/\beta_j)\,D_j]\,dD_j - \Sigma_{i=1}{}^{n}$
$[\int[v_i/\eta_i + (1/\eta_i)S_i]dS_i + FC_i] - \Sigma_{j=1}{}^{n}\Sigma_{i=1}{}^{n}\,(1/\beta_j)\,(r_{ij} + 1)$
$\int X_{ij}dX_{ij} - \Sigma_{j=1}{}^{n}\Sigma_{i=1}{}^{n}\,T_{ij}X_{ij},$

(82) subject to $D_j \leq \Sigma_{i=1}{}^{n}\,X_{ij},\ \forall j,$

(83) $\Sigma_{j=1}{}^{n}\,X_{ij} \leq S_i,\ \forall i,$ and

(84) $D_j \geq 0,\ S_i \geq 0,\ X_{ij} \geq 0,\ \forall i, j.$

The difference between *ANSP* in Equation (81) and the net social payoff (*NSP*) in the conventional spatial competitive-equilibrium model by Takayama and Judge (1971) is the term $-\Sigma_{j=1}{}^{n}\Sigma_{i=1}{}^{n}\,(1/\beta_j)\,(r_{ij} + 1)\int X_{ij}dX_{ij}$ (Kawaguchi, Suzuki, and Kaiser 1997, 2001; Matsuda and Kurokawa 1996; Nelson and McCarl 1984; Chen, McCarl, and Chang 2003). When the market is perfectly competitive ($r_{ij} = -1$), the term is zero and Equation (81) is equal to the original Takayama and Judge model. When Cournot behavior ($r_{ij} = 0$) is assumed, the term is equivalent to $-\Sigma_{j=1}{}^{n}\Sigma_{i=1}{}^{n}\,(1/\beta_j)\int X_{ij}dX_{ij}$, which is shown in Hashimoto's (1985) spatial

Nash equilibrium model. Cournot behavior means that region i believes that the other regions will not change their supply in response to the region's action.

Using the Lagrange function (L) with the multipliers λ_j and ω_i for constraints (82) and (83), respectively, the Kuhn-Tucker optimality conditions for the maximization problem can be expressed as follows:

(85) $\partial L / \partial D_j = \alpha_j / \beta_j - (1 / \beta_j)\, D_j - \lambda_j \leq 0,\ D_j\, (\partial L / \partial D_j) = 0,\ \forall j,$

(86) $\partial L / \partial S_i = -v_i / \eta_i - (1 / \eta_i) S_i + \omega_i \leq 0,\ S_i\, (\partial L / \partial S_i) = 0,\ \forall i,$

(87) $\partial L / \partial X_{ij} = -(1 / \beta_j)(r_{ij} + 1) X_{ij} - T_{ij} + \lambda_j - \omega_i \leq 0, X_{ij}(\partial L / \partial X_{ij}) = 0, \forall i, j,$

(88) $\partial L / \partial \lambda_j = -D_j + \Sigma_{i=1}{}^{n} X_{ij} \leq 0,\ \lambda_j\, (\partial L / \partial \lambda_j) = 0,\ \forall j,$ and

(89) $\partial L / \partial \omega_i = -\Sigma_{j=1}{}^{n} X_{ij} + S_i \leq 0,\ \omega_i\, (\partial L / \partial \omega_i) = 0,\ \forall i.$

The Lagrange multipliers (dual variables) λ_j and ω_i measure market price (PD_j) and marginal cost (MC_i), respectively. The condition represented by (87) indicates that each region must equalize marginal revenue net of transportation costs across all markets in which it sells products. Kawaguchi, Suzuki, and Kaiser (2001) found that U.S. regional fluid-milk price premiums can be approximated by the Cournot-Nash equilibrium but not by the competitive equilibrium.

Kawaguchi, Suzuki, and Kaiser's (1997; see Chapter 9) framework also can be used in international trade analyses, and the results of the simulation show that bilateral (two-way) dumping of homogeneous goods with hidden export subsidies can occur even in a free-trade market. In addition, Shono and Kawaguchi (1999) successfully introduced *ad valorem* tariffs as well as other trade policies such as tariff-rate quotas into the imperfectly competitive international spatial-equilibrium model by formulating the model as a linear complementarity problem (LCP). Maeda (2001) and Maeda et al. (2001a, 2001b) generalized this approach to a nonlinear model.

12. A Practical Application of NEIO to International Trade Policy Debates. Many international trade analyses have used conjectural variation, including Alston and Gray (2000); Baldwin and Krugman (1988); Buschena and Perloff

(1991); Chen, McCarl, and Chang (2003); Dixit (1988); Hwang (1984); Hwang and Mai (1988); Deodhar and Sheldon (1995, 1996, 1997); Delrome and van der Mensbrugghe (1990); Helpman and Krugman (1989); Karp and Perloff (1989, 1993); Love and Murniningtyas (1992); McCorriston, Sheldon, and Hirschberg (1993); McCorriston and Sheldon (1996); Suzuki, Kinoshita, and Kaiser (2001); Suzumura and Ishikawa (1996); Thursby and Thursby (1990); and Tiffin and Dawson (1997).

Suzuki, Kinoshita, and Kaiser (2001) developed a theoretical and practical measurement of the degree of market distortion due to price discrimination for hidden export subsidies by state trading enterprises (STEs) and also developed a practical, generalized definition of export subsidy equivalents (ESEs) applicable to "consumer financed," hidden export subsidies and to ordinal, "taxpayer financed" export subsidies.

The following model (Suzuki, Kinoshita, and Kaiser 2001) is useful for describing mechanisms by which to circumvent export subsidies by exporting STEs. Assume that an agency (an exporting STE) has exclusive authority to deal with the country's marketing of a certain product. Even if this agency is a monopoly within that country, it does not necessarily have monopoly power since it may be a price-taker through competition with international rivals if imports are not restricted. We assume here that there are some import-restricting measures that allow the agency to set the domestic price higher than the world price. The role of the agency is to allocate the country's supply of the commodity to domestic and export markets to maximize total sales revenue as the consignment seller of the commodity collected from farmers. The necessary condition for this to occur is to equate marginal revenues from domestic and export markets. The agency distributes the proceeds back to farmers by paying a weighted-average (blended) price from its sales to both the domestic and the export market. Farmers are assumed to be price-takers, so they produce at a level at which their marginal production costs are equal to the blended price they receive. The export price is determined by equating total world supply and demand. The agency is assumed to be a price-taker in the world market.

Based on these assumptions, Figure 1.3 illustrates this market situation conceptually. P_d is domestic price, Q_d is domestic supply, Q_s is total supply, P_b is the blended price, QR_d is foreign demand, and QR_s is foreign supply. The intersection (point C) of the marginal revenue line of domestic sales and a given P_w level (the horizontal marginal revenue line

Figure 1.3. Export Subsidy Equivalents (Two-Region Model)

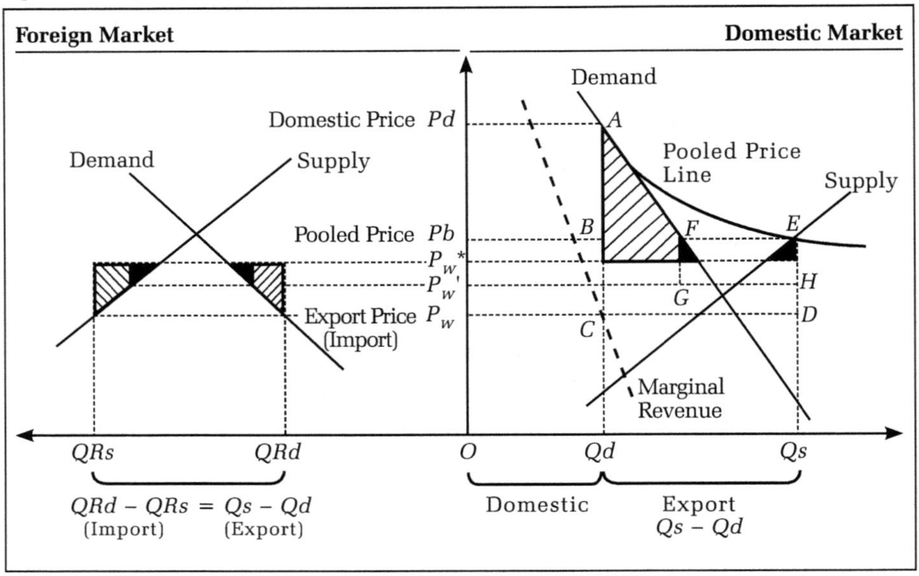

of export sales) in Figure 1.3 (right) is the point equalizing the perceived marginal revenues of domestic and export sales. Point *C* can be expressed mathematically as:

(90) $P_d + (\partial P_d / \partial Q_d)\, \theta Q_d = P_w$ or

(91) $P_d\,(1 - \theta/E) = P_w$

where θ $(0 \leq \theta \leq 1)$ is a degree-of-market-power parameter and E is the price elasticity of domestic demand in absolute value or $-(\partial Q_d / \partial P_d)(P_d / Q_d)$. The left-hand side of Equation (90) is the perceived marginal revenue from domestic sales. Because the agency is a price-setter in the domestic market, the domestic price is expected to decrease by $(\partial P_d / \partial Q_d)\theta$ when the domestic sales volume increases by one unit. Therefore, the incremental decrease in total revenue from domestic sales is $(\partial P_d / \partial Q_d)\theta Q_d$. Accordingly, perceived additional revenue from a one-unit volume increase in domestic sales is $P_d + (\partial P_d / \partial Q_d)\, \theta Q_d$ as shown in Equation (90). The right-hand side of Equation (90) is the perceived marginal revenue from export sales. It is expected to be constant regardless of changes in export volume because the model assumes that the agency is a price-taker in the export market. The value for θ is affected by the degree of border protection. The agency could have monopoly power, shown by $P_d\ (1 - 1/E) = P_w$, when imports are prohibited or it could become a price-taker in the domestic market (i.e., $P_d = P_w$) when there is no border protection.

The agency allocates its domestic sales up to the level Q_d with the domestic price determined at the intersection (point *A*) of the demand curve and Q_d on the x-axis. On the other hand, domestic production is determined at the intersection (point *E*) of the supply curve and the pooled (blended) price line. The pooled price line can be expressed mathematically as:

(92) $$P_b = [P_d Q_d + P_w (Q_s - Q_d)] / Q_s.$$

When equilibrium is achieved at a given world price (P_w), the agency's export volume ($Q_s - Q_d$) in Figure 1.3 (right) is equal to the import volume ($QR_d - QR_s$) in the rest of the world as shown in Figure 1.3 (left). When domestic and world markets are both perfectly competitive, Equation (90) becomes $P_d = P_w\ (= P_b)$ since $\theta = 0$, and $P_w{}^*$ in Figure 1.3 indicates the perfectly competitive level. Note that the competitive world price, $P_w{}^*$, is higher than the current world price, P_w, with price discrimination by the agency. The agency reduces the world price because it restricts domestic sales to achieve a higher domestic price and expands export sales, which generates a lower world equilibrium price.

Because producers receive the pooled price P_b and the export price P_w is lower than P_b, the amounts shown by the rectangle *BCDE* in Figure 1.3 (right) are subsidies. Under this scheme, consumers pay the entire amount of the subsidy, which is illustrated by the area of rectangle *BCDE* being equal to the area of P_dP_bBA. In the case of an ordinary export subsidy regulated by the WTO, the government, by way of taxpayers, pays the subsidy shown by area *BCDE*. Therefore, the areas P_dP_bBA and *BCDE* can be defined as a "consumer financed" ESE (Schluep 1999). In this case,

(93) $$\text{ESE} = Q_e (P_b - P_w) = Q_d (P_d - P_b)$$

where $Q_e = Q_s - Q_d$. The consumer-financed ESEs become larger as the value of θ grows, *ceteris paribus*, as shown by

(94) $$\text{ESE} = (\theta P_d Q_d Q_e) / (E Q_s)$$

from Equations (91) and (93). Therefore, θ is also an indicator of the magnitude of the ESEs.

From the viewpoint of economic welfare, the consumer-financed export subsidy may be worse than an ordinary subsidy (Alston and Gray 2000). As shown in Figure 1.3, when the producer price P_b is provided by an ordinary export subsidy, the export quantity is *FE*, the export price is P_w', and the government expenditure is the rectangle *FGHE*. In that case, welfare loss compared to a perfectly competitive free market is the sum of the four black triangles, which is smaller than the case of a "consumer financed" export subsidy by the sum of the three gray trapezoids.[15]

Although the model presented explains ESEs by price discrimination between domestic and export prices, the model also can be used for price discrimination among export markets. It can easily be extended in this direction by introducing cases where foreign consumers also pay the ESEs. Figure 1.4 illustrates several cases using three countries (domestic, country 1, and country 2). In case (a), domestic consumers finance the two foreign markets; in case (b), domestic consumers and consumers in country 1 finance the market in country 2; in case (c), consumers in country 1 finance the market in country 2;[16] and in case (d), consumers in country 1 finance the domestic market and the market in country 2. In all these cases, the black areas in the figure subsidize the gray areas. Therefore, a general consumer-financed ESE can be defined by

(95) $$\text{ESE} = \Sigma |P_j - P_b|\, Q_j / 2$$

15 Here no tax-collecting costs are assumed.

16 In case (c), the weighted-average price of export sales paid to farmers is equal to the price farmers receive from domestic sales. This is currently true for the Australian Wheat Board, which lost its control over domestic sales but is still a price-setter in export markets. Price-taking farmers must allocate their products between the domestic market and the Australian Wheat Board for the average export price to equal the price they receive from domestic sales.

Figure 1.4. Four Types of Export Subsidy Equivalents

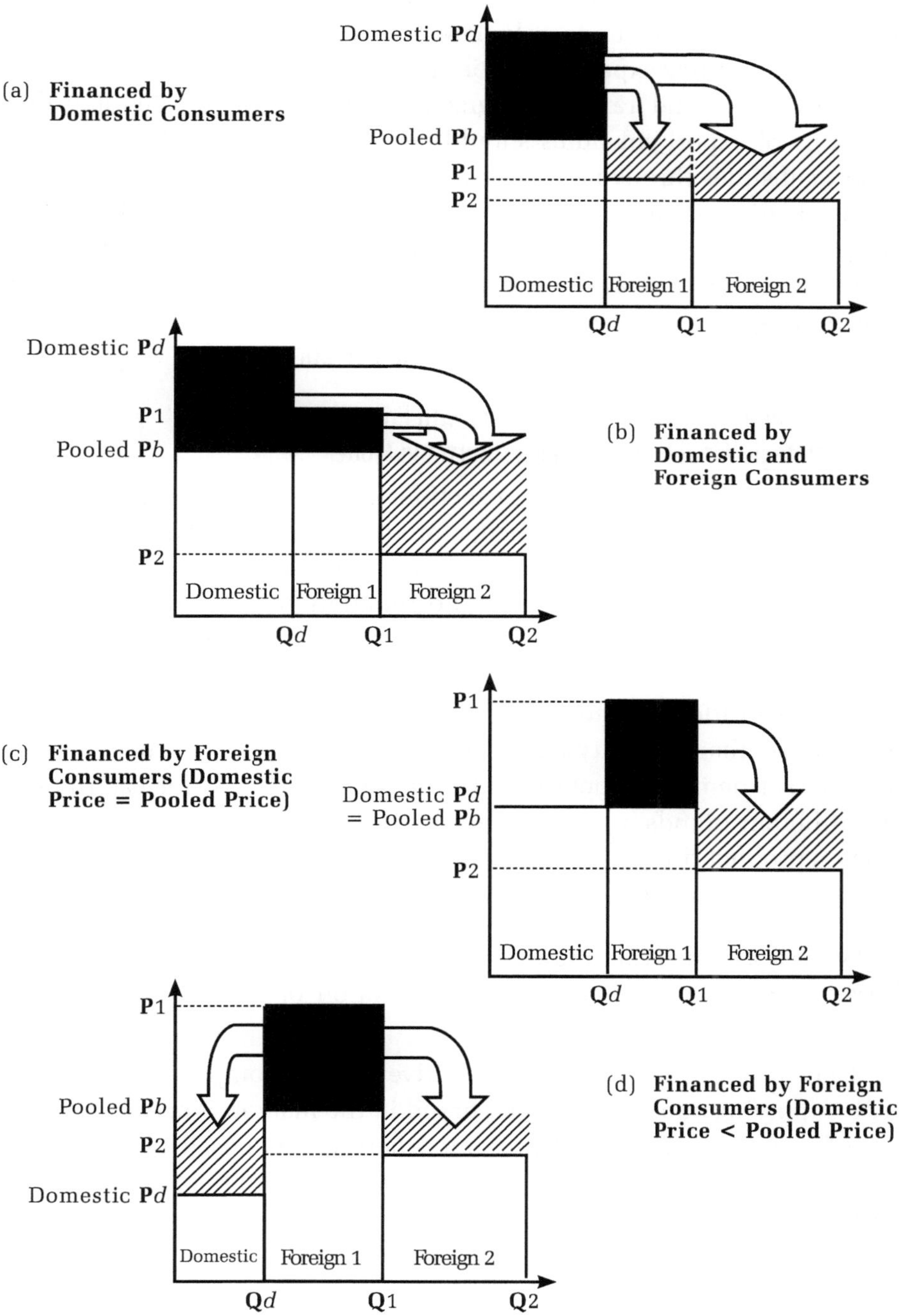

where P_j is the price received from the jth country net of transaction costs, P_b is the pooled price, and Q_j is the quantity sold in the jth country. This formula enables us to calculate any consumer-financed ESE with country-by-country export quantities and prices.

The Doha Round framework agreements established in 2004 state that all forms of export subsidies should be eliminated, including not only export subsidies through food aid and export credits but also consumer-financed "hidden" subsidies that come from exporting STEs. Therefore, one needs a theoretical definition and practical measurements for hidden export subsidies that are not regulated under current WTO rules. The ESE proposed here would provide a useful measurement of consumer-financed export subsidies (Suzuki and Kaiser 2005).

A Product-Differentiated Oligopoly Model with Price as the Strategic Variable: A Generalized Bertrand Model

The discussion in the preceding sections assumed that all firms within the industry supply a homogeneous product and use quantity as a strategic variable. The model discussed is a generalization of the Cournot model and there is only one demand function and one common price generated in the market.

However, most imperfectly competitive industries are characterized by multiple differentiated products that compete with each other with price rather than quantity being the strategic variable. As such, a separate market demand function and a unique price exist for each brand and different brands are incomplete substitutes for each other. In such cases, a product-differentiated oligopoly model with price as the strategic variable—a generalized Bertrand model—should be applied.

1. Basic Theory. Liang (1989) presented a basic generalized Bertrand model. Here, we explain an n-product generalization of Liang's original two-brand model.

Consider an imperfectly competitive market composed of n firms, each supplying one differentiated brand in the market. Since each brand is differentiated, the model presented here is composed of n demand functions for each brand i ($i = 1, \ldots, n$) and n price-reaction functions for each firm i ($i = 1, \ldots, n$).

To ensure empirical tractability, the demand function for *i* is expressed in a linear form as

(96) $Q_i = a_i - b_i P_i + \Sigma_{j \neq i} c_{ij} P_j$ for i and $j = 1, \ldots, n, i \neq j$

where Q_i is firm/brand *i*'s quantity supplied/demanded, P_i is brand *i*'s price, P_j is a substitute brand's price, and a_i, b_i, and c_{ij} are parameters. Other variables affecting Q_i are omitted here to simplify presentation of the conceptual model.

Firm *i*'s profit-maximization problem using price (P_i) as a strategic variable is represented by

(97) $\max_{P_i} [P_i Q_i - C_i]$

where C_i is firm *i*'s cost. The first-order condition is expressed as

(98) $Q_i + P_i (dQ_i / dP_i) - MC_i (dQ_i / dP_i) = 0$

where MC_i is firm *i*'s marginal cost ($dC_i / dQ_i = MC_i$). Firm *i*'s conjectural variation is defined as its perception of how rival firm *j* will respond to firm *i*'s price change. Mathematically, firm *i*'s conjectural variation, cv_{ji}, is defined as dP_j / dQ_i. Using this, dP_i / dQ_i in Equation (98) can be written as

(99) $dQ_i / dP_i = \partial Q_i / \partial P_i + \Sigma_{j \neq i} (\partial Q_i / \partial P_j)(dP_j / dP_i) = -b_i + \Sigma_{j \neq i} c_{ij} cv_{ij}.$

Using Equation (99), firm *i*'s perceived or conjectured own-price elasticity, e_i, can be defined as

(100) $e_i = (dQ_i / dP_i)(P_i / Q_i),$

$= [b_i + \Sigma_{j \neq i} (c_{ij} cv_{ji})] (P_i / Q_i),$

$= b_i (P_i / Q_i) + \Sigma_{j \neq i} [c_{ij} (P_j / Q_i) \, cv_{ji} (P_i / P_j)],$

$= e_i^{ow} + \Sigma_{j \neq i} (e_{ij}^{cr} e_{ji}^{cv}).$

Equation (100) shows that e_i is composed of three types of elasticities: (1) e_i^{ow} (partial own-price elasticity), (2) e_{ij}^{cr} (cross-price elasticity), and (3) e_{ji}^{cv} (conjectural-variation elasticity).

Substituting (96) and (99) into (98) and solving (98) for P_i yields the following price-reaction function, R_i, for firm i:

$$(101)\quad P_i = -a_i / (-2b_i + \Sigma_{j\neq i}\, c_{ij}\, cv_{ij}) - [1 / (-2b_i + \Sigma_{j\neq i}\, c_{ij}\, cv_{ij})]\, \Sigma_{j\neq i}\, c_{ij}P_j + [(-b_i + \Sigma_{j\neq i}\, c_{ij}\, cv_{ij}) / (-2b_i + \Sigma_{j\neq i}\, c_{ij}\, cv_{ij})]MC_i.$$

Values for cv_{ij} could be obtained through simultaneous estimation of (96) and (101). When $cv_{ij} = 0$, the firm determines its own price, taking its rivals' price as given, while $cv_{ij} = 1$ means that the firm believes that its rivals will, in response to i's price change, change their prices in an equal amount.

Studies that have developed and applied the Bertrand-type model to various industries include Cotterill (1994); Cotterill, Putsis, and Dhar (2000); Cotterill and Putsis (2000; see Chapter 12); Kinoshita, Suzuki, and Kaiser (2002; see Chapter 11); Liang (1989); Peterson and Cotterill (1998); and Vickner and Davies (1999). Most of these studies focused on processors' horizontal price competition, taking retailers' behavior as given. However, as in the homogeneous oligopoly cases, vertical competition between processors and retailers should be considered in product-differentiated oligopoly models (see Sudhir (2001) for this topic).

A difference between the generalized Cournot and Bertrand models is given by generalized Lerner Indices. The generalized Lerner Index for the generalized Cournot model is shown as

$$(102)\quad (P - mc_i) / P = \theta_i / e.$$

On the other hand, for the generalized Bertrand model it is:

$$(103)\quad (P_i - mc_i) / P_i = -1 / [(-b_i + \Sigma_{j\neq i}\, c_{ij}\, cv_{ij})\, P_i / Q_i].$$

2. Relationship between Price and Quantity Conjecture. As pointed out by Kawaguchi, a conjecture on price can be re-expressed by an equivalent conjecture on quantity that generates the same solution (Suzuki 2002). In this sense, price-setting competition should not be discussed separately from quantity-setting competition. Consider a two-firm (or two-brand) market that has the following two demand functions:

(104) $Q_1 = a_1 - b_1P_1 + c_{12}P_2$ and

(105) $Q_2 = a_2 + c_{21}P_1 - b_2P_2$.

Alternatively, and more compactly, by using the matrices

$$\mathbf{Q} = \begin{pmatrix} Q_1 \\ Q_2 \end{pmatrix}, \mathbf{a} = \begin{pmatrix} a_1 \\ a_2 \end{pmatrix}, \mathbf{B} = \begin{pmatrix} b_1 & -c_{12} \\ -c_{21} & b_2 \end{pmatrix}, \mathbf{P} = \begin{pmatrix} P_1 \\ P_2 \end{pmatrix},$$

Equations (104) and (105) can be expressed as

(106) $\mathbf{Q} = \mathbf{a} - \mathbf{BP}$.

Regarding firm 1's price conjectures, we use the following matrix expressions:

$$\begin{pmatrix} \frac{dP_1}{dP_1} \\ \frac{dP_2}{dP_1} \end{pmatrix} = \begin{pmatrix} cv_{11} \\ cv_{12} \end{pmatrix} \text{ or } \begin{pmatrix} dP_1 \\ dP_2 \end{pmatrix} = dP_1 \begin{pmatrix} 1 \\ cv_{12} \end{pmatrix}.$$

(107) $d\mathbf{P} = (dP_1)\,\mathbf{L}_1$.

Then, using (106), changes in quantity demanded associated with conjectured changes in price can be shown as:

(108) $d\mathbf{Q} = -\mathbf{B}d\mathbf{P} = -\mathbf{B}\,[(dP_1)\,\mathbf{L}_1] = -(dP_1)\,\mathbf{BL}_1$.

By using the expression B_i, which indicates the ith line of matrix **B**, we obtain:

(109) $dQ_1 = -(dP_1)B_1\mathbf{L}_1$ or $-(dP_1) = dQ_1 / B_1\mathbf{L}_1$.

Substituting (109) into (108) yields:

(110) $d\mathbf{Q} = (dQ_1)(1 / B_1\mathbf{L}_1)\,\mathbf{BL}_1$.

Equation (110) can be rewritten as

$$(111)\quad \begin{pmatrix} \dfrac{dQ_1}{dQ_1} \\ \dfrac{dQ_2}{dQ_1} \end{pmatrix} = \frac{1}{b_1 - c_{12}cv_{12}} = \begin{pmatrix} b_1 - c_{12}cv_{12} \\ -c_{21} + b_2cv_{12} \end{pmatrix} = \begin{pmatrix} 1 \\ \dfrac{-c_{21} + b_2cv_{12}}{b_1 - cv_{12}cv_{12}} \end{pmatrix}.$$

The vector shown by (111) provides firm 1's equivalent conjectures on quantity associated with firm 1's conjectures on price. Generally in an n-brand market, firm i's conjectures on price, shown by vector $\mathbf{L}_i$, are equivalent to firm i's conjectures on quantity shown by vector $(1 / B_i\mathbf{L}_i)\,\mathbf{BL}_i$. From (111), firm 1's conjectural variation on firm 2's quantity, $dQ_2 / dQ_1 \equiv v_{12}$, is expressed as

$$(112)\quad v_{12} = (-c_{21} + b_2cv_{12}) / (b_1 - c_{12}cv_{12}).$$

Inversely, firm 1's conjectural variation on firm 2's price, $dP_2 / dP_1 \equiv cv_{12}$, is expressed as

$$(113)\quad cv_{12} = (c_{21} + b_1v_{12}) / (b_2 + c_{12}v_{12}).$$

From (113), firm 1's conjecture on price corresponding to firm 1's perfectly collusive conjecture on quantity ($v_{12} = 1$) is shown as

$$(114)\quad cv_{12} = (c_{21} + b_1) / (b_2 + c_{12}).$$

Similarly, from (113), firm 1's conjecture on price corresponding to firm 1's perfectly competitive conjecture on quantity ($v_{12} = -1$) is shown as

$$(115)\quad cv_{12} = (c_{21} - b_1) / (b_2 - c_{12}).$$

See Liang (1989) for a similar discussion.

In connection with the preceding discussion, we can define such conjecture terms as dP_1 / dQ_1. Dixit (1988) and Baldwin and Krugman (1988) introduced this conjecture term into their product-differentiated oligopoly models for trade in automobiles and random access memory, respectively. The derivative dP_1 / dQ_1 represents the firm's belief about the effect on its price if it sells another unit, taking into account its conjectures about the output responses of other firms (Dixit 1988). McCorriston and Sheldon (1996) applied Dixit's type of model to the banana market in Germany. Tiffin and Dawson (1997; see Chapter 13) applied Dixit's type of model to the United Kingdom's frozen-potato-product sector. Dixit's modeling is another approach to differentiated oligopoly.

3. Identification Problems. There are two conjectural variations (conduct parameters) in a two-brand market, but in general there are $n\ (n - 1)$ conjectural variations in an n-brand market. Therefore, we have an identification problem in estimating conjectural variations in a market with more than two brands.[17] More specifically, we should simultaneously estimate n demand functions expressed by $Q_i = a_i - b_iP_i + \Sigma_{j\neq i}c_{ij}P_j$ and n price-reaction functions expressed by

$$(116)\quad P_i = -a_i / (-2b_i + \Sigma_{j\neq i}c_{ij}cv_{ij}) - [1 / (-2b_i + \Sigma_{j\neq i}c_{ij}cv_{ij})]\ \Sigma_{j\neq i}c_{ij}P_j + [(-b_i + \Sigma_{j\neq i}c_{ij}cv_{ij}) / (-2b_i + \Sigma_{j\neq i}c_{ij}cv_{ij}j)]MC_i.$$

Liang (1989) and Kinoshita, Suzuki, and Kaiser (2002) did not have an identification problem because they used two-brand models. Kadiyali, Vilcassim, and Chintagunta (1999), on the other hand, employed a three-brand model and solved the problem by adding the following constraint:

$$(117)\quad cv_{13} / cv_{31} = c_{13} / c_{31}.$$

Peterson and Cotterill (1998) reviewed possibilities for identifying conjectural variations in many alternative demand-system models with n brands.

Another approach to this problem is given by Cotterill (1994, 2001); Cotterill and Putsis (2000); Cotterill, Putsis, and Dhar (2000; see Chapter 12); and Vickner and Davies (1999). They separated demand functions from price-reaction functions and then estimated n demand functions as one independent system and n price-reaction functions as another system. Thus, they avoided estimation of conjectural variations, and the "actual" price-reaction parameters obtained from estimation of price-reaction

17 The identification problem means that the number of unknown conjectural-variation parameters exceeds the number of equations.

functions were put into the conjectural-variations terms in the following relationship for the "perceived" own-price elasticity:

(118) $dQ_i / dP_i = \partial Q_i/\partial P_i + \Sigma_{j\neq i} (\partial Q_i / \partial P_j) (dP_j / dP_i) = -b_i + \Sigma_{j\neq i} c_{ij} cv_{ij}$.

This approach considers the firm's conjecture to be equal to the rival's actual price reaction. However, this approach does not satisfy the *ccv* (consistent conjectural variation) condition. For instance, in the case of a two-brand market, the following two relationships must simultaneously hold to realize a *ccv*:

(119) $cv_{12} = -c_{21} / (-2b_2 + c_{21} cv_{21})$ and

(120) $cv_{21} = -c_{12} / (-2b_1 + c_{12} cv_{12})$.

The solution (Liang 1989) is:

(121) $\c{c}cv_{ij} = [-b_i b_j + \{(b_i b_j) (b_i b_j - c_{ij} c_{ji})\}^{1/2}] / (-b_j c_{ij})$.

Therefore, treating the price-reaction parameters as equal to relevant *cv* parameters is not a sufficient condition for a *ccv*. In addition, because price-reaction functions are derived from the necessary condition for profit maximization using certain demand functions, there should be parameter constraints between the two functions. More specifically, consider the following demand and price-reaction functions again.

(122) $Q_i = a_i - b_i P_i + \Sigma_{j\neq i} c_{ij} P_j$.

(123) $P_i = -a_i/(-2b_i + \Sigma_{j\neq i} c_{ij} cv_{ij}) - [1 / (-2b_i + \Sigma_{j\neq i} c_{ij} cv_{ij})] \Sigma_{j\neq i} c_{ij} P_j$
$+ [(-b_i + \Sigma_{j\neq i} c_{ij} cv_{ij}) / (-2b_i + \Sigma_{j\neq i} c_{ij} cv_{ij})] MC_i$.

The separate-estimation approach ignores these parameter constraints. See Nevo (1998) for a more complete discussion of the identification problem in product-differentiated oligopoly models.

Concluding Remarks

Although mergers and absorptions are frequent occurrences today in many industries, including the food industry, the importance of concentration ratios is diminishing when determining whether competition is precluded. As the world economy becomes more global, monopoly within a country is no longer a monopoly in a free-trade market. IO analyses increasingly rely on case-by-case reviews rather than on concentration ratios to determine whether welfare losses have been or will be generated. Therefore, NEIO's role is becoming increasingly important because it can provide a framework and tools with which to estimate actual degrees of market power.

NEIO's theoretical framework and analytical tools have not been systematically explained in any previous literature. The primary function of the NEIO framework is a generalization of conventional perfect-competition (monopoly) models to intermediate imperfect-competition models that can be empirically estimated. We divided the NEIO models into two categories: quantity-setting homogeneous oligopoly (a generalization of the Cournot model) and price-setting product-differentiated oligopoly (a generalization of the Bertrand model). In addition, we presented a number of types of generalization methods for conventional perfect-competition (monopoly) models for many topics.

The NEIO approach presents several challenges. The biggest challenge is how to develop an improved empirical model of bilateral oligopoly that incorporates both horizontal and vertical competition. Because there are many bilateral oligopolies in the real world, this is an important problem to address even though finding a solution may be difficult. Related to bilateral oligopoly, as each stage of the food system becomes more oligopolistic, how to evaluate countervailing power is also an important topic because such power can create a near-competitive price even in bilateral monopoly. A subsequent challenge is harmonization of NEIO with game-theoretic NIO approaches. Efforts to incorporate dynamic aspects into NEIO models (e.g., Karp and Perloff 1989, 1993; Deodhar and Sheldon 1996; Katchova, Sheldon, and Miranda 2005) should be intensified. A third challenge is how to develop a more theoretically acceptable and empirically feasible model for analyzing price-setting product-differentiated oligopoly. Chapter 14 addresses these and other challenges for NEIO in more detail.

References

Abbott, P.C., P.M. Patterson, and A. Reca. "Imperfect Competition and Exchange Rate Pass-through in the Food Processing Sector." *American Journal of Agricultural Economics* 75 (1993):1226–1230.

Alston, J.M., and R. Gray. "State Trading versus Export Subsidies: The Case of Canadian Wheat." *Journal of Agricultural and Resource Economics* 25 (2000):51–67.

Alvarez, A.M., E.G. Fidalgo, R.J. Sexton, and M. Zhang. "Oligopsony Power with Uniform Spatial Pricing: Theory and Application to Milk Processing in Spain." *European Review of Agricultural Economics* 26 (1999):229–243.

Amos, G., L.S. Karp, and J.M. Perloff. "Estimating Coke and Pepsi's Price and Advertising Strategies." *Journal of Business and Economic Statistics* 18 (2000):398–409.

Appelbaum, E. "The Estimation on the Degree of Oligopoly Power." *Journal of Econometrics* 19 (1982):287–299.

Azzam, A.M. "Testing the Competitiveness of Food Price Spreads." *Journal of Agricultural Economics* 43 (1992):248–256.

———. "Estimating the Degree of Dominance in a Bilateral Oligopoly." *Applied Economics Letters* 3 (1996):209–211.

———. "Market Transparency and Market Structure: The Livestock Mandatory Reporting Act of 1999." *American Journal of Agricultural Economics* 85 (2003):387–395.

Azzam, A.M., and E. Pagoulatos. "Testing Oligopolistic and Oligopsonistic Behavior: An Application to the U.S. Meat-Packing Industry." *Journal of Agricultural Economics* 41 (1990):362–370.

———. "Measurement of Technological and Market Structure-Alternative Methodological Perspectives: Discussion." *American Journal of Agricultural Economics* 81 (1999):644–646.

Azzam, A.M., and J.R. Schroeter. "Implications of Increased Regional Concentration and Oligopsonistic Coordination in the Beef Packing Industry." *Western Journal of Agricultural Economics* 16 (1991):374–381.

———. "The Tradeoff between Oligopsony Power and Cost Efficiency in Horizontal Consolidation: An Example from Beef Packing." *American Journal of Agricultural Economics* 77 (1995):825–836.

Bain, J.S. *Barriers to New Competition*. Cambridge MA: Harvard University Press, 1956.

Baldwin, R.E., and P.R. Krugman. "Market Access and International Competition: A Simulation Study of 16K Random Access Memories." *Empirical Methods for International Trade*. R. Feenstra, ed., pp. 171–197. Boston MA: MIT Press, 1988.

Baumer, D.L., R.T. Masson, and R.A. Masson. "Curdling the Competition: An Economic and Legal Analysis of the Antitrust Exemption for Agriculture." *Villanova Law Review* 31 (1986):183–252.

Baumol, W.J. "Contestable Markets—An Uprising in the Theory of Industry Structure." *American Economic Review* 72 (1982):1–15.

Baumol, W.J., J.C. Panzar, and R.D. Willig. *Contestable Markets and the Theory of Industry Structure*. Oxford England: Oxford University Press, 1982.

Bettendorf, L., and F. Verboven. "Incomplete Transmission of Coffee Bean Prices: Evidence from the Netherlands." *European Review of Agricultural Economics* 27 2000):1–16.

Bhuyan, S., and R.A. Lopez. "Welfare Losses under Oligopoly Regimes: The U.S. Food and Tobacco Manufacturing Industries." *Journal of Agricultural and Applied Economics* 27 (1995):577–587.

———. "Oligopoly Power in the Food and Tobacco Industries." *American Journal of Agricultural Economics* 79 (1997):1035–1043.

———. "Oligopoly Power and Allocative Efficiency in U.S. Food and Tobacco Industries." *Journal of Agricultural Economics* 49 (1998a):434–442.

———. "What Determines Welfare Losses from Oligopoly Power in the Food and Tobacco Industries." *Agricultural and Resource Economics Review* 27 (1998b):258–265.

Bowley, A.L. *The Mathematical Groundwork of Economics.* Oxford England: Clarendon Oxford University Press, 1924.

Brennan, T.J. "The Economics of Competition Policy: Recent Developments and Cautionary Notes in Antitrust and Regulation." Washington DC: Resources for the Future Discussion Paper 00-07, 2000.

Bresnahan, T.F. "Duopoly Models with Consistent Conjectures." *American Economic Review* 71 (1981):934–945.

———. "The Oligopoly Solution Concept Is Identified." *Economics Letters* 10 (1982):87–92.

———. "Empirical Studies of Industries with Market Power." *Handbook of Industrial Organization*, vol. 2. R. Schmalensee and R.D. Willig, ed., pp. 1011–1157. Amsterdam Netherlands: Elsevier, 1989.

Buschena, D.V., and J.M. Perloff. "The Creation of Dominant Firm Market Power in the Coconut Oil Export Market." *American Journal of Agricultural Economics* 73 (1991):1000–1008.

Cabral, L.M.B. "Conjectural Variations as a Reduced Form." *Economic Letters* 49 (1995):397–402.

Carlton, D.W., and J.M. Perloff. *Modern Industrial Organization*, fourth edition. Boston MA: Addison-Wesley, 2004.

Chen, C., B.A. McCarl, and C. Chang. "The Potential Impacts of Government Intervention in the International Rice Market." Unpublished, 2003.

Chen, K., and J. Weerahewa. "Regulation, Market Power, and Advertising Effectiveness." *Agricultural and Resource Economics Review* 27 (1998):209–217.

Chen, Z., and R. Lent. "Supply Analysis in an Oligopsony Model." *American Journal of Agricultural Economics* 74 (1992):973–979.

Clarke, R. *Industrial Economics.* Oxford England, Malden MA: Blackwell, 1985.

Clarke, R., and S.W. Davies. "Market Structure and Price-Cost Margins." *Economica* 49 (1982):277–287.

Connor, J.M. "Empirical Challenges in Analyzing Market Performance in the U.S. Food System." *American Journal of Agricultural Economics* 72 (1990):1219–1226.

Corts, K.S. "Conduct Parameters and the Measurement of Market Power." *Journal of Econometrics* 88 (1999):227–250.

Cotterill, R.W. "Scanner Data: New Opportunities for Demand and Competitive Strategy Analysis." *Agricultural and Resource Economics Review* 23 (1994):125–139.

———. "Neoclassical Explanations of Vertical Organization and Performance of Food Industries." *Agribusiness: An International Journal* 17 (2001):33–57.

Cotterill, R.W., and W.P. Putsis. "Market Share and Price Setting Behavior for Private Labels and National Brands." *Review of Industrial Organization* 17 (2000):17–39.

Cotterill, R.W., W.P. Putsis, and R. Dhar. "Assessing the Competitive Interaction between Private Labels and National Brands." *Journal of Business* 73 (2000):109–137.

Cranfield, J.A.L., and W. Goddard. "Open Economy and Processor Oligopoly Power Effects of Beef Advertising in Canada." *Canadian Journal of Agricultural Economics* 47 (1999):1–19.

Delrome, F., and D. van der Mensbrugghe. "Assessing the Role of Scale Economies and Imperfect Competition in the Context of Agricultural Trade Liberalization: A Canadian Case Study." *OECD Economic Studies* (Organisation for Economic Co-operation and Development) (1990):205–236.

Deodhar, S.Y., and I.M. Sheldon. "Is Foreign Trade (Im)perfectly Competitive? An Analysis of the German Market for Banana Imports." *Journal of Agricultural Economics* 46 (1995):336–348.

———. "Estimation of Imperfect Competition in Food Marketing: A Dynamic Analysis of the German Banana Market." *Journal of Food Distribution Research* 27 (1996):1–10.

———. "Market Power in the World Market for Soymeal Exports." *Journal of Agricultural and Resource Economics* 22 (1997):78–86.

Dhar, T., and R.W. Cotterill. *Price Transmission in Differentiated Product Market Channels: A Study of the Boston Fluid Milk Market and the North East Dairy Compact.* Storrs CT: University of Connecticut Food Marketing Policy Center Research Report 67, 2002.

Dixit, A. "Optimal Trade and Industrial Policies for the U.S. Automobile Industry." *Empirical Methods for International Trade*. R. Feenstra, ed., pp. 141–165. Boston MA: MIT Press, 1988.

Driskill, R., and S. McCafferty. "Dynamic Duopoly with Output Adjustment Costs: A Differential Game Approach." *Journal of Economic Theory* 49 (1989):324–338.

Durham, C.A., and R.J. Sexton. "Oligopsony Potential in Agriculture: Residual Supply Estimation in California's Processing Tomato Market." *American Journal of Agricultural Economics* 74 (1992):962–972.

Egaitsu, F., and Higuchi, T., ed. *Industrial Organization of Agribusiness*. Tokyo Japan: Tokyo University Press, 1995 (in Japanese).

Frisch, R. "Monopoly-Polypoly: The Concept of Force in the Economy." *International Economic Papers* 1 (1933):23–26.

Fulton, M., and Y. Tang. "Testing the Competitiveness of a Multistage Food Marketing System: The Canadian Chicken Industry." *Canadian Journal of Agricultural Economics* 47 (1999):225–250.

Galbraith, J.K. *American Capitalism: The Concept of Countervailing Power*. Boston MA: Houghton-Mifflin, 1952.

Goddard, E.W., and M.L. McCutcheon. "Optimal Producer Investment in Generic Advertising: The Case of Fluid Milk in Ontario and Quebec." *Canadian Journal of Agricultural Economics* 41 (1993):329–347.

Gohin, A., and H. Guyomard. "Measuring Market Power for Food Retail Activities: French Evidence." *Journal of Agricultural Economics* 51 (2000):181–195.

Gollop, F.M., and M.J. Roberts. "Firm Interdependence in Oligopolistic Markets." *Journal of Econometrics* 10 (1979):313–331.

Gopinath, M., and J. Wu. "Environmental Externality and the Optimal Level of Market Power." *American Journal of Agricultural Economics* 81 (1999):825–833.

Hashimoto, H. "A Spatial Nash Equilibrium Model." *Spatial Price Equilibrium: Advances in Theory, Computation and Application*. P.T. Harker, ed., pp. 20–40. New York NY: Springer-Verlag, 1985.

Helpman, E., and P.R. Krugman. *Trade Policy and Market Structure*. Boston MA: MIT Press, 1989.

Hicks, J.R. "Annual Survey of Economic Theory: The Theory of Monopoly." *Econometrica* 3 (1935):1–20.

Higaki, Y., K. Gunjal, and G. Coffin. "Competition in the Japanese Wholesale Potato Market." *Agribusiness: An International Journal* 17 (2001):319–332.

Holloway, G.J. "The Farm-Retail Prices Spread in an Imperfectly Competitive Food Industry." *American Journal of Agricultural Economics* 73 (1991):979–989.

Hosoe, M., ed. *Applied Microeconomics Analyses*. Tokyo Japan: Yuhikaku, 1990 (in Japanese).

Huang, S., and R.J. Sexton. "Measuring Returns to Innovation in an Imperfectly Competitive Market: Application to Mechanical Harvesting of Processing Tomatoes in Taiwan." *American Journal of Agricultural Economics* 78 (1996):558–571.

Hwang, H. "Intra-Industry Trade and Oligopoly: A Conjectural Variations Approach." *Canadian Journal of Economics* 17 (1984):126–137.

Hwang, H., and C. Mai. "On the Equivalence of Tariffs and Quotas under Duopoly: A Conjectural Variation Approach." *Journal of International Economics* 24 (1988):373–380.

Hyde, C.E., and J.M. Perloff. "Can Market Power Be Estimated?" *Review of Industrial Organization* 10 (1995):465–485.

Irie, M., N. Suzuki, and K. Maeda. "Dynamic Oligopoly with Conjectural Variations." *Science Bulletin of the Faculty of Agriculture, Kyushu University* 59 (2004):247–254 (in Japanese).

Itaya, J., and K. Shimomura. "A Dynamic Conjectural Variations Model in the Private Provision of Public Goods: A Differential Game Approach." *Journal of Public Economics* 81 (2001):153–172.

Ito, M., K. Kiyono, M. Okuno, and K. Suzumura. *Economic Analyses of Industry Policy*. Tokyo Japan: Tokyo University Press, 1988 (in Japanese).

Iwata, G. "Measurement of Conjectural Variations in Oligopoly." *Econometrica* 42 (1974):947–966.

Kadiyali, V., N. Vilcassim, and P. Chintagunta. "Product Line Extensions and Competitive Market Interactions: An Empirical Analysis." *Journal of Econometrics* 89 (1999):339–363.

Kano, H. "Estimating the Effectiveness of Controlling Production and Classified Marketing of Mandarin Oranges in Japan." Masters thesis, Kyushu University, 1999 (in Japanese).

Karp, L.S., and J.M. Perloff. "Dynamic Oligopoly in the Rice Export Market." *Review of Economics and Statistics* 71 (1989):462–470.

———. "A Dynamic Model of Oligopoly in the Coffee Export Market." *American Journal of Agricultural Economics* 75 (1993):448–457.

Katchova, A.L., I.M. Sheldon, and M.J. Miranda. "A Dynamic Model of Oligopoly and Oligopsony in the U.S. Potato-Processing Industry." *Agribusiness: An International Journal* 21 (2005):409–428.

Kawaguchi, T., N. Suzuki, and H.M. Kaiser. "A Spatial Equilibrium Model for Imperfectly Competitive Milk Markets." *American Journal of Agricultural Economics* 79 (1997):851–859.

———. "Evaluating Class I Differentials in the New Federal Milk Marketing Order System." *Agribusiness: An International Journal* 17 (2001):527–538.

Kawamura, T. "Theoretical Models of Cooperatives' Collective Marketing." *Theories and Analytical Frameworks for the Food System*, vol. 6. M. Takahashi, ed., pp. 61–79. Tokyo Japan: Norintokeikyokai, 2001 (in Japanese).

Kinnucan, H.W. "Optimal Generic Advertising Decisions in Supply-Managed Industries: Clarification and Some Further Results." *Canadian Journal of Agricultural Economics* 47 (1999):57–66.

———. "Optimal Generic Advertising in an Imperfectly Competitive Food Industry with Variable Proportions." *Agricultural Economics* 29 (2003):143–158.

Kinoshita, J., N. Suzuki, and H.M. Kaiser. "Explaining Pricing Conduct in a Product-Differentiated Oligopolistic Market: An Empirical Application of a Price Conjectural Variations Model." *Agribusiness: An International Journal* 18 (2002):427–436.

———. "An Economic Evaluation of Recombinant Bovine Somatotropin Approval in Japan." *Journal of Dairy Science* 87 (2004a):1565–1577.

———. "The Degree of Vertical and Horizontal Competition among Dairy Cooperatives, Processors, and Retailers in Japanese Milk Markets." Unpublished, Cornell University, 2004b.

Kinoshita, J., N. Suzuki, T. Kawamura, Y. Watanabe, and H.M. Kaiser. "Estimating Own and Cross Brand Price Elasticities and Price-Cost Margin Ratios using Store-Level Daily Scanner Data." *Agribusiness: An International Journal* 17 (2001):515–525.

Kojima, Y. "Competitiveness of the Japanese Bread Baking Industry and Downward Rigidity of Price: A Conjectural Variations Approach." *Journal of Rural Economics* 77 (2005):23–35 (in Japanese).

Konishi, T., ed. *New Trends in Industrial Organization and Competition Policy*, Tokyo Japan: Koyoshobo, 1994 (in Japanese).

Koontz, S.R., and P. Garcia. "Meat-Packer Conduct in Fed Cattle Pricing: Multiple-Market Oligopsony Power." *Journal of Agricultural and Resource Economics* 22 (1997):87–103.

Koontz, S.R., P. Garcia, and M.A. Hudson. "Meatpacker Conduct in Fed Cattle Pricing: An Investigation of Oligopsony Power." *American Journal of Agricultural Economics* 75 (1993):537–548.

Liang, J.N. "Price Reaction Function and Conjectural Variations: An Application to the Breakfast Cereal Industry." *Review of Industrial Organization* 4 (1989):31–58.

Liu, D.J., C. Sun, and H.M. Kaiser. "Market Conduct under Government Price Intervention in the U.S. Dairy Industry." *Journal of Agricultural and Resource Economics* 20 (1995):301–315.

Lopez, R.A., and Z. You. "Determinants of Oligopsony Power: The Haitian Coffee Case." *Journal of Development Economics* 41 (1993):275–284.

Lopez, R.E. "Measuring Oligopoly Power and Production Responses of the Canadian Food Processing Industry." *Journal of Agricultural Economics* 35 (1984):219–230.

Love, H.A., and E. Murniningtyas. "Measuring the Degree of Market Power Exerted by Government Trade Agencies." *American Journal of Agricultural Economics* 74 (1992):546–555.

Madhavan, A.N., R.T. Masson, and W.H. Lesser. "Cooperation for Monopolization?: An Empirical Analysis of Cartelization." *Review of Economics and Statistics* 76 (1994):161–175.

Maeda, K. "A Policy Simulation Model of International Trade under Imperfect Competition: A Spatial Equilibrium Analysis of International Wheat Trade by the Use of a Mixed Complementarity Problem." *Journal of Rural Economics* 73 (2001):119–132 (in Japanese).

Maeda, K., N. Suzuki, and H.M. Kaiser. "An Economic Evaluation of the New Agricultural Trade Negotiations: A Nonlinear Imperfectly Competitive Spatial Equilibrium Approach." Ithaca NY: Cornell University Working Paper 2001-14, 2001a.

———. "Measuring the Effects of Eliminating State Trading Enterprises on the World Wheat Sector." Ithaca NY: Cornell University Working Paper 2001-17, 2001b.

Maier, L. "Policy Analysis in an Imperfectly Competitive Market: A Conjectural Variations Model for the Food Manufacturing Industry." Ithaca NY: Cornell University Working Paper in Agricultural Economics 90-10, 1990.

———. "The Costs and Benefits of U.S. Agricultural Policies with Imperfect Competition in Food Manufacturing." Cornell University Ph.D. thesis, 1991.

Martin, S. *Advanced Industrial Economics*. Oxford England, Cambridge MA: Blackwell, 1993.

———. *Advanced Industrial Economics*, second edition. Oxford England, Cambridge MA: Blackwell, 1993.

Matsuda, T., and I. Kurokawa. "Measurement of Oligopoly Behavior in the Japanese Onion Markets." *Journal of Rural Economics* 68 (1996):129–142.

McCorriston, S., C.W. Morgan, and A.J. Rayner. "Processing Technology, Market Power, and Price Transmission." *Journal of Agricultural Economics* 49 (1998):185–201.

———. "Price Transmission: The Interaction between Market Power and Return to Scale." *European Review of Agricultural Economics* 28 (2001):143–159.

McCorriston, S., and I.M. Sheldon. "Incorporating Industrial Organization into Agricultural Trade Modelling." *Agricultural Markets: Mechanisms, Failures and Regulations*. D. Martimort, ed., pp. 313–330. Amsterdam Netherlands: North-Holland Publishing, 1996.

McCorriston, S., I.M. Sheldon, and J.G. Hirschberg. "Simulating Changes in Conjectural Variations: Tariff and Quota Nonequivalence in Food Processing." *American Journal of Agricultural Economics* 75 (1993):1221–1225.

Milgrom, P., and J. Roberts. "Predation, Reputation, and Entry Deterrence." *Journal of Economic Theory* 27 (1982):280–312.

Millan, J.A. "Short-Run Equilibrium and Market Power in the Spanish Food, Drink and Tobacco Industries." *European Review of Agricultural Economics* 26 (1999):229–243.

Muth, M.K., and M.K. Wohlgenant. "Measuring the Degree of Oligopsony Power in the Beef Packing Industry in the Absence of Marketing Input Quantity Data." *Journal of Agricultural and Resource Economics* 24 (1999a):299–312.

———. "A Test for Market Power Using Marginal Input and Output Prices with Application to the U.S. Beef Processing Industry." *American Journal of Agricultural Economics* 81 (1999b):638–643.

Nakashima, Y. "Industrial Organization Analyses of the Food System." *Theories and Analytical Frameworks for the Food System*, vol. 1. M. Takahashi, ed. Tokyo Japan: Norintokeikyokai, 2002 (in Japanese).

Nelson, C.H., and B.A. McCarl. "Including Imperfect Competition in Spatial Equilibrium Models." *Canadian Journal of Agricultural Economics* 32 (1984):55–70.

Nevo, A. "Identification of the Oligopoly Solution Concept in a Differentiated-Products Industry." *Economics Letters* 59 (1998):391–395.

Niiyama, Y. "Food System Structures and Vertical Coordination." *Theories and Analytical Frameworks for the Food System*, vol. 1. M. Takahashi, ed., pp. 313–328. Tokyo Japan: Norintokeikyokai, 2002 (in Japanese).

Nishimura, K. *Microeconomics*. Tokyo Japan: Toyokeizaishinposha, 1990 (in Japanese).

Odagiri, H. *New Industrial Organization: Theory, Empirical Analyses, and Policy*. Tokyo Japan: Yuhikaku, 2001(in Japanese).

Park, T., and A. Weliwita. "Competitive Behavior in the U.S. Food Retailing Industry." *Canadian Journal of Agricultural Economics* 47 (1999):45–55.

Paul, C.J.M. "Production Structure and Trends in the U.S. Meat and Poultry Products Industries." *Journal of Agricultural and Resource Economics* 24 (1999a):281–298.

———. "Aggregation and the Measurement of Technological and Market Structure: The Case of the U.S. Meatpacking Industry." *American Journal of Agricultural Economics* 81 (1999b):624–629.

———. "Market and Cost Structure in the U.S. Beef Packing Industry: A Plant-Level Analysis." *American Journal of Agricultural Economics* 83 (2001):64–76.

Peterson, E.B., and J.M. Connor. "A Comparison of Oligopoly Welfare Loss Estimates for U.S. Food Manufacturing." *American Journal of Agricultural Economics* 77 (1995):300–308.

Peterson, E.B., and R.W. Cotterill. *Incorporating Flexible Demand Systems in Empirical Models of Market Power*. Storrs CT: University of Connecticut Food Marketing Policy Center Research Report 43, 1998.

Pfaffermayer, M. "Conjectural Variation Models and Supergames with Price-Competition in a Differentiated Product Oligopoly." *Journal of Economics* 70 (1999):209–326.

Powers, N.J. "Evaluating Orange Growers' Exercise of Market Power with Marketing Order Volume Control Regulations." *Journal of Agricultural Economics Research* 44 (1993):33–42.

Purcell, W.D. "Economics of Consolidation in the Beef Sector: Research Challenges." *American Journal of Agricultural Economics* 72 (1990):1210–1218.

Richards, T.J., A. Kagan, P. Mischen, and R. Adu-Asamoah. "Marketing Order Suspensions and Fresh Lemon Retail-FOB Margins." *Journal of Agricultural and Applied Economics* 28 (1996):263–277.

Rogers, R.T. "The Role of Marketing Cooperatives in Increasingly Concentrated Agricultural Markets: Reaction." *Cooperatives: Their Importance in the Future Food and Agricultural System*. M. Cook, R. Torgerson, T. Sporleder, and D. Padberg, eds. Washington DC: National Council of Farmer Cooperatives, 1997.

Rogers, R.T., and R.J. Sexton. "Assessing the Importance of Oligopsony Power in Agricultural Markets." *American Journal of Agricultural Economics* 76 (1994):1143–1150.

Saito, O. *Innovation and Firm Conduct in the Food System.* Tokyo Japan: Norintokei-kyokai, 1999 (in Japanese).

Schluep, Isabelle. "The Law and Economics of 'Consumer Only' Financed Export Subsidies: A Context for the WTO Panel on Canadian Dairy Pricing Policy." Masters thesis, Cornell University, 1999.

Schmalensee, R. "The New Industrial Organization and the Economic Analysis of Modern Markets." *Advances in Economic Theory.* W. Hildenbrand, ed., pp. 253–285. Cambridge England: Cambridge University Press, 1982.

Schroeter, J.R. "Estimating the Degree of Market Power in the Beef Packing Industry." *Review of Economics and Statistics* 70 (1988):158–162.

———. "Measuring Market Power in Food-Processing Industries: Discussion." *American Journal of Agricultural Economics* 72 (1990):1227–1229.

Schroeter, J.R., and A.M. Azzam. "Measuring Market Power in Multiproduct Oligopolies: The U.S. Meat Industry." *Applied Economics* 22 (1990):1365–1376.

———. "Marketing Margins, Market Power, and Price Uncertainty." *American Journal of Agricultural Economics* 73 (1991):990–999.

Schroeter, J.R., A.M. Azzam, and M. Zhang. "Measuring Market Power in Bilateral Oligopoly: The Wholesale Market for Beef." *Southern Economic Journal* 66 (2000):526–547.

Schumpeter, J.A. *Capitalism, Socialism and Democracy.* New York NY: Harper, 1942.

Sexton, R.J. "Imperfect Competition in Agricultural Markets and the Role of Cooperatives: A Spatial Analysis." *American Journal of Agricultural Economics* 72 (1990):709–720.

———. "Industrialization and Consolidation in the U.S. Food Sector: Implications for Competition and Welfare." *American Journal of Agricultural Economics* 82 (2000):1087–1104.

Sexton, R.J., and N. Lavoie. "Food Processing and Distribution: An Industrial Organization Approach." *Handbook of Agricultural Economics,* vol. 1. B.L. Gardner and G.C. Rausser, eds., pp. 865–932. Amsterdam Netherlands: Elsevier, 2001.

Sexton, R.J, and M. Zhang. "An Assessment of the Impact of Food Industry Market Power on U.S. Consumers." *Agribusiness: An International Journal* 17 (2001):59–79.

Sheldon, I., and R. Sperling "Estimating the Extent of Imperfect Competition in the Food Industry: What Have We Learned?" *Journal of Agricultural Economics* 54 (2003):89–109.

Shogenji, S. "Competition Policy and Social Regulation for the Food System." *Journal of Food System Research* 8 (2001):12–22 (in Japanese).

Shono, C., and T. Kawaguchi. "Studies on the Spatial Equilibrium Model of International Trade under the Tariff Quota System with Specific and Ad Valorem Duties: The Case of Oligopolistic International Trade." *Science Bulletin of the Faculty of Agriculture, Kyushu University* 54 (1999):85–96 (in Japanese).

Slade, M.E. "Conjectures, Firm Characteristics, and Market Structure: An Empirical Assessment." *International Journal of Industrial Organization* 4 (1986):347–369.

Stiegert, K.W., A. Azzam, and B.W. Brorsen. "Markdown Pricing and Cattle Supply in the Beef Packing Industry." *American Journal of Agricultural Economics* 75 (1993):549–558.

Stigler, G.J. "The Theory of Economic Regulation." *Bell Journal of Economics* 2 (1971):3–21.

Sudhir, K. "Structural Analysis of Manufacturer Pricing in the Presence of a Strategic Retailer." *Marketing Science* 20 (2001):244–264.

Sullivan, D. "Testing Hypotheses about Firm Behavior in the Cigarette Industry." *Journal of Political Economy* 93 (1985):586–598.

Suzuki, N. *Econometric Analyses of Imperfectly Competitive Milk Markets*. Tokyo Japan: Norintokeikyokai, 1994 (in Japanese).

———. *Econometric Analyses of Imperfectly Competitive Food Systems*. Tokyo Japan: Norintokeikyokai, 2002 (in Japanese).

Suzuki, N., and D.H. Judson. "An Assessment of Deficiency Payment to Milk Producers in Japan." *Western Journal of Agricultural Economics* 16 (1991):119–131.

Suzuki, N., and H.M. Kaiser. "Basic Mechanisms of Japanese Dairy Policy and Milk Market Models: A Comparison with U.S. Dairy Policy." *Journal of Dairy Science* 77 (1994):1746–1754.

———. "Imperfect Competition Models and Commodity Promotion: The Case of U.S. Generic Milk Advertising." *Journal of Agricultural and Applied Economics* 29 (1997):315–325.

———. "Market Impacts of Japanese Rice Policies with and without Supply Control." *Agribusiness: An International Journal* 14 (1998):355–362.

———. "Impacts of the Doha Round Framework Agreements on Dairy Policies." *Journal of Dairy Science* 88 (2005):1901–1908.

Suzuki, N., H.M. Kaiser, J.E. Lenz, and O.D. Forker. "An Analysis of U.S. Dairy Policy Deregulation Using an Imperfect Competition Model." *Agricultural and Resource Economic Review* 23 (1994a):84–93.

Suzuki, N., H.M. Kaiser, J.E. Lenz, O.D. Forker, and K. Kobayashi. "Evaluating Generic Milk Promotion Effectiveness with an Imperfect Competition Model." *American Journal of Agricultural Economics* 76 (1994b):296–302.

Suzuki, N., J. Kinoshita, and H.M. Kaiser. "Measuring the Degree of Market Distortion through Price Discrimination for Hidden Export Subsidies Generated by State Trading Enterprises." *Report on Economic Analysis and Investigation of State Trading Enterprises*. Japan International Agricultural Council, ed. Tokyo Japan: Japan International Agricultural Council, 2001.

Suzuki, N., J.E. Lenz, and O.D. Forker. "A Conjectural Variations Model of Reduced Japanese Milk Price Supports." *American Journal of Agricultural Economics* 75 (1993):210–218.

Suzumura, K., and J. Ishikawa. "Voluntary Export Restraints and Economic Welfare." *Japanese Economic Review* 48 (1996):176–186.

Takayama, T., and G.G. Judge. *Spatial and Temporal Price and Allocation Models*. Amsterdam Netherlands: North-Holland Publishing, 1971.

Thursby, M.C., and J.G. Thursby. "Strategic Trade Theory and Agricultural Markets: An Application to Canadian and U.S. Wheat Exports to Japan." *Imperfect Competition and Political Economy: The New Trade Theory in Agricultural Trade Research*. C.A. Carter, A.F. McCalla, and J.A. Sharples, ed., pp. 87–106. Boulder CO: Westview Press, 1990.

Tiffin, A.L., and P.J. Dawson. "Measuring Oligopolistic Distortion in the UK Frozen Potato Product Sector: A Calibration Modelling Approach." *Journal of Agricultural Economics* 48 (1997):300–312.

Tirole, J. *The Theory of Industrial Organization*. Boston MA: MIT Press, 1988.

Tsujimura, K. *Econometrics*. Tokyo Japan: Iwanamizensho, 1981 (in Japanese).

Uekusa, M., ed. *Japan's Industrial Organization*. Tokyo Japan: Yuhikaku, 1995 (in Japanese).
Varian, H.R. *Microeconomic Analysis*, second edition. New York NY: W.W. Norton, 1984.
Vickner, S.S., and S.P. Davies. "Estimating Market Power and Pricing Conduct in a Product-Differentiated Oligopoly: The Case of the Domestic Spaghetti Sauce Industry." *Journal of Agricultural and Applied Economics* 31 (1999):1–13.
Wann, J.J., and R.J. Sexton. "Imperfect Competition in Multiproduct Food Industries with Application to Pear Processing." *American Journal of Agricultural Economics* 74 (1992):980–990.
Weldegebriel, H.T. "Imperfect Price Transmission: Is Market Power Really to Blame?" *Journal of Agricultural Economics* 55 (2004):101–114.
Weliwita, A., and A. Azzam. "Identifying Implicit Collusion under Declining Output Demand." *Journal of Agricultural and Resource Economics* 21 (1996):235–246.
Williamson, O.E. "The Economics of Organization: The Transaction Cost Approach." *American Journal of Sociology* 87 (1981):548–577.
Wilson, W.W., and X.L. Casavant. "Some Market Power Implications of the Shipping Act of 1984: A Case Study of the U.S. to Pacific Rim Transportation Markets." *Western Journal of Agricultural Economics* 16 (1991):427–434.
Yanagawa, N. *Strategic Trade Policy: Application of Game Theory to Policy Analyses*. Tokyo Japan: Yuhikaku, 1998 (in Japanese).
Yano, M. *Applied Microeconomics*. Tokyo Japan: Iwanamishoten, 2001 (in Japanese).
Zhang, M., and R.J. Sexton. "Optimal Commodity Promotion when Downstream Markets Are Imperfectly Competitive." *American Journal of Agricultural Economics* 84 (2002):352–365.

PART II

FARMER OLIGOPOLY STUDIES

2

An Analysis of U.S. Dairy Policy Deregulation Using an Imperfect-Competition Model

Nobuhiro Suzuki, Harry M. Kaiser,
John E. Lenz, and Olan D. Forker

Since the General Agreement on Tariffs and Trade (GATT) was successfully reached in December 1993, it has been useful for the U.S. dairy industry to be proactive and consider the impact of deregulation of the U.S. dairy industry. There are three main federal programs that form the basis of U.S. dairy policy and will be affected by trade liberalization: the dairy price-support program, federal milk marketing orders, and import quotas. Domestic pressures also bring various programs into question, such as regional concerns over federal milk marketing orders and budgetary pressure on the dairy price-support program.

Most previous dairy-policy models (e.g., Kaiser, Streeter, and Liu 1988; Liu et al. 1990) have assumed an exogenous fluid differential because the dairy industry is regulated by federal milk marketing orders that obligate milk handlers to pay the minimum Class I (fluid) differential.[1] However, the fluid differential actually is not exogenous because, in addition to

---

This chapter originally appeared as an article titled "An Analysis of U.S. Dairy Policy Deregulation Using an Imperfect Competition Model" by Suzuki, Kaiser, Lenz, and Forker in *Agricultural and Resource Economics Review* in 1994. This material is reprinted by permission of the Northeastern Agricultural and Resource Economics Association and *Agricultural and Resource Economics Review*.

1 Federal marketing orders also set the minimum Class II and Class III prices. The M-W price has been used as the minimum Class III price for most orders since 1960. The M-W price is an estimate of the average price paid for manufacturing-grade (Grade B) milk by manufacturing plants in Minnesota and Wisconsin. Grade B milk is now a very small portion of the national market (approximately 6 percent), although it is still used to set M-W prices.

the premiums associated with the federal minimum prices, there are over-order payments that result from negotiations[2] between dairy cooperatives and processors (or manufacturers).[3] Most previous models have not accounted for over-order payments. An imperfect-competition model with an endogenous fluid price differential is necessary for estimating how large the fluid price differential might be without existing regulations.

While imperfect-competition models of the Japanese milk market have been developed by Suzuki, Lenz, and Forker (1993) and by Suzuki et al. (1994), no imperfect-competition models have been developed for the U.S. milk market.[4] In this chapter we present an imperfect-competition model with an endogenous fluid price differential to evaluate the market effects of deregulating the U.S. dairy industry. The usefulness of the model is demonstrated by comparing the results of dynamic simulations of this model with the results of a conventional exogenous fluid differential model.

We estimated the effective fluid-milk price (the M-W price + the minimum fluid price differential + any over-order payment) to measure the degree of competition in the U.S. milk market and incorporated a degree-of-competition measure into the U.S. dairy-policy model. The degree-of-competition measure, which is based on the effective fluid price differential (the minimum fluid price differential + any over-order payment), is considered to be an aggregate indicator of the degree of imperfection created by federal policies and the market power of cooperatives, processors, and manufacturers in the U.S. milk market. While over-order payments exist not only for Class I milk but also for Class II and Class III milk, over-order payments for Class II and Class III milk are included in the fluid differential in this analysis because of data limitations.

---

2 The Capper-Volstead Act accords cooperatives special treatment regarding collective bargaining.

3 In this chapter, the term "processor" refers to a processor of fluid milk while the term "manufacturer" refers to a manufacturer of nonfluid dairy products.

4 The American Agricultural Economics Association Task Force Report (1986) stated that "since the 1930s, agricultural economists have emphasized that some model besides pure competition is needed. But no one has yet proposed such a model in a form capable of generating comparative-static results concerning the effects of marketing orders as compared to no orders." (p. 34)

Theoretical Model

To measure the degree of imperfection, we defined a perfectly competitive market as a basis of comparison. One would expect a relatively uniform manufacturing-milk price nationwide. According to Robinson,

> Class II or manufacturing milk prices are approximately the same in all markets and are linked to the M-W price. Uniform pricing of manufacturing milk is necessary because products derived from surplus milk are easily transported between regions. Cheese, butter, and skim-milk powder produced in federal-order markets must compete with similar products manufactured from grade B milk in Minnesota and Wisconsin. Handlers operating in federal-order markets will not purchase surplus milk if it is priced higher than what unregulated plants pay for manufacturing milk in the Midwest." (Robinson 1989, p. 116)

Without cooperative market power and revenue pooling, individual farmers would compete with each other until the price difference between fluid and manufacturing milk disappeared except for modest locational differences. If a market did not have enough milk to meet local fluid uses, there would be some locational or transportation differentials paid for fluid milk even without marketing orders and cooperatives because fluid-milk plants would have to bring milk from further distances. Fluid-milk plants tend to be located near population centers while manufacturing plants tend to be located near farms because dairy products are less bulky to ship than raw or fluid milk.

We do not consider such possibilities in the current analysis because the number of deficit areas and the magnitude of fluid differentials in a perfectly competitive market is difficult to predict.[5] Several previous studies that tried to estimate welfare losses caused by marketing orders also assumed no differentials as a benchmark for comparison (Buxton 1977; Dahlgran 1980; Ippolito and Masson 1978; Masson and Eisenstat 1980).

If one specifies that, under imperfect competition, the role of dairy cooperatives is to allocate their raw-milk supply to fluid and manufacturing markets so as to maximize total milk sales revenues, the first-order condition is to equate marginal revenues from fluid and manufacturing milk. When cooperatives undertake processing themselves, manufacturing

---

5 McDowell, Fleming, and Spinelli (1990) looked at differentials that would be obtained in different markets under a multiple-base-point pricing system.

costs should be taken into account. For simplicity, our model does not incorporate these costs. Under perfect competition, the first-order condition is expressed as

(1) $$P_f = P_m$$

where P_f is the fluid-milk price and P_m is the manufacturing-milk price.

At the opposite extreme, the first-order condition for monopoly or collusion is

(2) $$P_f(1 - 1/\varepsilon_f) = P_m(1 - 1/\varepsilon_m)$$

where $\varepsilon_f = |(\partial Q_f / \partial P_f)(P_f / Q_f)|$ and $\varepsilon_m = |(\partial Q_m / \partial P_m)(P_m / Q_m)|$ are price elasticities of fluid- and manufacturing-milk demand in absolute value terms, Q_f is the aggregate quantity of fluid-milk demand, and Q_m is the aggregate quantity of manufacturing-milk demand.

To express an intermediate degree of imperfect competition, a degree-of-competition parameter, θ ($0 \leq \theta \leq 1$), is introduced. Then, equality across markets of "perceived" marginal revenue is expressed as

(3) $$P_f(1 - \theta_f/\varepsilon_f) = P_m(1 - \theta_m/\varepsilon_m), \text{ or}$$

(4) $$P_f + \theta_f\, Q_f/(\partial Q_f / \partial P_f) = P_m + \theta_m\, Q_m/(\partial Q_m / \partial P_m).$$

The parameter θ is considered an aggregate indicator of the degree of competition in the milk market. Marginal milk production costs do not enter Equation (3) or (4) because milk production is almost never controlled by cooperatives; rather, it is determined by individual farmer's responses to the blend prices they receive.

Although the degree-of-competition parameter is specified by cooperatives' revenue-maximizing behavior, it also is affected by the countervailing power of processors and manufacturers. Therefore, we note that the degree-of-competition parameter reflects both cooperatives' and processors' and manufacturers' market power. We do not have a way to explicitly incorporate processors' and manufacturers' oligopsonistic power into the current model.

If θ can be assumed to be the same for both fluid-milk and manufacturing-milk markets, one can identify a value of θ that satisfies Equation (3) or (4) with the values of the milk price elasticities estimated by demand functions and observations of P_f, P_m, Q_f, and Q_m. However, θ_m will probably be lower than θ_f because cooperatives face more competition in the manufacturing-milk market than in the fluid-milk market. Fluid-milk products are costlier to transport than manufactured milk products and, therefore, the geographical scope of markets for manufactured milk products generally will exceed that for fluid-milk products. Since there likely is more competition in the manufactured-product market than in the fluid market, the manufacturing-milk market is probably more competitive.

Instead of deriving θ with the assumption that $\theta_f = \theta_m$ by estimating both the fluid-milk and the manufacturing-milk demand equations, one could estimate the fluid-milk (or manufacturing-milk) demand equation and Equation (3) or (4), into which the manufacturing-milk (or fluid-milk) demand equation is substituted. The parameter θ is directly estimated as a coefficient of (3) or (4) using this method (Bresnahan 1982) and θ_f and θ_m can be separately identified. However, the coefficients for the manufacturing-milk (or fluid-milk) demand equation cannot be identified (See the Appendix).

We solve this problem by assuming that $\theta_m = 0$ and then solve for θ_f. The assumption that $\theta_m = 0$ is plausible because the manufacturing-milk price for each market is given as the M-W price and the M-W price is indirectly supported by government purchases of dairy products. In this analysis, we do not consider the fact that some orders, such as the one established for Chicago, obtain significant over-order Class III payments (Babb 1989) because we have no national average data on over-order payments for Class III milk. We use the assumption $\theta_m = 0$ and identify a value of θ_f that satisfies (3) or (4) by assuming that θ_f is constant in each time period and that cooperatives approximately realize the condition expressed by (3) or (4). This means that (3) and (4) can be replaced with

(5) $$P_f(1 - \theta_f/\varepsilon_f) = P_m \text{ or}$$

(6) $$P_f + \theta_f\ Q_f/(\partial Q_f / \partial P_f) = P_m.$$

The full imperfect-competition model is expressed as

Milk production: (7) $Q = f(BP)$

Fluid-milk demand: (8) $Q_f = g(P_f)$

Manufacturing-milk demand: (9) $Q_m = h(P_m)$

Milk sales-maximizing allocation: (10) $P_f + \theta_f Q_f/(\partial Q_f / \partial P_f) = P_m$

Milk uses identity: (11) $Q \equiv Q_f + Q_m + FUSE$

Blend price: (12) $BP = (P_f Q_f + P_m Q_m) / (Q - FUSE)$

where Q is aggregate milk production, BP is the blend price, and $FUSE$ is on-farm use of milk produced (assumed to be exogenous) with all other variables as previously defined. The other exogenous variables (feed price, income, advertising expenditures, and trend) are not included in the preceding simplified expressions. With six endogenous variables (Q, Q_f, Q_m, P_f, P_m, BP) and six equations, the model is complete. Because this model expresses farmers' supply and processors' and manufacturers' demand for raw milk, government purchases of dairy products and changes in commercial inventories are not treated separately; in other words, manufacturing-milk demand (Q_m) includes commercial manufacturing demand, government purchases of dairy products, and changes in commercial inventories on a milk-equivalent basis.

Empirical Model Estimation

Over-Order Payment Data

The effective fluid-milk price is equal to the M-W price plus the minimum Class I differential plus any over-order payments. Since the available data on over-order payments pertain only to over-order payments "announced" by the U.S. Department of Agriculture (USDA) (USDA annual issues) in thirty-five markets, it was difficult to collect the over-order payment data for all cooperatives over time and to construct a national average time-series data set. Instead, we estimated the effective fluid-milk price (P_f) by solving the blend price equation for P_f:

(13) $P_f = [BP(Q - FUSE) - P_m Q_m] / Q_f$.

The difference between the Class II and Class III price is minor and our model, like most previous ones, neglects it. However, since over-order payments exist not only for Class I milk but also for Class II and Class III milk, Equation (13) assigns the over-order payments for Class II and Class III milk to the fluid differential. Consequently, the estimated effective fluid prices may be higher than the actual ones because they include all premiums over the M-W price. To check for possible bias from using

Table 2.1. Differences between Effective Fluid-Milk Prices Estimated by Equation (13) and the "Announced" Cooperative Class I Prices Reported by the USDA (Simple Annual Average)

	Effective Fluid-Milk (Class I) Price		
Year	Estimates by Equation (13) (A) dollars/cwt	"Announced" Cooperative Prices in Thirty-Five Markets (B) dollars/cwt	Difference (A) – (B) dollars/cwt
1976	11.13	10.91	+0.22
1977	11.17	10.81	+0.36
1978	11.89	11.61	+0.28
1979	13.52	13.25	+0.27
1980	14.75	14.26	+0.49
1981	15.62	15.22	+0.40
1982	15.45	15.09	+0.36
1983	15.45	15.12	+0.33
1984	15.22	14.88	+0.34
1985	14.85	14.89	–0.04
1986	14.51	14.52	–0.01
1987	14.67	14.52	+0.15
1988	14.10	14.22	–0.12
1989	15.24	15.37	–0.13
1990	16.03	16.41	–0.38

Source: U.S. Department of Agriculture, Agricultural Marketing Service, *Dairy Market Statistics*.

the estimates from Equation (13), we compared the estimated prices with simple annual average values of "announced" cooperative Class I prices in the thirty-five markets reported by USDA. As shown in Table 2.1, as expected, the estimated effective prices are larger than the cooperative prices for many years, but the differences are relatively small and in some recent years the announced cooperative prices are slightly higher. This comparison implies that over-order premiums for Class II and Class III milk are not large on average and, therefore, that the effective fluid-milk prices estimated by Equation (13) probably do not generate a serious bias in estimating the degree-of-competition parameter.

Supply Function

The aggregate milk supply (Q) was estimated using quarterly data from 1975 through 1990 as a function of the current and lagged milk-feed price ratio (MF = blend price / feed price), a time trend ($TREND$) representing technical progress, intercept dummy variables for the Milk Diversion Program (MDP) and the Dairy Termination Program (DTP), and harmonic seasonality variables ($SIN1$, $COS1$, and $COS2$) (see Table 2.3). We present the econometric results in Table 2.2 along with all of the estimated equations.

A polynomial distributed lag was imposed to account for lagged effects of the milk-feed price ratio.[6] Among the many alternative forms, the second-degree polynomial distributed lag with both endpoints constrained to lie close to zero and a six-quarter lag provided the best results. This lag length seems reasonable considering the biological reproduction cycle of dairy cattle.

The estimated long-run price elasticity of milk supply is 0.224, which is similar to Chavas and Klemme's (1986) estimated two-year price elasticity of 0.20 and Weersink's (1990) estimate of 0.29. To overcome significant first-order autocorrelation in the disturbance term, the Cochrance-Orcutt procedure was employed. Two-stage least squares (TSLS) estimation was used because both milk production and the blend price are endogenous variables in the model.

---

6 Because long-run milk-feed price effects are considered by imposing a polynomial distributed lag, the number of cows is not included as an explanatory variable.

Table 2.2. The Estimated Equations

Milk Supply (1975.2–1990.4)

$$\ln(Q) = 3.899 + 0.019 \ln(MF) + 0.032 \ln(MF)_{-1} + 0.040 \ln(MF)_{-2}$$

(24.75)[a] *(3.86)* *(3.86)* *(3.86)*

$$+ 0.043 \ln(MF)_{-3} + 0.040 \ln(MF)_{-4} + 0.032 \ln(MF)_{-5}$$

(3.86) *(3.86)* *(3.86)*

$$+ 0.019 \ln(MF)_{-6} + 0.0039\, TREND - 0.024\, MDP - 0.041\, DTP$$

(3.86) *(8.17)* *(–1.67)* *(–2.94)*

$$- 0.0053\, SIN1 - 0.052\, COS1 + 0.0071\, COS2 + 0.734\, (U^{Q})_{-1}$$

(–1.94) *(–19.57)* *(5.40)* *(7.57)*

Adjusted $R^2 = 0.95$, Durbin Watson = 1.79

Fluid Demand (1976.3–1990.4)

$$Q_f / N = -0.077 - 0.105\, (P_f / \mathrm{CPI}) + 0.0011\, (INC / \mathrm{CPI})$$

(–2.49) *(–3.16)* *(2.70)*

$$+ 1.0 \times 10^{-7} (GA_f) + 1.7 \times 10^{-7} (GA_f)_{-1} + 2.0 \times 10^{-7} (GA_f)_{-2}$$

(3.10) *(3.10)* *(3.10)*

$$+ 2.0 \times 10^{-7} (\mathrm{GA}_f)_{-3} + 1.7 \times 10^{-7} (\mathrm{GA}_f)_{-4} + 1.0 \times 10^{-7} (\mathrm{GA}_f)_{-5}$$

(3.10) *(3.10)* *(3.10)*

$$+ 6.8 \times 10^{-7} (BA_f) + 0.387\, AU19 + 0.0016\, SIN1 + 0.0023\, COS1$$

(2.60) *(4.85)* *(8.28)* *(10.15)*

$$+ 0.00018\, COS2 + 0.788\, (U^{Q_f / N})_{-1}$$

(3.70) *(4.94)*

Adjusted $R^2 = 0.92$, Durbin Watson = 2.02

Manufacturing Demand (1976.3 – 1990.4)

$$Q_m / N = 0.378 - 1.113\, (P_m / \mathrm{CPI}) - 0.0069\, (INC / \mathrm{CPI})$$

(4.66) *(–3.96)* *(–3.55)*

$$+ 3.6 \times 10^{-7} (BA_m) + 5.4 \times 10^{-7} (BA_m)_{-1} + 5.4 \times 10^{-7} (BA_m)_{-2}$$

(2.34) *(2.34)* *(2.34)*

$$+ 3.6 \times 10^{-7} (BA_m)_{-3} - 0.0059\, DTP + 0.018\, D89.4$$

(2.34) *(–1.71)* *(2.80)*

$$- 0.022\, D90.4 - 0.0013\, SIN1 - 0.0074\, COS1 + 0.00074\, COS2$$

(–3.16) *(–1.98)* *(–9.08)* *(2.12)*

$$+ 0.670\, (U^{Q_m / N})_{-1}$$

(3.78)

Adjusted $R^2 = 0.78$, Durbin Watson = 1.74

[a] Figures in italics and parentheses are t–values.

Table 2.3. Definitions for the Variables Used in the Equations Presented in Table 2.2

Q	Milk production (billion pounds).
MF	(blend price) / (feed price) where the blend price is the all-milk price (dollars/hundredweight) and the feed price is the U.S. average price of 16-percent-protein dairy feed (dollars/ton).
$TREND$	A time-trend variable equal to one for quarter 1, 1970, ...
MDP	Intercept dummy variable for the Milk Diversion Program equal to one for quarter 1 of 1984 through quarter 2 of 1985 and zero otherwise.
DTP	Intercept dummy variable for the Dairy Termination Program equal to one for quarter 2 of 1986 through quarter 3 of 1987 and zero otherwise.
$SIN1$, $COS1$, and $COS2$	Harmonic seasonality variables representing the first wave of the sine function (1, 0, −1, 0), the first wave of the cosine function (0, −1, 0, 1), and the second wave of the cosine function (−1, 1, −1, 1), respectively. (1, 0, −1, 0), etc. are values for each quarter where the first quarter means $\pi / 2$, the second is π, the third is $3\pi / 2$, and the fourth is 2π.
U_{-1}	Lagged residual.
Q_f	Fluid milk marketed (billion pounds).
N	U.S. population (million persons).
P_f	Effective Class I price estimated using Equation (13) (dollars/hundredweight).
CPI	Consumer Price Index for all items (1982–84 = 100).
INC	Disposable personal income per capita (in thousands of dollars).
GA_f, BA_f	Generic and branded fluid advertising expenditures deflated by the Media Price Index (in thousands of dollars).
$AU19$	Ratio of persons under nineteen years of age to the total population (total = 1).
Q_m	Manufacturing milk marketed (billion pounds).
P_m	M-W price (dollars/hundredweight).
BA_m	Branded manufacturing advertising expenditures (including branded butter, ice cream, and cheese advertising) deflated by the Media Price Index (in thousands of dollars).
$D89.4$	Intercept dummy variable equal to one for quarter 4 of 1989 and zero otherwise.
$D90.4$	Intercept dummy variable equal to one for quarter 4 of 1990 and zero otherwise.

Note: The data sources are listed in Kaiser et al. (1994).

Fluid-Milk Demand Function

The fluid-milk demand function is the processors' "derived" demand for raw milk.[7] To insure that all identities were meaningful, all quantities in the model were measured on a milk-fat-equivalent basis. Per capita fluid-milk demand (Q_f / N) is explained by the effective fluid-milk price (P_f), per capita income (*INC*), the ratio of persons under nineteen years of age to the total population (*AU*19), current and lagged fluid-milk advertising expenditures (branded fluid advertising – BA_f and generic fluid-milk advertising – GA_f), and harmonic seasonality variables (*SIN*1, *COS*1, and *COS*2). The variables P_f and *INC* are deflated by the Consumer Price Index (CPI) and BA_f and GA_f are deflated by the Media Price Index.

A polynomial distributed lag was imposed to account for lagged generic fluid-milk advertising effects. The second-degree polynomial distributed lag with both endpoints constrained to lie close to zero and a five-quarter lag provided the best results. The effects were largest four to six months later and eroded in about a year. No lagged effects of branded fluid-milk advertising were found to be significant but the current effect was significant.

Calculated at mean data points, the estimated elasticities of fluid-milk demand with respect to price, income, and branded fluid-milk advertising are –0.293, 0.483, and 0.0089, respectively. Liu et al.'s (1990) estimated elasticities for retail fluid-milk demand with respect to price and income were –0.282 and 0.154, respectively. Our estimated long-run generic advertising elasticity is 0.054, which is similar to Kinnucan and Forker's (1988) estimate of 0.051 in New York City but larger than Liu et al.'s (1990) estimate of 0.0175 for retail-level national fluid-milk demand.

The fluid-milk demand function was estimated using a linear form because other functional forms (double-log, semi-log, log-inverse, and inverse) resulted in negative marginal revenue estimates and were rejected because negative fluid-milk marginal revenue precludes discussion of the collusion case expressed by Equation (2).[8] TSLS was used to estimate this equation because both quantity and price are endogenous in the model.

---

7 It is assumed that changes in retail fluid and manufactured-product prices correspond to changes in raw-milk prices.

8 The manufacturing demand function is also estimated using a linear form to make it consistent with the fluid demand function.

Manufacturing-Milk Demand Function

Because this is the manufacturers' "derived" demand for raw milk, government purchases of dairy products and changes in commercial inventories are not treated separately. Per capita manufacturing-milk demand (Q_m / N) was estimated as a function of the manufacturing-milk price (P_m) deflated by the CPI, per capita income (*INC*) deflated by the CPI, the ratio of persons under nineteen years of age to the total population (*AU*19), current and lagged manufacturing-milk advertising expenditures (branded manufactured product – BA_m and generic manufactured product – GA_m), intercept dummy variables for the *DTP* and the *MDP*, and harmonic seasonality variables (*SIN*1, *COS*1, and *COS*2).

The *DTP* and *MDP* were included in this demand function because this derived demand includes government purchases of dairy products as well as commercial use. Since these two programs significantly affect the government component of manufacturer demand, they are included in this equation. The federal dairy price-support program is considered in this equation in that the manufacturing-milk price (P_m = M-W price) is indirectly supported through government purchases of dairy products. Intercept dummy variables are also included for the fourth quarters of 1989 and 1990 because regression residuals for both periods are very large. The outlier for the fourth quarter of 1989 is likely due to unusually strong demand for nonfat dry milk during that quarter but we have no explanation for the 1990 fourth-quarter outlier.[9]

A polynomial distributed lag is imposed to account for lagged branded manufacturing advertising effects. The second-degree polynomial distributed lag with both endpoints constrained to lie close to zero and a three-quarter lag provided the best results. We could not estimate any significant effects of generic manufacturing advertising (a negative coefficient with very small t-value was found). The variable *AU*19 and the intercept dummy variable for the *MDP* also were not significant. Consequently, these variables were dropped from the model. The estimated coefficient on the income variable was negative and significant, which is not consistent with expectations. Because each dairy product has a very different demand trend and structure, disaggregated estimation would

9 Inventory-building by cheese wholesalers and manufacturers in the face of uncertainties in milk supply could be another factor explaining unprecedented increases in M-W prices.

likely produce better results; however, this is beyond the scope of the present analysis.

Calculated at mean data points, the estimated elasticities of manufacturing demand with respect to price and long-run branded advertising are −1.575 and 0.234, respectively. The estimated price elasticity is relatively large compared to estimates generated by previous studies, such as −0.928 by Liu et al. (1990). Again, TSLS was used to estimate this equation because both the manufacturing demand and the price are endogenous in the model.

Degree-of-Competition Parameter

The degree-of-competition parameter is equal to one under monopoly or collusion and zero under perfect competition or price-taking behavior. As shown in Table 2.4, derived annual average estimates of θ_fs using Equation (5) or (6) indicate that the U.S. milk market is neither perfectly competitive nor purely monopolistic. The estimates imply some degree of market imperfection that has been declining over time.

Competitive pressures in the fluid-milk market have increased over time due to improvements in transportation technology and increasing reserves of milk in areas other than Minnesota and Wisconsin. Dairy farmers have tried to reduce competitive pressures by merging cooperatives and milk marketing orders while the size of manufacturing plants has become larger. The gradually decreasing degree-of-competition

Table 2.4. Estimated Degree-of-Competition Parameters (Annual Average)

Year	Degree-of-Competition Parameter (θ_f)	Year	Degree-of-Competition Parameter (θ_f)
1977	0.077 (0.024)[a]	1984	0.056 (0.018)
1978	0.065 (0.021)	1985	0.061 (0.019)
1979	0.066 (0.021)	1986	0.057 (0.018)
1980	0.066 (0.021)	1987	0.058 (0.018)
1981	0.065 (0.020)	1988	0.050 (0.016)
1982	0.061 (0.019)	1989	0.044 (0.014)
1983	0.059 (0.019)	1990	0.055 (0.017)

[a] Figures in parentheses are standard errors defined by $((P_f - P_m) / \text{CPI}) / (Q_f / N) \times$ [standard error of the fluid demand function's estimated slope].

parameters could be a consequence of a power balance caused by these developments.

Simulations

To determine the validity of the estimated model for analyses of deregulation, values for the endogenous variables, given the values for the exogenous variables, were determined in a fully dynamic simulation by the Gauss-Seidel technique[10] for the historical period 1980 through 1990. As illustrated by the mean absolute percent errors[11] shown in Table 2.5, the largest error is less than 4 percent, which is quite reasonable for a dynamic simulation.

It is not clear how deregulation might be implemented. Because our focal point is to examine the relative differences between model results, we do not consider any gradual deregulation following some time schedule.

For simulating cases without import quotas where the manufacturing-milk market is open to foreign imports, the manufacturing-milk price (P_m) in Equation (10) is interpreted as a given imported-product price measured on a milk-fat-equivalent basis. It is assumed that farmers would consider the demand for manufacturing milk to be perfectly elastic at the import-product (world) price. Because ours is a single-country model, we cannot determine the import-product price level after eliminating import quotas endogenously by solving a multicountry model such as the Ministerial Trade Mandate (MTM) model of the Organization for Economic Cooperation and Development (OECD) and USDA's Static World Policy Simulation (SWOPSIM) model (OECD 1991; Roningen 1985). Because our focal point is to examine the differences between the two model results, we assume a 20 percent decline from the current P_m level for each year as an example.

In addition, Q_m is replaced by Q_{Mi} (total manufacturing-milk demand, including dairy imports, in milk equivalents) in (9); in other words:

(14) $Q_{Mi} = h(P_m)$.

---

10 This is a numerical technique used in the *TSP-Micro* econometric software. It is similar to the Newton and the Fletcher-Powell methods but is more powerful in dealing with large models that have block components.

11 The formula is $(1 / n)\Sigma|(P - A) / A| \times 100$ where P is the predicted value and A is the actual value.

Table 2.5. Mean Absolute Percent Errors[a] (1980.1–1990.4)

Endogenous Variables	Mean Absolute Percent Error
Fluid-Milk Price (P_f)	3.10
Manufacturing-Milk Price (P_m)	3.70
Blend Price (BP)	3.54
Fluid-Milk Demand (Q_f)	1.60
Manufacturing-Milk Demand (Q_m)	2.91
Milk Production (Q)	1.67

[a] The formula is $(1 / n)\Sigma|(P - A) / A| \times 100$ where P is the predicted value and A is the actual value.

Also, the following definitional identity for imports is added:

(15) $$Q_I \equiv Q_{Mi} - Q_m$$

where Q_I is dairy imports on a milk-fat-equivalent basis.

Imperfect-Competition Model

The imperfect-competition model for simulation in the case where there are no price supports, no import quotas, and no marketing orders is

(16) $$Q = f(BP),$$

(17) $$Q_f = g(P_f),$$

(18) $$Q_{Mi} = h(P_m),$$

(19) $$P_f + \theta_f Q_f/(\partial Q_f / \partial P_f) = P_m,$$

(20) $$Q \equiv Q_f + Q_m + FUSE,$$

(21) $$BP = (P_f Q_f + P_m Q_m) / (Q - FUSE), \text{ and}$$

(22) $$Q_I \equiv Q_{Mi} - Q_m.$$

Exogenous Fluid Differential Model

The difference between the current imperfect-competition model and the conventional exogenous fluid differential model is found in Equation (19). The imperfect-competition model is transformed into the exogenous fluid differential model used in most previous analyses by replacing Equation (19) with

$$(23) \qquad P_f = P_m + DIFF$$

where *DIFF* is the exogenous fluid price differential. Note that *DIFF* is not the minimum Class I differential but rather is the effective fluid differential treated as an exogenous variable.

Comparison of Simulation Results

We used historical time periods for our simulations to avoid having to estimate future values for exogenous variables. The dynamic simulation results from 1980 through 1990 are shown in Table 2.6.

Comparison of the two models illustrates how the imperfect-competition model yields different estimates of deregulation effects than does the conventional exogenous fluid differential model.

The differences in results between the imperfect-competition and the exogenous fluid differential models tend to be rather small. For instance, results of the imperfect-competition model show that, compared to the current level,[12] the fluid-milk price would be 15.2 percent lower on average for the simulation period, the blend price would be 17.2 percent lower, and milk production would be 4.2 percent less. Results of the exogenous fluid differential model show that the fluid-milk price would be 16.1 percent lower on average for the simulation period, the blend price would be 17.6 percent lower, and milk production would be 4.3 percent less.

In the imperfect-competition model, $\partial P_f / \partial P_m$ is expressed by

$$(24) \qquad \partial P_f / \partial P_m = 1 / (1 + \theta_f) = 0.946$$

12 The current values are not observations but are values solved by fully dynamic simulation of Equations (7) to (12).

Table 2.6. Estimated Effects of Deregulation (Average of 1980–1990)

Percent Change in	Fluid-Milk Price	All-Milk Price	Milk Production
Imperfect-Competition Model	–15.2	–17.2	–4.2
Exogenous Differential Model	–16.1	–17.6	–4.3

where 0.946 is an average of simulation periods. This means that fluid-milk prices decline by 95 percent of the magnitude of manufacturing-milk price declines. In the exogenous fluid differential model,

$$(25) \quad \partial P_f / \partial P_m = 1.$$

Therefore, the exogenous fluid differential model tends to overestimate the negative effects of deregulation compared to the imperfect-competition model, but the differences are relatively small. As Equation (24) shows, the larger θ_f is, the smaller $\partial P_f / \partial P_m$. In other words, more fluid market imperfection could cause less direct translation of decreases in the manufacturing price into decreases in the fluid price.

Summary and Conclusions

The effective fluid-milk price differential (the minimum differential plus any over-order payment) is considered to reflect the degree of competition in the U.S. milk market. An imperfect-competition model with the fluid price differential endogenously explained by the degree of competitiveness is theoretically preferred to the exogenous fluid differential model because the fluid price differential is not exogenous due to over-order payments.

This chapter presents the first imperfect-competition model with an endogenous fluid differential for the U.S. dairy market using degree-of-competition parameter estimates. The estimated parameters imply that there is a degree of imperfection in the U.S. milk market that has been declining over time. Competitive pressures in the fluid-milk market have increased over time due to improvements in transportation technology and increasing reserves of milk in areas other than Minnesota and Wisconsin. Dairy farmers have tried to reduce competitive pressures by merging cooperatives and milk marketing orders while the size of manufacturing plants has become larger. The gradually decreasing degree-of-competition parameters could be the consequence of a balance of power due to these developments.

The usefulness of the model was demonstrated by showing the relative differences in dynamic simulation results between the imperfect-competition model and an exogenous fluid differential model.

Because it is not clear whether the current degree of market competition would remain unchanged after deregulation, we cannot determine that the imperfect-competition model provides more likely estimates of key milk-market parameters after deregulation than does the exogenous fluid differential model. At a minimum, the exogenous fluid differential model tends to overestimate the negative effects of deregulation compared to the imperfect-competition model if the current degree of competitiveness in the U.S. milk market is maintained after deregulation. Decreases in the manufacturing-milk price exactly correspond to decreases in the fluid-milk price in the exogenous fluid differential model. However, some market imperfection could cause less direct translation of decreases in the manufacturing price into decreases in the fluid price.

Our simulation results show that the effects of deregulation are indeed larger with the exogenous fluid price differential model than with the imperfect-competition model but that the differences are relatively small. This provides some justification for using the exogenous fluid price differential model as a simplifying but acceptable alternative. There is a trade-off between realism and tractability in model-building. If the losses associated with simplifying assumptions are small, it is often beneficial to make them and then concentrate on the parameters that impact the analysis most significantly. Since the degree-of-competition parameter has been declining in the U.S. dairy industry, researchers may not be guilty of a capital offense by treating the fluid differential as exogenous in some cases.

The model presented here is a basic framework, and improvements to its shortcomings would be valuable in future research. In particular, processors' and manufacturers' market power is not explicitly incorporated into the current model and over-order payments for Class II and Class III milk are aggregated in the effective fluid differential. Also, cooperatives with manufacturing plants, a characteristic of the U.S. dairy market that does exist to an important degree, are not considered.

Appendix: An Alternative Solution

For simplicity, fluid and manufacturing demand equations are specified as follows:

(A1) $Q_f = a + bP_f$ and

(A2) $Q_m = c + dP_m$.

Then, equality across markets of "perceived" marginal revenue is

(A3) $P_f + \theta_f Q_f/b = P_m + \theta_m Q_m/d$.

In this chapter we attempt to estimate θs from (A3) using estimates of (A1) and (A2).

Alternatively, substituting (A2) into (A3) yields

(A4) $P_f = -\theta_f/b\, Q_f + (1 + \theta_m)P_m + \theta_m\, c/d$.

If (A1) and (A4) are estimated without estimating (A2), both θ_f and θ_m are identified but *c* and *d* cannot be identified separately.

References

American Agricultural Economics Association. *Federal Milk Marketing Orders: A Review of Research on Their Economic Consequences*. Task Force on Dairy Marketing Orders Occasional Paper 3, Ames IA, June 1986.

Babb, E.M. *Case Studies of Over-Order Payments in Federal Milk Marketing Orders*. Gainesville FL: University of Florida Food and Resource Department Staff Paper 557, 1989.

Bresnahan, T.F. "The Oligopoly Solution Concept Is Identified." *Economics Letters* 10 (1982):87–92.

Buxton, B.M. "Welfare Implications of Alternative Classified Pricing Policies for Milk." *American Journal of Agricultural Economics* 59 (1977):525–529.

Chavas, J-P., and R.M. Klemme. "Aggregate Milk Supply Response and Investment Behavior on U.S. Dairy Farms." *American Journal of Agricultural Economics* 68 (1986):55–66.

Dahlgran, R.A. "Welfare Costs and Interregional Income Transfers due to Regulation of Dairy Markets." *American Journal of Agricultural Economics* 62 (1980):288–296.

Ippolito, R.A., and R.T. Masson. "The Social Cost of Government Regulation of Milk." *Journal of Law and Economics* 21 (1978):33–65.

Kaiser, H.M., O.D. Forker, J.E. Lenz, and C. Sun. "Evaluating Generic Dairy Advertising Impacts on Retail, Wholesale, and Farm Milk Markets." *Journal of Agricultural Economic Research* 44 (1994):3–17.

Kaiser, H.M., D.H. Streeter, and D.J. Liu. "Welfare Comparison of U.S. Dairy Policies with and without Mandatory Supply Control." *American Journal of Agricultural Economics* 70 (1988):848–858.

Kinnucan, H.W., and O.D. Forker. "Allocation of Generic Advertising Funds among Products: A Sales Maximization Approach." *Northeastern Journal of Agricultural and Resource Economics* 17 (1988):64–71.

Liu, D.J., H.M. Kaiser, O.D. Forker, and T.D. Mount. "An Economic Analysis of the U.S. Generic Dairy Advertising Program Using an Industry Model." *Northeastern Journal of Agricultural and Resource Economics* 19 (1990):37–48.

Masson, R.T., and P.M. Eisenstat. "Welfare Impacts of Milk Orders and the Antitrust Immunities for Cooperatives." *American Journal of Agricultural Economics* 62 (1980):270–278.

McDowell, H., A. Fleming, and F. Spinelli. *U.S. Milk Markets under Alternative Federal Order Pricing Policies*. Washington DC: U.S. Department of Agriculture, Economic Research Service Staff Report AGES 9068, 1990.

Organization for Economic Cooperation and Development (OECD). "The OECD Ministerial Trade Mandate Model for Policy Analysis: Documentation." OECD/GD(91) 159, Paris, 1991.

Robinson, K.L. *Farm and Food Policies and Their Consequences*. Upper Saddle River NJ: Prentice-Hall Inc., 1989.

Roningen, V.O. *A Static World Policy Simulation (SWOPSIM) Modeling Framework*. Washington DC: U.S. Department of Agriculture, Economic Research Service Staff Report AGES 860625, 1985.

Suzuki, N., H.M. Kaiser, J.E. Lenz, K. Kobayashi, and O.D. Forker. "Evaluating Generic Promotion Effectiveness with an Imperfect Competition Model." *American Journal of Agricultural Economics* 76 (1994):296–302.

Suzuki, N., J.E. Lenz, and O.D. Forker. "A Conjectural Variations Model of Reduced Japanese Milk Price Supports." *American Journal of Agricultural Economics* 75 (1993):210–218.

U.S. Department of Agriculture (USDA), Agricultural Marketing Service. *Dairy Market Statistics*. Washington, DC, annual issues.

U.S. Department of Labor, Bureau of Labor Statistics. *Consumer Price Index*. Washington DC, selected issues.

Weersink, A. "Nonrobustness of Dynamic Dual Models of Northeastern and U.S. Dairy Industries." *Northeastern Journal of Agricultural and Resource Economics* 19 (1990):12–16.

3

Evaluating Generic Milk Promotion Effectiveness with an Imperfect-Competition Model

Nobuhiro Suzuki, Harry M. Kaiser, John E. Lenz, Kohei Kobayashi, and Olan D. Forker

Although raw milk is essentially a homogeneous input in the production of fluid milk and manufactured dairy products, in many countries the price received for fluid-milk usage is higher than the price received for manufactured-product usage. Such differences indicate that prices are not competitively determined. This certainly is the case in the U.S., where federal and state milk marketing order programs establish minimum Class I price differentials (premiums) for most milk marketed. There are also over-order fluid premium payments resulting from negotiations between cooperatives and fluid processors.

Changes in milk advertising expenditures should produce changes in milk prices as well as in milk demand. Thus, an advertising program's effectiveness should account for changes in both price and quantity; each should be treated as endogenous. Most studies of U.S. dairy markets assume either an exogenous fluid-milk price (Thompson, Eiler, and Forker 1976; Liu and Forker 1988, 1990; Ward and Dixon 1989; Blisard, Sun, and Blaylock 1991; Forker and Ward 1993) or an exogenous fluid (Class I) price differential (Kaiser, Streeter, and Liu 1988; Kaiser et al. 1994; Liu et al. 1990). No studies known to the authors have used a

---

This chapter originally appeared as an article titled "Evaluating Generic Milk Promotion Effectiveness with an Imperfect Competition Model" by Suzuki, Kaiser, Lenz, Kobayashi, and Forker in the *American Journal of Agricultural Economics* in 1994. This material is reprinted by permission of the *American Journal of Agricultural Economics*.

degree-of-competition measure or an endogenously determined fluid price differential in U.S. dairy models.

In the Japanese market, the assumption of an exogenous fluid price or premium is even less appropriate because milk prices are determined in individual negotiations between prefectural milk marketing boards (designated dairy cooperatives) and the processors they supply. The market power of prefectural boards and processors is crucial in milk price determination. In addition, because generic promotion expenditures are defrayed by assessments on milk marketings, assessment increases affect fluid-milk marketings, which in turn increase assessments. Promotion expenditures should therefore be treated as endogenous.

Using Japanese generic milk promotion data, we develop a framework for measuring the effectiveness of generic milk advertising incorporating the degree of market competition. Results assuming endogenously determined fluid premiums are then compared with a conventional model assuming an exogenous fluid premium.

Our imperfect-competition model employs a type of conjectural variation that is similar to that used by Suzuki, Lenz, and Forker (1993) to examine market impacts of decreasing Japanese milk-support prices. However, the present conceptual model differs from their model in its consideration of both manufacturers' (buyer) and producer marketing boards' (seller) market power.

Japanese Milk-Promotion System

The National Milk Promotion Association of Japan (NMPAJ) has been the sole agency responsible for Japanese generic milk promotion since 1978. Each prefectural milk marketing board contributes to NMPAJ in addition to promoting its prefectural brand. In the present analysis, we consider only NMPAJ's nationwide generic program.[1]

1 While it would be desirable to analyze the total advertising program for fluid milk and manufactured dairy products, the proprietary nature of the data pertaining to the prefectural boards' and dairy-product manufacturers' branded advertising programs precludes incorporation of the data into our analysis. However, as with most brand advertising, these programs are aimed primarily at increasing the advertisers' market shares. NMPAJ's generic program, on the other hand, is primarily a demand expansion program for fluid milk and is the focus of our study.

For fiscal year 1990 (April–March), the total budget of NMPAJ was eight billion yen (equivalent to $61.5 million). This revenue came from three sources: voluntary assessments (2.8 billion yen), government subsidies (4.0 billion yen), and carryover from fiscal year 1989 (1.2 billion yen). Unlike the U.S., where only dairy farmers pay mandatory promotion assessments, the Japanese program is voluntary and retailers and wholesalers are included in addition to farmers. Assessments on farmers and wholesalers were 0.24 yen per kilogram (8.4 cents per hundredweight) of fluid milk and 0.10 yen per kilogram (3.4 cents per hundredweight) of manufacturing milk marketed. Retailers were assessed 0.24 yen per kilogram (8.4 cents per hundredweight) of fluid milk purchased from wholesale processors.

The eight billion yen from NMPAJ's 1990 budget was divided among three primary expenditure categories: promotion (6.4 billion yen), administration (0.4 billion yen), and carryover to 1991 (1.2 billion yen).[2] NMPAJ's promotion activities encompass media advertising and a wide variety of nonadvertising activities.[3] With the exception of a cheese fair, NMPAJ's promotional activities are primarily focused on expanding fluid-milk consumption.

Model

Because there are many fluid processors in the Japanese fluid sector and none has a large share of the market, fluid processors do not have much influence on price negotiations with the prefectural marketing board. On the other hand, there are few nonfluid dairy manufacturing firms in Japan and the market share of some of these firms is larger than their fluid counterparts. As a result, some dairy manufacturing-product firms may have some market power and could influence the determination of the manufacturing-milk price. Since Japan operates a deficiency payment system with two-tiered pricing for manufacturing milk, dairy manufacturing firms are required to pay the standard purchase price to farmers

2 Carryover to next year = (actual expense) – (budget).

3 The nonadvertising activities encompass booklet distribution (on topics such as milk and health, cooking with milk, sports nutrition, guidelines for health professionals, and information for teachers and young children), special events (including a milk home-delivery campaign, a milk and dairy-product fair, and a cooking contest), cooking classes, seminars for opinion leaders, research on market trends, and special projects for promoting milk consumption in schools.

for within-quota milk marketings.[4] The degree of manufacturing firms' market power becomes important in negotiating the over-quota milk price with the prefectural marketing board. Dominance of the manufacturers' power would be reflected by a rapidly declining over-quota milk price as quantity increases while dominance of the boards' power would be reflected by an over-quota milk price that is close to the standard purchase price even as quantity increases.

Mathematically, the *i*th board's maximization problem is comparable to the following constrained maximization problem.[5]

$$\max NR^i = P_f q_f^i + GP\, QT^i + P_m(q^i - q_f^i - QT^i) - a^i$$

subject to

(1) $Q_f = f(P_f, M)$,

(2) $Q_f = q_f^i + \sum_{j \neq i} q_f^j$,

(3) $OQT = \sum q^i - Q_f - \sum QT^i$,

(4) $\sum_{j \neq i} q_f^j = g(q_f^i)$,

(5) $P_m = h(OQT)$,

(6) $M = A / c$,

(7) $A = a^i + \{0.24 \sum_{j \neq i} q_f^j + 0.10 \left(\sum_{j \neq i} q^j - \sum_{j \neq i} q_f^j \right) + 0.24 \times 2Q_f + 0.10 (Q - Q_f)\} L + S$, and

(8) $a^i = \{0.24\, q_f^i + 0.10\, (q^i - q_f^i)\} L$

4 The government pays farmers a deficiency payment for all within-quota milk marketings equal to the difference between the standard purchase price and the guaranteed price.

5 If $q^i - q_f^i < QT^i$, the objective function is replaced by $\max NR^i = P_f q_f^i + GP\,(q^i - q_f^i) - a^i$. This is not the case in our analyses.

where

NR = total revenue net of promotion costs,

P_f = fluid-milk price,

q_f^j = ith board's fluid-milk supply,

GP = guaranteed price for manufacturing milk,

QT^i = ith board's quota,

OQT = total over-quota manufacturing milk,

P_m = price paid for over-quota manufacturing milk,

q^i = ith board's total milk supply,

a^i = ith board's assessment expenditures,

Q_f = total fluid-milk supply,

M = total promotion messages,

A = total promotion expenditures,

c = unit price of promotion messages (the Consumer Price Index is used as a proxy for c in our empirical model),

Q = total milk supply,[6]

L = collection ratio (the ratio of actual to expected assessments), and

S = government subsidies.

---

6 Because the interest of this chapter is on the prefectural boards' decision-making, it is assumed that total milk supply is given and that each board allocates its raw-milk supply to fluid and manufacturing uses to maximize its total milk sales revenues net of promotion assessments.

Equation (1) is the aggregate fluid-milk demand function while Equation (2) requires that aggregate fluid-milk demand equal the sum of all marketing boards' fluid-milk supplies. Total over-quota manufacturing-milk marketings are measured by Equation (3). Equation (4) is the ith board's conjecture of the other boards' aggregate reaction function. The relationship between the price for over-quota manufacturing milk and the quantity of over-quota marketings is captured by Equation (5) (note that the coefficient $\partial P_m / \partial(OQT)$ is a measure of the degree of dominance by the farmers' boards and manufacturing companies). Total promotion messages are defined by Equation (6). Equation (7) measures total promotion expenditures, which is composed of the voluntary assessments paid by the ith board, all other boards, wholesale fluid processors, fluid-milk retailers, and wholesale manufacturing-milk processors plus government subsidies. The ith board's contribution to total promotion expenditures is defined by Equation (8). With aggregate assessment collection losses ranging from 32 to 55 percent during 1981–1990, the collection ratio L must be included in Equations (7) and (8) to avoid overstating assessment income. Although L should be endogenous, it is treated as exogenous in this model because of the difficulty of incorporating the factors that affect collection loss.

Taking q^i, QT^i, GP, c, Q, L, and S as given, the first-order conditions for this maximization problem can be reduced to the following expression by solving for variables P_f, P_m, Q_f, OQT, q_f^i, a^i, M, $\sum_{j \neq i} q_f^j$, and A:

$$(9) \quad P_f - 0.24\,L + q_f^i(\partial P_f / \partial Q_f)\ [1 - 0.52\,L\,(\partial Q_f / \partial M) / c]$$
$$[1 + \partial(\sum_{j \neq i} q_f^j) / \partial q_f^i] = P_m - 0.10\,L +$$
$$(q^i - q_f^i - QT^i)(\partial P_m / \partial(OQT))\ [1 + \partial(\sum_{j \neq i} q_f^j) / \partial q_f^i]$$

where the coefficient 0.52 is derived from $[0.24 \times 3 - 0.10 \times 2]$. The term

$$\partial(\sum_{j \neq i} q_f^j) / \partial q_f^i$$

is conjectural variation. Equation (9) expresses equality across markets of perceived marginal revenue net of perceived marginal assessments. This equation poses a problem for empirical analysis as there are no observations available for P_m. Consequently, it is assumed that $P_m = SP$ (the standard purchase price) and $\partial P_m / \partial(OQT) = 0$, which means that the boards' power is dominant and the following equation replaces Equation (9) in the empirical model:

$$(10) \quad P_f - 0.24\,L + [1 - 0.52\,L\,(\partial Q_f / \partial M) / c]\,[1 + \partial(\sum_{j \neq i} q_f^j)/\partial q_f^i] \cdot q_f^i (\partial P_f / \partial Q_f) = SP - 0.10\,L.$$

Equation (10) can be expressed in elasticity terms as

$$(11) \quad P_f - 0.24\,L - (1 - 0.52\,L\,\eta\,Q_f/A)\,P_f\,\theta^i/\varepsilon = SP - 0.10\,L$$

where $\varepsilon = |(\partial Q_f / \partial P_f)\,(P_f / Q_f)|$, which is the absolute value of the fluid-milk own-price elasticity; $\eta = (\partial Q_f / \partial M)\,(M / Q_f)$, which is the fluid-milk promotion elasticity; and $\theta^i = (\partial Q_f / \partial q_f^i)\,(q_f^i / Q_f)$, which is conjectural elasticity (Appelbaum 1982) or Tsujimura's (1981) "market response elasticity."[7] By definition of conjectural elasticity, $\theta = 1$ implies monopoly or collusion and $\theta = 0$ implies price-taking behavior.

Since P_f and ε are common to all boards, Equation (11) requires that θ^i be the same for all boards (Holloway 1991). We consider $\theta^i \equiv \theta$ with the assumption that the boards approximately realize the condition expressed by (11).

The market-power parameter, θ, is difficult to directly estimate as a function of explanatory variables because it is not derived from demand or costs but rather depends on behavior (Helpman and Krugman 1989). Another approach for obtaining a value for θ is to solve Equation (11) for θ in each time period (we assume θ is constant for each time period) given actual observations for P_f, SP, L, Q_f, and A, as well as estimates of ε and η. Using the derived values for θ, the full model is:

Fluid-milk demand: $(12)\ Q_f = f(P_f, M)$

Milk sales-maximizing allocation: $(13)\ P_f - 0.24\,L + [1 - 0.52\,L\,(\partial Q_f / \partial M) / c] \cdot \theta\,Q_f/(\partial Q_f / \partial P_f) = SP - 0.10\,L$

Milk promotion messages: $(14)\ M = A / c$

---

7 The term c is not in Equation (11) because $(\partial Q_f / \partial M) / c = [(\partial Q_f / \partial M)(M / Q_f)] \cdot Q_f / (M c) = \eta\,Q_f / A$, which is because $M c = A$.

Total milk-promotion expenditures: (15) $A = AS + S$

Total assessments: (16) $AS = \{0.24 \times 3\, Q_f + 0.10 \times 2\, (Q - Q_f)\}\, L$

Milk sales revenue: (17) $GR = P_f Q_f + GP \times QT + SP\,(Q - Q_f - QT)$

where *AS* is total assessments, *GR* is milk sales revenue, *QT* is total quota manufacturing milk, and all other variables are as previously defined. From this model, equilibrium values are obtained for Q_f, P_f, *M*, *A*, *AS*, and *GR*, conditional on *L*, *c*, θ, *S*, *Q*, *GP*, *QT*, and *SP*.

Fluid-Milk Demand Function

The fluid-milk demand function (12) is estimated in linear form using monthly data from April 1981 through December 1990. Two-stage least squares is used because the fluid-milk price and quantity, as well as promotion expenditures, are endogenous. To overcome significant first-order autocorrelation in the disturbance term, the Cochrance-Orcutt procedure is employed. The fluid-milk demand function is initially specified as a function of the following variables: fluid-milk price, consumption expenditures, milk-promotion expenditures, temperature, the ratio of persons under fourteen years old to the total population, price of soft drinks, and eleven monthly intercept dummy variables. Temperature is included because Japanese consumers drink more milk during warmer months. The monthly intercept dummy variables are included to capture monthly seasonality in milk consumption. Contrary to *a priori* expectations, neither the price of soft drinks nor the ratio of persons under fourteen years old to the total population is a statistically significant explanatory variable. Consequently, these variables are dropped from the final model. The estimated fluid-milk demand function used is

$$
\begin{aligned}
Q_f / N = {} & \underset{(6.31)}{3.444} - \underset{(-5.18)}{1.004}\, P_f / CPIF + \underset{(2.56)}{4.524}\, FEXP/CPIF + \underset{(3.45)}{0.015}\, TEMP \\
& + \underset{(1.91)}{0.224}\, (A/N/CPI)_t + \underset{(1.91)}{0.418}\, (A/N/CPI)_{t-1} + \underset{(1.91)}{0.583}\, (A/N/CPI)_{t-2} \\
& + \underset{(1.91)}{0.717}\, (A/N/CPI)_{t-3} + \underset{(1.91)}{0.822}\, (A/N/CPI)_{t-4} + \underset{(1.91)}{0.896}\, (A/N/CPI)_{t-5}
\end{aligned}
$$

$$+ \underset{(1.91)}{0.941} (A/N/CPI)_{t-6} + \underset{(1.91)}{0.956} (A/N/CPI)_{t-7} + \underset{(1.91)}{0.941} (A/N/CPI)_{t-8}$$

$$+ \underset{(1.91)}{0.896} (A/N/CPI)_{t-9} + \underset{(1.91)}{0.822} (A/N/CPI)_{t-10} + \underset{(1.91)}{0.717} (A/N/CPI)_{t-11}$$

$$+ \underset{(1.91)}{0.583} (A/N/CPI)_{t-12} + \underset{(1.91)}{0.418} (A/N/CPI)_{t-13} + \underset{(1.91)}{0.224} (A/N/CPI)_{t-14}$$

$$+ \underset{(1.81)}{0.388} D1 + \underset{(2.17)}{0.459} D2 + \underset{(2.37)}{0.394} D3 + \underset{(2.32)}{0.418} D4 + \underset{(3.46)}{0.570} D5 + \underset{(3.39)}{0.606} D6$$

$$+ \underset{(3.06)}{0.470} D7 + \underset{(1.67)}{0.252} D8 + \underset{(3.35)}{0.609} D9 + \underset{(3.36)}{0.571} D10 + \underset{(2.66)}{0.478} D11$$

$$+ \underset{(10.84)}{0.825} (U^{Q_f/N})_{t-1}$$

$R^2 = 0.970$, Durbin Watson = 2.447
(italicized t-values given in parentheses)

where:

$Q_f =$	Total fluid-milk demand in thousand metric tons ("Milk and Milk Products Statistics," Ministry of Agriculture, Forestry, and Fisheries).
$N =$	Population in millions ("Japan Statistical Monthly Report," Japan Prime Minister's Office).
$P_f =$	Retail fluid-milk price in yen per kilogram ("Household Survey," Japan Prime Minister's Office).
CPIF =	Consumer Price Index for food (1985 = 100) ("Household Survey," Japan Prime Minister's Office).
FEXP =	Average per capita food expenditures in thousand yen ("Household Survey," Japan Prime Minister's Office).
TEMP =	Average temperature in Tokyo in degrees Celsius (data provided by the Japanese Meteorological Agency).

A = Generic fluid-milk promotion expenditures in million yen (NMPAJ).

CPI = Consumer Price Index for all commodities (1985 = 100) ("Household Survey," Japan Prime Minister's Office).

$D1$–$D11$ = Monthly intercept dummy variables equal to one for respective months and zero otherwise.

$(U^{Q_f/N})_{t-1}$ = Lagged residual.

R^2 = Adjusted coefficient of determination.

Existing theory suggests that, at some level of promotion expenditures, diminishing marginal returns occur. To incorporate this effect into the model, the demand function is estimated using several forms with diminishing marginal promotion effects (linear expenditure system, double-log, semi-log, log-inverse, inverse). In terms of statistical significance of coefficients and R^2, the linear form has the best results.

As a proxy for M in Equation (12), milk-promotion expenditures deflated by the Consumer Price Index (CPI) are used. Previous empirical studies indicate that consumer reaction to promotion messages is larger in later periods due to accumulated effects. These effects then decay as consumers become satiated with the promotion messages (Thompson, Eiler, and Forker 1976; Ward and Dixon 1989; Liu et al. 1990). To account for lagged promotion effects, a polynomial distributed lag is specified. A second-degree polynomial with both endpoints restricted to lie close to zero and a fourteen-month lag is used because it results in the best statistical fit.

Calculated at mean data points, the own-price elasticity of fluid demand is –0.697, the expenditure elasticity is 0.319, and the long-run promotion elasticity is 0.058. Of the twenty-seven U.S., United Kingdom, and Canadian studies reviewed by the International Dairy Federation (1991), only six reported a fluid advertising elasticity greater than 0.06 (International Dairy Federation 1991, p. 21). Thus, compared to results from previous fluid-milk studies, our estimated promotion response for Japan is large. Japan's relatively short history of milk consumption compared to other countries may be contributing to the relatively large estimated promotion response.

Simulations

The Japanese marketing boards actually make promotion decisions on an annual rather than a monthly basis. Because the fluid demand function is estimated using monthly data, the equation is converted into an annual equation by summing all variables into annual values and assuming that the estimated parameters are valid for the annual relationship.[8]

Although we would have preferred to use the wholesale price the boards receive from manufacturers, due to data availability, retail prices are used in estimating the demand equation. Consequently, it is assumed that the wholesale price parameter equals the retail price parameter. In the aggregated annual model, the distributed lag coefficients for promotion are summed into a single, long-run promotion-response coefficient. While other considerations, such as optimal timing of promotion expenditures, would dictate a more detailed analysis of carryover effects and discounted present values of net returns to promotion, such considerations are beyond the scope of this analysis. Previous studies also have used the sum of advertising coefficients and did not consider discounted present values (Thompson, Eiler, and Forker 1976; Thompson and Eiler 1977). Moreover, Case and Shamblin (1972) showed that the optimal advertising level remains surprisingly constant over a wide range of advertising carryovers.

The derived annual market-response elasticity (θ) is shown in the last column of Table 3.1. This elasticity declines monotonically from 0.16 in 1981 to 0.10 in 1989. Based on our estimates, the Japanese milk market is far from collusive and is becoming more competitive over time. This result is consistent with findings by Suzuki, Lenz, and Forker (1993).

Marginal Rates of Return to Promotion

To estimate marginal rates of return to promotion, Equations (12) through (17) are solved for annual equilibrium values of Q_f, P_f, and A with the collection ratio (L) increased 1 percent above actual values. Because promotion expenditures are endogenous, an exogenous shock is applied

8 Because the fluid-milk price variable cannot be summed into an annual value, its parameter is multiplied by twelve.

to increase promotion expenditures using L. Marginal rates of return to promotion are calculated as the increase in wholesale fluid-milk revenue divided by the increase in promotion expenditures. Since total milk supply is given and only the allocations to fluid and manufacturing uses are in question, the only costs associated with the increased wholesale-level revenues are the increased promotion expenditures.

The second column of Table 3.1 contains estimated marginal rates of return to promotion for 1981 through 1989 using the imperfect-competition model. As the market becomes more competitive over time, the marginal rate of return to promotion declines, going from 6.04 in 1981 to 4.33 in 1989. Although the marginal rate of return to promotion does decrease over time, the results still imply substantial opportunities for revenue enhancement through increased promotion expenditures since the rates of return are always larger than one.

To compare these results with a conventional model, the present model is simulated assuming an exogenous fluid-milk premium (or an exogenous fluid-milk price because the manufacturing-milk price is given). The third column of Table 3.1 contains estimated marginal rates of return to promotion assuming an exogenous fluid-milk premium. Since the fluid-milk price declines over time, the estimated rates of return to fluid-milk promotion also decline throughout the simulation period. The estimated exogenous-price marginal rates of return range from 73 to 77

Table 3.1. Marginal Rate of Return to Promotion, $\partial P_f / \partial L$, and θ

Fiscal Year	Marginal Rate of Return: Imperfect-Competition Model	Marginal Rate of Return: Exogenous-Premium Model	$\partial P_f / \partial L$	θ
1981	6.04	4.66	1.41	0.16
1982	6.13	4.71	1.44	0.16
1983	5.68	4.31	1.39	0.15
1984	5.26	3.96	1.30	0.13
1985	5.17	3.87	1.31	0.13
1986	4.46	3.30	1.13	0.11
1987	3.87	2.85	0.98	0.09
1988	4.22	3.12	1.05	0.10
1989	4.33	3.19	1.10	0.10

percent of the estimates in the imperfect-competition model. The rate of return is higher in the imperfect-competition model because increased promotion causes increases in both price and quantity, whereas only the quantity increases in the exogenous-premium model. As long as fluid demand is price-inelastic, revenue must increase more in the imperfect-competition model than in the exogenous-price model.

Equations (12) through (16) can be manipulated to derive the following relationship:

(18) $$\partial P_f / \partial L = [0.14 - 0.2\ \theta\ (\partial Q_f / \partial M) / c / (\partial Q_f / \partial P_f)\ Q] / (1 + \theta).$$

Equation (18) gives the change in the fluid-milk price associated with a marginal change in promotion expenditures (caused by a marginal change in the collection ratio). This equation shows how increases in the fluid-milk price become larger as θ and $\partial Q_f / \partial M$ become larger and $\partial Q_f / \partial P_f$ becomes smaller. Because θ has declined over time, $\partial P_f / \partial L$ has also become smaller, as shown in the fourth column of Table 3.1.

Conclusion

We have developed a framework incorporating the degree of market competition in the evaluation of milk-promotion effectiveness. The fluid-milk premium is endogenously explained by the degree of market power. Using Japanese generic milk promotion data, we demonstrated how our framework improves the estimates over a conventional exogenous-premium model. In Japan, estimated marginal rates of return with the exogenous-premium model are 23 to 27 percent smaller than those with our imperfect-competition model. Such results indicate that analyses with an exogenous-premium model may underestimate the rate of return to milk promotion when imperfect competition is present.

An equation was also derived to show how several parameters affect the magnitude of increase in the fluid-milk price as promotion expenditures increase. The increase in milk price associated with an increase in promotion becomes larger as the market-power parameter and promotion elasticity of fluid demand become larger and as the price elasticity of fluid demand becomes smaller.

References

Appelbaum, E. "The Estimation on the Degree of Oligopoly Power." *Journal of Econometrics* 19 (1982):287–299.

Blisard, W.N., T. Sun, and J.M. Blaylock. *Effects of Advertising on the Demand for Cheese and Fluid Milk.* Washington DC: U.S. Department of Agriculture, Economic Research Service Staff Report AGES 9154, 1991.

Case, K.E., and J.E. Shamblin. "The Effects of Advertising Carry-over." *Journal of Advertising Research* 12 (1972):37–40.

Forker, O.D., and R.W. Ward. *Commodity Advertising: The Economics and Measurement of Generic Programs.* New York NY: Lexington Books, 1993.

Helpman, E., and P.K. Krugman. *Trade Policy and Market Structure.* Cambridge England: MIT Press, 1989.

Holloway, G.J. "The Farm-Retail Price Spread in an Imperfectly Competitive Food Industry." *American Journal of Agricultural Economics* 73 (1991):979–989.

International Dairy Federation. *Econometric Measurement of Generic Advertising.* Special Issue 9202, Brussels Belgium, 1991.

Japan Ministry of Agriculture, Forestry, and Fisheries. "Milk and Milk Products Statistics." Tokyo Japan, 1993.

Japan Prime Minister's Office. "Japan Statistical Monthly Report." Tokyo Japan, 1993.

———. "Household Survey." Tokyo Japan, 1993.

Kaiser, H.M., O.D. Forker, J.E. Lenz, and C. Sun. "Evaluating Generic Dairy Advertising Impacts on Retail, Wholesale, and Farm Milk Markets." *Journal of Agricultural Economic Research* 44 (1994):3–17.

Kaiser, H.M., D.H. Streeter, and D.J. Liu. "Welfare Comparison of U.S. Dairy Policies with and without Mandatory Supply Control." *American Journal of Agricultural Economics* 70 (1988):848–858.

Liu, D.J., and O.D. Forker. "Generic Fluid Milk Advertising, Demand Expansion, and Supply Response: The Case of New York City." *American Journal of Agricultural Economics* 70 (1988):229–236.

———. "Optimal Control of Generic Fluid Milk Advertising Expenditures." *American Journal of Agricultural Economics* 72 (1990):1047–1055.

Liu, D.J., H.M. Kaiser, O.D. Forker, and T.D. Mount. "An Economic Analysis of the U. S. Generic Dairy Advertising Program Using an Industry Model." *Northeastern Journal of Agricultural and Resource Economics* 19 (1990):37–48.

Suzuki, N., J.E. Lenz, and O.D. Forker. "A Conjectural Variations Model of Reduced Japanese Milk Price Supports." *American Journal of Agricultural Economics* 75 (1993): 210–218.

Thompson, S.R., and D.A. Eiler. "Determinants of Milk Advertising Effectiveness." *American Journal of Agricultural Economics* 59 (1977):330–335.

Thompson, S.R., D.A. Eiler, and O.D. Forker. "An Econometric Analysis of Sales Response to Generic Milk Advertising in New York State." *Search Agriculture,* vol. 6, no. 3. Ithaca NY: Cornell University, New York State College of Agriculture and Life Sciences, 1976.

Tsujimura, K. *Econometrics.* Tokyo: Iwanamishoten, 1981.

U.S. Department of Labor, Bureau of Labor Statistics. *Consumer Price Index.* Washington DC, selected issues.

Ward, R.W., and B.L. Dixon. "Effectiveness of Fluid Milk Advertising since the Dairy and Tobacco Adjustment Act of 1983." *American Journal of Agricultural Economics* 71 (1989):730–740.

Part III

Processor Oligopoly Studies

4

Market Conduct under Government Price Intervention in the U.S. Dairy Industry

Donald J. Liu, Chin-Hwa Sun, and
Harry M. Kaiser

The U.S. dairy industry has become more concentrated over the last several decades. For example, between 1963 and 1987, the twenty-firm concentration ratios for wholesale butter, cheese, and fluid-milk companies increased from 31 percent to 94 percent, 59 percent to 68 percent, and 48 percent to 67 percent, respectively (U.S. Department of Labor, Bureau of the Census). These concentration ratios suggest that models of the dairy industry should account for the market power of processors.

A framework that became popular in the 1980s for assessing the degree of market power was developed by Appelbaum (1982). Rather than assuming a certain market conduct, the Appelbaum procedure uses the concept of conjectural variation, which is estimated endogenously as a measure of the degree of market power.[1] There have been several applications of this technique to agricultural industries (Schroeter 1988; Schroeter and Azzam 1990, 1991; Azzam and Pagoulatos 1990; Buschena and Perloff 1991; Durham and Sexton 1992; Wann and Sexton 1992;

This chapter originally appeared as an article titled "Market Conduct under Government Price Intervention in the U.S. Dairy Industry" by Liu, Sun, and Kaiser in the *Journal of Agricultural and Resource Economics* in 1995. This material is reprinted by permission of the *Journal of Agricultural and Resource Economics*.

1 Appelbaum's procedure involves deriving the first-order condition of the profit-maximizing oligopolist using a dual framework. The markup term of price over marginal cost appearing in the first-order condition contains both the slope of the demand curve and the so-called "conjectural variation" to be discussed shortly. The first-order condition is then estimated in conjunction with the market demand equation.

Azzam and Park 1993). However, with few exceptions, models of the U.S. dairy industry have assumed that the market is perfectly competitive (e.g., Kaiser, Streeter, and Liu (1988) and La France and de Gorter (1985)). To our knowledge, Suzuki et al. (1994) is the only U.S. dairy study that incorporated a market-power parameter of cooperatives and fluid processors. However, the role of government intervention was ignored in Suzuki et al. In the U.S. dairy industry, government intervention through the dairy price-support program causes prices to be determined under two different structural regimes: a "market equilibrium" regime where the market price is above the support price and a "government supported" regime where the support price is the effective price. In a recent study, Liu et al. (1991) presented an econometric model that allows for endogenous switching between the two market regimes. Under this framework, government intervention becomes part of the market structure since the reduced-form equations for each regime are different. However, Liu et al. assumed no market power on the part of industry participants.[2]

Given the importance of government programs in many of the U.S. and foreign agricultural industries, an obvious question is how to estimate the market power of industry participants when there is government intervention in the price formation process. The task calls for merging the literature on market power with that on switching-regime estimations. This chapter presents a framework for that purpose and then estimates the degree of selling power exercised by U.S. fluid and manufacturing dairy processors using that framework.[3] This study also examines whether the conduct of processors is different between the two market regimes, a relevant public policy issue.

A Conceptual Framework

Consider an oligopolistic industry in which individual firms face a downward-sloping aggregate demand curve and there is a government

2 Also see Shonkwiler and Maddala (1985) for a treatment of modeling agricultural markets with government price-support programs.

3 Ideally, a complete model would include the selling and buying power of both processors. However, since this would greatly increase the complexity of the switching-regime estimation, this study focuses solely on the selling power of processors.

price-support program for the product. Denote the product by m. The demand equation is specified in inverse form as

$$(1) \qquad P^m = P^m(Q^{md}, Z^m)$$

where P^m is the price of product m, Q^{md} is the aggregate demand quantity, and Z^m is a vector of demand shifters.

Due to government price supports, the observed price (P^{m*}) depends on whether the government support price (P^g) is binding. In the market-equilibrium regime, the observed price is higher than the government-support price and hence is equal to P^m plus a shock (ε^m) to the demand equation in (1):

$$(2a) \qquad P^{m^*} = P^m + \varepsilon^m \text{ if } P^m + \varepsilon^m > P^g.$$

Under the government-supported regime, $P^m + \varepsilon^m \le P^g$ and, hence, the observed price is equal to P^g:

$$(2b) \qquad P^{m^*} = P^g \text{ if } P^m + \varepsilon^m \le P^g.$$

There is a distinction made between supply and demand quantities (Q^{ms} versus Q^{md}) because they need not be equal due to possible government purchases (Q^g). More explicitly, one has

$$(3) \qquad Q^{ms} = \begin{cases} Q^{md} + Q^g & \text{in the government-supported regime, and} \\ Q^{md} & \text{in the market-equilibrium regime.} \end{cases}$$

The supply relation for an individual producer or processor (henceforth processor) is given by the first-order condition of her maximization problem. Facing the demand equation in (1) and the switching scheme in (2), the ith processor maximizes profit by choosing the optimal supply quantity,

$$(4) \qquad \max_{\{q_i^m\}} \pi_i^m = E[P^{m^*}]\, q_i^m - C_i^m(q_i^m, W^m),$$

where $E[P^{m^*}]$ is the expected output price, q_i^m is the ith processor's output supply quantity (with $\Sigma_i q_i^m = Q^{ms}$), and C_i^m is the cost function, which depends on q_i^m and a vector of parametric variable costs (W^m).

The first-order condition for the optimization problem is

$$(5) \qquad E[P^{m^*}] - \partial C_i^m / \partial q_i^m + (\partial E[P^{m^*}] / \partial q_i^m)\, q_i^m = 0.$$

The last term in (5), $\partial E[P^{m^*}] / \partial q_i^m$, captures the ith processor's perceived effect of a change in her q_i^m on the manufactured price, $E[P^{m^*}]$. This term is analogous to the conventional market-power term in a monopolist's problem, except in this case the price effect also accounts for the perceived quantity change (arising from a change in q_i^m) of other suppliers.

The expected price $E[P^{m*}]$ depends, in part, on the associated probabilities of the market-equilibrium and government-supported regimes. Define the probability that the government-supported regime occurs as $\Phi(\alpha)$ and the probability that the market-equilibrium regime occurs as $1 - \Phi(\alpha)$:

$$(6a) \qquad \text{Prob}(P^m + \varepsilon^m \leq P^g) = \Phi(\alpha) \text{ and}$$

$$(6b) \qquad \text{Prob}(P^m + \varepsilon^m > P^g) = 1 - \Phi(\alpha)$$

where $\Phi(\alpha)$ is the cumulative standard normal of ε^m evaluated at $\alpha \equiv (P^g - P^m) / \sigma^m$ and σ^m is the standard deviation of ε^m. Hence, the unconditional expectation for P^{m^*} is

$$(7) \qquad E[P^{m^*}] = (1 - \Phi)\, E[P^{m^*} \mid P^m + \varepsilon^m > P^g] + \Phi\, P^g.$$

Given (2a), the first term on the right-hand side of (7) can be expressed as (Maddala 1983, pp. 158–159)

$$(8) \qquad E[P^{m^*} \mid P^m + \varepsilon^m > P^g] = P^m + \sigma^m \phi / (1 - \Phi)$$

where ϕ is the standard normal density, again evaluated at α. The last term in (8) is similar to Heckman's bias correction term for selectivity bias arising from price censoring caused by the price-support program. Upon substituting (8) into (7), the unconditional expectation for P^{m^*} is

$$(9)\qquad E[P^{m*}] = (1-\Phi)\left\{P^m + \sigma^m \phi / (1-\Phi)\right\} + \Phi P^g.$$

The expression in (9) can be substituted for the first term in the first-order condition in (5). In addition, (9) can be used to derive the last term in (5), $\partial E[P^{m*}] / \partial q_i^m$. Differentiate (9) with respect to q_i^m and make use of the fact that Φ and ϕ are both evaluated at $\alpha \equiv (P^g - P^m) / \sigma^m$:

$$(10)\qquad \frac{\partial E[P^{m*}]}{\partial q_i^m} = (1-\Phi)\frac{\partial P^m}{\partial q_i^m} + \frac{\partial(1-\Phi)}{\partial P^m}\frac{\partial P^m}{\partial q_i^m}P^m + \sigma^m \frac{\partial \phi}{\partial P^m}\frac{\partial P^m}{\partial q_i^m} + \frac{\partial \Phi}{\partial P^m}\frac{\partial P^m}{\partial q_i^m}P^g .$$

Let $\xi \equiv P^g - P^m$ and, hence, $\alpha \equiv \xi / \sigma^m$. Using the derivative rules that $\partial\Phi / \partial\xi = \phi / \sigma^m$ and $\partial\phi / \partial\xi = -\phi\xi / \sigma^{m2}$ (Maddala 1983, p. 365), (10) can be rewritten as

$$(11)\qquad \frac{\partial E[P^{m*}]}{\partial q_i^m} q_i^m = \left\{(1-\Phi) + \frac{\phi}{\sigma^m}P^m + \frac{P^g - P^m}{\sigma^m}\phi - \frac{\phi}{\sigma^m}P^g\right\}\frac{\partial P^m}{\partial q_i^m} q_i^m = (1-\Phi)\frac{\partial P^m}{\partial q_i^m} q_i^m .$$

Since the terms associated with P^g cancel out, it is clear that the remaining term, $(1-\Phi)\, \partial P^m / \partial q_i^m q_i^m$, in the second line of (11) pertains to the case where the market-equilibrium regime prevails. In this regime, Equation (3) is simply $Q^{md} = Q^{ms}$. To further manipulate (11), one seeks an expression for $(\partial P^m / \partial q_i^m) q_i^m$. Denote $\partial Q^{ms} / \partial q_i^m$ as the ith processor's conjectural variation pertaining to the aggregate supply quantity of the product. Applying the chain rule to (1) and making use of the fact that, in the market-equilibrium regime, $Q^{md} = Q^{ms}$ (and, hence, $\partial Q^{md} / \partial Q^{ms}$ and Q^{ms} / Q^{md} are both equal to one), one obtains

$$(12)\qquad \frac{\partial P^m}{\partial q_i^m} q_i^m = \left\{\frac{\partial P^m}{\partial Q^{md}}\frac{\partial Q^{md}}{\partial Q^{ms}}\frac{\partial Q^{ms}}{\partial q_i^m}\right\} q_i^m = -P^m \eta^{mm}\frac{Q^{ms}}{Q^{md}}\frac{\partial Q^{md}}{\partial Q^{ms}}\lambda_i^m = -P^m \eta^{mm} \lambda_i^m$$

where $\eta^{mm} \equiv -\partial(\ln P^m) / \partial(\ln Q^{md})$, which is the price flexibility with respect to commercial quantity, and $\lambda_i^m \equiv \partial(\ln Q^{ms}) / \partial(\ln q_i^m)$, which is the ith processor's conjecture elasticity of aggregate supply with respect to a change in q_i^m.

If the individual processor behaves competitively, she would conjecture that, as she changes her output, other firms will adjust their quantities in such a way that the price the individual faces will remain unchanged. That is, $\lambda_i^m = 0$. In contrast, if the individual processor has monopoly power, any change in her supply will perfectly coincide with the change in the aggregate supply; that is, $\lambda_i^m = 1$. In general, $\lambda_i^m \in [0,1]$. Substituting (12) into (11) and then the resulting expression into (5) for $\partial E[P^{m*}] / \partial q_i^m q_i^m$, the first-order condition can finally be expressed as

$$(13) \quad \frac{[(1 - \Phi)\, P^m + \sigma^m \phi + \Phi\, P^g] - \partial C_i^m / \partial q_i^m}{P^m} = (1 - \Phi)\, \eta^{mm}\, \lambda_i^m.$$

Notice that the bracketed terms in the numerator in (13) are the expected output price, $E[P^{m*}]$, and, hence, the left-hand side of (13) is the Lerner Index measuring the price-cost margin. The right-hand side of (13) indicates that the price-cost margin is a function of market regime probability $(1 - \Phi)$, conjectural elasticity of the processor (λ_i^m), and the price flexibility of the demand function (η^{mm}).

To summarize, the model includes a demand function with government price intervention, (1) and (2); a supply relation, (13); and a market equilibrium condition with possible government purchases, (3). Therefore, there are four equations in the model. In the case of the market-equilibrium regime, the endogenous variables contained in the four-equation system are P^m, P^{m*}, Q^{md}, and Q^{ms}. In the case of the government-supported regime, Q^g replaces $P^{m*}(\because P^{m*} \equiv P^g)$ as an endogenous variable.

The Switching Dairy Model

The dairy model used in this study includes a manufactured dairy product subsector and a fluid milk subsector. The model focuses on the wholesale processing level of the dairy industry because government price intervention occurs at this level. Under the dairy price-support program, the government supports the farm milk price indirectly by agreeing to buy unlimited quantities of manufactured dairy products (cheese, butter, and nonfat dry milk) in the wholesale market at announced "purchase prices."

With minor modifications, the framework presented in the previous section is appropriate for the manufactured dairy product subsector.

The retailer's demand equation for manufactured dairy products in the wholesale market is specified in inverse form as

$$(14) \qquad P^m = P^m(Q^{md}, Q^f, Z^m)$$

where the superscript *m* is used to denote the manufactured dairy product and *f* the fluid milk product. Equation (14) is the same as (1) except now the fluid quantity (Q^f) is also included as a right-hand-side variable accounting for the cross-quantity effect on price. Notice that no distinction is made between fluid-milk supply and demand (i.e., $Q^{fs} = Q^{fd} \equiv Q^f$) because there is no direct government intervention in this market. With the additional cross-quantity term, (12) is modified[4] as follows:

$$(15) \qquad \frac{\partial P^m}{\partial q_i^m} q_i^m = -P^m \left\{ \eta^{mm} - \eta^{mf} \frac{Q^{ms}}{Q^f} \right\} \lambda_i^m$$

where

$$\eta^{mf} \equiv -\, \partial(\ln P^m) \,/\, \partial(\ln Q^f).$$

To simplify the estimation, it is assumed that milk processing at the wholesale level follows a Leontief-type technology of fixed proportions between farm milk and other inputs. With (12) being replaced by (15) and the fixed-proportions assumption, the first-order condition in (13) becomes

$$(16) \qquad \frac{\left[(1 - \Phi)\, P^m + \sigma^m \,\phi + \Phi\, P^g\right] - P^{\mathrm{II}} - \partial C_i^m \,/\, \partial q_i^m}{P^m} = \left(1 - \Phi\right) \left(\eta^{mm} - \eta^{mf}\, Q^{ms} \,/\, Q^f\right) \lambda_i^m$$

4 With Q^f appearing as an additional argument for P^m, one adds $(\partial P^m / \partial Q^f)\,(\partial Q^f / \partial q_i^m)$ to the curly bracketed term in the first line of (12). The term $\partial Q^f / \partial q_i^m$ is interpreted as the *i*th manufactured-product processor's conjectural variation pertaining to the aggregate fluid supply quantity. Details on the derivation can be obtained from the authors upon request.

where P^{II} is the Class II price that manufactured dairy processors must pay for their farm milk input[5] and C_i^m is now defined as the cost function associated with other variable inputs.

The Fluid-Milk Subsector

The retailer's demand equation for fluid milk in the wholesale market is specified in inverse form as

$$(17) \quad P^f = P^f(Q^f, Q^{md}, Z^f)$$

where Z^f is a vector of fluid demand shifters. Denoting ε^f as a shock to the fluid demand equation, the observed fluid price is

$$(18) \quad P^{f^*} = P^f + \varepsilon^f.$$

Since there is no direct government price intervention in the fluid-milk market, $E[P^{f^*}] = P^f$.

Regarding the supply side of the fluid submodel, consider the following profit-maximization problem for individual fluid processor j:

$$(19) \quad \max_{\{q_j^f\}} \pi_j^f = (P^f - P^{II} - d)\, q_j^f - C_j^f(q_j^f, W^f)$$

where $P^{II} + d$ is the Class I farm milk price, which is equal to the Class II price plus the exogenous Class I differential (d); q_j^f is the jth fluid processor's supply quantity (with $\Sigma_j\, q_j^f = Q^f$); and C_j^f is the processing cost function, which depends on q_j^f and a vector of parametric variable processing costs (W^f).

Upon manipulation, the fluid processor's first-order condition can be written as the following price-cost margin expression:[6]

$$(20) \quad \frac{P^f - P^{II} - d - \partial C_j^f / \partial q_j^f}{P^f} = \left[\eta^{ff} - (1 - \Phi)\, \eta^{fm}\, Q^f / Q^{md}\right] \lambda_j^f$$

---

5 Under the federal milk marketing order system, manufactured-dairy-product processors pay the Class II price for their milk while fluid-milk processors pay the Class I price, which is equal to the Class II price plus a fixed Class I differential.

6 Similar to the manufactured-dairy-products case, the derivation begins with applying the chain rule to the demand equation in (17). Details of the derivation can be obtained from the authors upon request.

where

$$\eta^{ff} \equiv -\partial(\ln P^f) / \partial(\ln Q^f),$$

$$\eta^{fm} \equiv -\partial(\ln P^f) / \partial(\ln Q^{md}), \text{ and}$$

$$\lambda_j^f \equiv \partial(\ln Q^f) / \partial(\ln q_j^f).$$

To facilitate time-series data estimation, the individual processors' first-order conditions in (16) and (20) are aggregated. Assuming a generalized Leontief technology, the aggregate processing cost function for each subsector k ($k = m$ and f) can be expressed as

$$C^k(Q^k, W^k) = Q^k \Sigma_i \Sigma_j \beta_{ij}^k (W_i^k W_j^k)^{1/2} + \Sigma_i \beta_i^k (W_i^k)^{1/2}$$

where the subscripts i and j now denote processing inputs rather than firms. The aggregate marginal cost for the kth subsector is

$$\Sigma_i \Sigma_j \beta_{ij}^k (W_i^k W_j^k)^{1/2}.$$

The aggregated first-order conditions can then be written as

$$(21) \quad P^{\mathrm{II}} = (1 - \Phi) P^m + \sigma^m \phi + \Phi P^g - \Sigma_i \Sigma_j \beta_{ij}^m (W_i^m W_j^m)^{1/2} - P^m (1 - \Phi) (\eta^{mm} - \eta^{mf} Q^{ms} / Q^f) \lambda^m, \text{ and}$$

$$(22) \quad P^{\mathrm{II}} + d = P^f - \Sigma_i \Sigma_j \beta_{ij}^f (W_i^f W_j^f)^{1/2} - P^f \left[\eta^{ff} - (1 - \Phi) \eta^{fm} Q^f / Q^{md}\right] \lambda^f$$

where, as discussed in Appelbaum (1982), λ^k is the aggregate conjectural elasticity (measuring the average industry conduct) for processors in the kth subsector.

To give some structure to the average industry-conduct parameters in (21) and (22), it is hypothesized that λ^k ($k = m$ and f) is a function of the probability of the market-equilibrium regime occurring $(1 - \Phi)$. Further, since λ^k lies between zero and one, the following logistic function is specified:

$$(23) \quad \lambda^k = 1 / \left\{1 + \exp\left[\gamma^k - \delta^k (1 - \Phi)\right]\right\}.$$

Including $1 - \Phi$ as an explanatory variable for λ^k can provide insight into the issue of whether competition is more pervasive in market-equilibrium or government-supported regimes. For example, according to Rotemberg and Saloner (1986), one might expect to find a negative relationship between λ^k and $1 - \Phi$ because individual dairy processors, in an attempt to capture a larger share of the "boom" market, are more inclined to behave more competitively in the market-equilibrium regime.

Substituting (23) into (21) and (22), the aggregate first-order conditions for manufactured and fluid processors become

$$(24) \quad P^{II} = (1 - \Phi)\, P^m + \sigma^m \phi + \Phi P^g - \Sigma_i \Sigma_j \beta^m_{ij} (W^m_i W^m_j)^{1/2} - \frac{P^m (1 - \Phi) \left(\eta^{mm} - \eta^{mf} Q^{ms} / Q^f\right)}{1 + \exp\left[\gamma^m - \delta^m (1 - \Phi)\right]}, \text{ and}$$

$$(25) \quad P^{II} + d = P^f - \Sigma_i \Sigma_j \beta^f_{ij} (W^f_i W^f_j)^{1/2} - \frac{P^f \left[\eta^{ff} - (1 - \Phi)\, \eta^{fm} Q^f / Q^{md}\right]}{1 + \exp\left[\gamma^f - \delta^f (1 - \Phi)\right]}.$$

The Closure

To close the model, the farm component is briefly introduced. Given the Leontief fixed-proportions assumption between farm milk and other processing inputs, the quantities of wholesale fluid and manufactured products can be expressed on a farm-milk-equivalent basis. Then, Q^{ms} and Q^f also can be used to denote the derived demand for farm milk of dairy processors at the farm level. The linkage between the farm and wholesale markets can be written as

$$(26) \quad \bar{Q} = Q^{ms} + Q^f$$

where $\bar{Q}$ is the farm milk supply, which is assumed to be predetermined due to lags in farm milk production.

To summarize, the wholesale component of the model includes the following: a wholesale manufactured-product demand function with government price intervention, (14) and (2); a wholesale manufactured-product supply relation, (24); a wholesale fluid demand function, (17) and (18); a wholesale fluid supply relation, (25); a wholesale manufactured-product equilibrium condition with possible government purchases, (3); and a wholesale fluid equilibrium condition (imposed by using a common notation; i.e., $Q^{fd} = Q^{fs} \equiv Q^{f}$). The farm component of the model includes a farm milk demand function, a farm milk supply function, and a farm equilibrium condition. The predetermined farm milk supply assumption yields farm milk supply equals $\bar{Q}$ Given that all quantity variables are expressed on an equivalent basis, farm milk demand equals $Q^{ms} + Q^{f}$. Thus, the farm component of the model is concisely captured by the farm-wholesale linkage, (26). There are eight equations in the model containing eight endogenous variables. In the case of the market-equilibrium regime, the endogenous variables are P^{m}, P^{m^*}, P^{f}, P^{f^*}, P^{II}, Q^{md}, Q^{ms}, and Q^{f}. In the case of the government-supported regime, Q^{g} replaces P^{m^*} ($\because P^{m^*} \equiv P^{g}$) as an endogenous variable.

The Estimation

The estimation procedure is similar to conventional two-stage (nonlinear) least squares, with several exceptions. The structural equations to be estimated are the wholesale manufactured and fluid demand functions and supply relations. Similar to the two-stage least squares procedure, the first stage involves estimating instruments for the endogenous variables in the right-hand side of the structural equations, and the second stage consists of substituting the instruments into the structural equations, which are then estimated.

Instruments for the quantity variables appearing on the right-hand side of the structural equations are first obtained by regressing the quantity variables on all the exogenous variables and their one-period to four-period lags. Given the quantity instruments, the two inverse demand equations are estimated. Specifically, the manufactured-product inverse demand function is estimated by applying a maximum likelihood Tobit procedure to (14) and (2) and the fluid inverse demand function is estimated by using ordinary least squares on (17) and (18). The Tobit procedure is needed for the manufactured demand function because of the limited dependent variable problem associated with the manufactured price; as indicated by (2), the manufactured price is constrained to be no less than the government purchase price.

In addition to the quantity variables, the right-hand sides of the supply relations in (24) and (25) involve other endogenous variables for which instruments also must be obtained. From the Tobit estimation of the manufactured demand function, one obtains instruments for Φ, ϕ, and P^{m}, as well as estimates of η^{mm}, η^{mf}, and σ^{m}. From the ordinary least squares estimation of the fluid demand function, one obtains an instrument for P^f, as well as estimates of η^{ff} and η^{fm}. Upon substituting the obtained instruments (for quantity variables P^m, P^f, Φ, and ϕ) and estimates (of η^{mm}, η^{mf}, η^{ff}, η^{fm}, and σ^m) into (24) and (25), the two first-order conditions can then be estimated. Rather than using single-equation estimation, the two first-order conditions are estimated as a system of nonlinear, seemingly unrelated equations because P^{II} and P^{I} are related by an exogenous Class I price differential. From the system estimation, one obtains estimates of the remaining parameters (β^k_{ij}, γ^k, and δ^k; $k = m$ and f).

While the two-stage procedure on a structural equation system with limited dependent variables is asymptotically equivalent to a maximum likelihood estimation of the system, the conventionally computed second-stage standard errors on the structural parameters are biased (Maddala 1983). The asymptotic theory for the preceding two-stage estimation method has been derived by Lee (1981) and may be used to correctly compute standard errors for the second-stage coefficients. However, as pointed out by Cornick and Cox (1994), such theory is both complicated and not very general (i.e., the asymptotic covariance matrices have to be derived for each permutation of the model). Hence, a bootstrapping procedure after the fashion of Cornick and Cox (1994) is adopted to compute the second-stage standard errors of the structural coefficients.[7]

Empirical Results

Quarterly time-series data from 1975 through 1992 are used to estimate the model. Variable definitions and sources of data are given in Table 4.1. Since time-series data are used in the estimation, all the price variables in the model are deflated by the Consumer Price Index for all items. Table 4.2 presents the empirical results for the two inverse demand equations. Both inverse demand equations are estimated in double-logarithmic form and as a function of commercial manufactured and fluid demand quantities (Q^{md} and Q^f); the consumer price indices for nonalcoholic beverages (*CPIBEV*), fats and oils (*CPIFAT*), and away-from-home food (*CPIAFH*); quarterly dummy variables (*Quarter-1*, *Quarter-2*, and *Quarter-3*); and generic manufactured/fluid advertising expenditures (*GMA/GFA*). The

7 The procedure involves re-estimating the model for each bootstrap data set. The number of replications is 350.

Table 4.1. Variable Definitions and Data Sources

Variable	Definition	Unit of Measurement	Source[a]
Q^{ms}	Wholesale manufactured supply	billion pounds of of milk fat equivalent	DSO
Q^{md}	Wholesale manufactured demand	billion pounds of milk fat equivalent	DSO
Q^{f}	Wholesale fluid supply and demand	billion pounds of milk fat equivalent	DSO
P^{m}	Wholesale manufactured-product price	dollars per hundredweight	DSO
P^{f}	Whole fluid-milk price	dollars per hundredweight	DSO
P^{g}	Government manufactured-product purchase price	dollars per hundredweight	DSO
P^{II}	Class II price	dollars per hundredweight	FMOMS
d	Class I price differential	dollars per hundredweight	FMOMS
CPIFAT	Consumer Price Index for fats and oil	1967=100	CPI
CPIBEV	Consumer Price Index for nonalcoholic beverages	1967=100	CPI
CPIAFH	Consumer Price Index for away-from-home food	1967=100	CPI
PPIFE	Producer price index for fuel and energy	1967=100	EE
WAGE	Average hourly wage in food-manufacturing sector	dollars per hour	EE
GMA	Generic manufactured-product advertising expenditures	thousand dollars	LNA
GFA	Generic fluid advertising expenditures	thousand dollars	LNA

[a] DSO: USDA, *Dairy Situation and Outlook*. FMOMS: USDA, *Federal Milk Order Market Statistics*. CPI: Consumer Price Index. EE: U.S. Department of Labor, *Employment and Earnings*. LNA: Leading National Advertisers. See the references for detailed citations.

Table 4.2. Estimated Manufactured and Fluid Inverse Demand Equations (Double-Log)

	Manufactured Equation		Fluid Equation	
Variable	Estimated Coefficient	t-Value	Estimated Coefficient	t-Value
Intercept	5.799	7.3	1.405	4.0
Quarter-1	–0.104	–2.9	–0.043	–3.3
Quarter-2	–0.170	–4.2	–0.040	–2.4
Quarter-3	–0.138	–3.7	–0.033	–2.0
ln (Q^{md})	–0.136	–0.7	–0.296	–3.9
ln (Q^{f})	–2.818	–7.9	–0.841	–5.4
ln (*CPIBEV*)	0.081	0.6	–0.160	–3.1
ln (*CPIAFH*)	1.470	2.4	0.081	0.4
ln (*CPIFAT*)	0.731	1.8	0.724	6.4
GFA[a]			0.013	3.8
GMA1[a]	–0.001	–0.1		
GMA2[a]	0.002	0.1		
GMA3[a]	–0.003	–0.3		
AR(1)	0.552	5.0	0.497	4.6
AR(2)	0.455	4.1		
σ^m	0.073	11.1		
Log-Likelihood	69.9			
Adjusted R^2			0.92	
Durbin Watson	2.4		1.86	

[a] *GMA* and *GFA* are specified as a second-order polynomial distributed lag function of the previous four quarters' advertising expenditures. End-point restrictions are imposed for *GFA* (in the ordinary least squares fluid equation) but not for *GMA* (in the Tobit manufactured equation).

quarterly dummies capture demand seasonality while the advertising expenditures account for the impact on demand of generic dairy promotion activities.[8] Autoregressive terms (*AR*) for the residuals are added to the demand equations to correct for serial correlation.

The estimated manufactured and fluid own-price flexibility coefficients are both negative, confirming that the demand curves are downward

8 To capture the carryover effect of advertising, *GMA* and *GFA* are specified as a second-order polynomial distributed lag function of the previous four quarters' advertising expenditures with end-point restrictions imposed for *GFA* but not for *GMA*. Imposition of the end-point restrictions for *GMA* is difficult because the manufactured demand equation is estimated by the Tobit procedure.

sloping. The estimated cross-price flexibility coefficients in both demand equations are negative, indicating that the two dairy products are gross substitutes. Except for *CPIBEV* in the fluid demand equation, the coefficients for the three price-index variables are positive in both equations, suggesting a substitution relationship between the dairy product in question and the food groups represented by the price indices. The coefficients for generic advertising expenditures are positive in both the fluid and manufactured demand equations but statistically significant only in the fluid case.[9] Finally, the estimated σ^m in the Tobit equation is significantly different from zero, corroborating the importance of correcting for selectivity bias arising from the dairy price-support program.

The estimated equations for the first-order conditions in (24) and (25) are presented in Table 4.3. In this table, *PPIFE* is the producer price index for fuel and energy and *WAGE* is the average hourly wage in the manufacturing sector of the general economy. These two prices are included to reflect the variable processing costs (W) appearing in (24) and (25). As in the demand equations, an autoregressive term for the residuals is added to each of the two first-order conditions to correct for serial correlation.

The coefficients of interest to this study are the ones associated with the average industry-conduct parameters in Equation (23). As mentioned, the relationship between $(1 - \Phi)$ and λ^k is expected to be negative because individual dairy processors, in an attempt to capture a larger share of the boom market, may be more inclined to behave more competitively in the market-equilibrium regime. This hypothesis is not rejected by the empirical results, as the estimated coefficients for fluid and manufactured milk markets are negative and statistically significant at the 1 percent level. The implications of this result are rather interesting. If the government continues to deregulate the dairy price-support program in the future, then the probability of a market-equilibrium regime occurring will increase over time. Since λ^k and $(1 - \Phi)$ are negatively related, the result implies that deregulation will have a procompetitive effect on the market conduct of fluid and manufactured processors.

To gain insight into the magnitude of market power in both markets, the conjectural elasticities for manufactured and fluid processors are

---

9 Deleting the advertising variable from the manufactured demand equation does not change in any significant way the estimated coefficients of the remaining variables. Hence, it is left in the equation to be consistent with the fluid demand equation.

Table 4.3. Estimated Manufactured and Fluid Processor First-Order Conditions

Variable		Estimated Coefficient	t-Value
A. Manufactured First-Order Condition			
PPIFE	(β^m_{ii})	–0.222	–1.4
(*PPIFE** *WAGE*)$^{1/2}$	(β^m_{ij})	0.704	1.4
WAGE	(β^m_{jj})	–2.117	–1.4
Intercept	(γ^m)	3.330	14.8
$1 - \Phi$	(δ^m)	–2.660	–10.9
AR(1)		0.609	6.2
Adjusted R^2 = 0.91, Durbin Watson = 1.9			
B. Fluid First-Order Condition			
PPIFE	(β^f_{ii})	0.318	59.3
(*PPIFE** *WAGE*)$^{1/2}$	(β^f_{ij})	–1.002	–64.3
WAGE	(β^f_{jj})	4.655	77.8
Intercept	(γ^f)	1.778	26.5
$1 - \Phi$	(δ^f)	–0.630	–22.4
AR(1)		0.647	6.9
Adjusted R^2 = 0.88, Durbin Watson = 2.1			

Note: The system of first-order conditions is estimated by the seemingly unrelated regression procedure.

computed from (23) using the estimates of γ and δ. Table 4.4 presents the simulated conjectural elasticities and their t-ratios over the period 1976–1992.[10] For most periods of the sample, the conjectural elasticities of manufactured processors are found to be smaller than those of fluid processors; the mean values of λ^m and λ^f are 0.100 and 0.176, respectively. The finding that fluid processors behave in a less competitive manner than manufactured processors is intuitive because markets for fluid milk

10 The variances of the simulated conjectural elasticities are obtained through bootstrapping.

Table 4.4. Simulated Manufactured and Fluid Conjectural Elasticities

Year	Quarter	λ^m	t-Value	λ^f	t-Value
1976	I	0.092	11.3	0.178	21.7
	II	0.172	21.5	0.204	25.5
	III	0.080	9.8	0.173	21.1
	IV	0.153	19.0	0.198	24.7
1977	I	0.101	12.4	0.181	22.2
	II	0.096	11.7	0.179	21.9
	III	0.050	6.1	0.156	19.1
	IV	0.057	6.9	0.160	19.5
1978	I	0.045	5.5	0.152	18.6
	II	0.072	8.7	0.168	20.5
	III	0.041	5.1	0.150	18.4
	IV	0.078	9.5	0.172	20.9
1979	I	0.103	12.7	0.182	22.3
	II	0.154	19.1	0.199	24.7
	III	0.053	6.5	0.158	19.3
	IV	0.186	23.4	0.208	26.1
1980	I	0.313	41.1	0.236	31.0
	II	0.249	32.0	0.223	28.6
	III	0.207	26.2	0.213	26.9
	IV	0.249	32.0	0.223	28.6
1981	I	0.224	28.5	0.217	27.6
	II	0.148	18.4	0.197	24.5
	III	0.128	15.8	0.191	23.5
	IV	0.149	18.4	0.197	24.5
1982	I	0.207	26.2	0.213	26.9
	II	0.135	16.7	0.193	23.9
	III	0.119	14.6	0.188	23.1
	IV	0.137	17.0	0.194	24.0
1983	I	0.143	17.8	0.196	24.3
	II	0.150	18.6	0.198	24.5
	III	0.164	20.4	0.202	25.2
	IV	0.072	8.8	0.169	20.6
1984	I	0.110	13.6	0.185	22.7
	II	0.086	10.5	0.175	21.4
	III	0.050	6.1	0.156	19.0
	IV	0.063	7.7	0.164	20.0

Continued on following page

Table 4.4 (continued)

Year	Quarter	λ^m	t-Value	λ^f	t-Value
1985	I	0.080	9.7	0.172	21.0
	II	0.056	6.8	0.160	19.5
	III	0.059	7.2	0.162	19.7
	IV	0.071	8.7	0.168	20.5
1986	I	0.139	17.2	0.194	24.1
	II	0.071	8.7	0.168	20.5
	III	0.102	12.5	0.182	22.3
	IV	0.062	7.6	0.163	19.9
1987	I	0.156	19.4	0.199	24.8
	II	0.121	14.8	0.188	23.2
	III	0.064	7.8	0.165	20.1
	IV	0.075	9.2	0.170	20.8
1988	I	0.108	13.3	0.184	22.6
	II	0.062	7.5	0.163	19.9
	III	0.058	7.0	0.161	19.6
	IV	0.043	5.2	0.151	18.5
1989	I	0.090	11.0	0.177	21.6
	II	0.092	11.3	0.178	21.7
	III	0.050	6.0	0.156	19.0
	IV	0.053	6.5	0.158	19.3
1990	I	0.047	5.7	0.154	18.8
	II	0.036	4.4	0.145	17.8
	III	0.036	4.4	0.145	17.8
	IV	0.046	5.6	0.153	18.7
1991	I	0.037	4.6	0.147	18.0
	II	0.035	4.3	0.145	17.8
	III	0.052	6.3	0.157	19.2
	IV	0.040	5.0	0.149	18.3
1992	I	0.084	10.2	0.174	21.3
	II	0.036	4.5	0.146	17.9
	III	0.061	7.4	0.163	19.8
	IV	0.035	4.3	0.145	17.8
Mean		**0.100**		**0.176**	

are less national in scope due to the perishability and relatively high transportation costs of the product. While the average industry-conduct parameters of manufactured and fluid processors are statistically different from zero for all the quarters, the magnitudes of these parameters are not alarming as they are still closer to zero (perfect competition) than to one (monopoly). Furthermore, neither parameter exhibits a strong pattern of increasing over time, a finding that is reassuring given that the industry became more concentrated over the sample period.

Summary

Bridging the market-conduct and switching-regime literatures, this chapter presents a framework by which to estimate the market power of an oligopolistic industry where there is a government price-support program impacting firms' output price. The proposed framework was then used to estimate the degree of market power exercised by manufactured and fluid processors in the U.S. dairy industry. The study also examined whether government price intervention in the dairy industry has a procompetitive or anticompetitive influence on market conduct.

The results indicated that the average industry-conduct parameters of manufactured and fluid processors were statistically different from zero (perfect competition) for all quarters during the period 1976–1992. However, the magnitudes of these estimated parameters were not alarming as they were still closer to zero than to one (monopoly). The results also indicated that manufactured and fluid processors tend to behave in a more competitive manner in the market-equilibrium regime than in the government-supported regime. This result suggests that further deregulation of the dairy price-support program will have a procompetitive impact on market conduct.

Though the oligopolistic switching-regime estimation framework was specifically applied to the dairy processing industry, it also can be applied to a farm-level problem. For example, the procedure can be invoked to examine the selling power of a group of large-scale farmers whose output price is under the control of a government price-support program. Further, the framework can be modified to derive a procedure for estimating the buying power of processors (e.g., flour processors buying wheat) and large-scale farmers (e.g., large-scale hog and poultry producers buying corn) whose input price is the subject of government price interventions.

References

Appelbaum, E. "The Estimation of the Degree of Oligopoly Power." *Journal of Econometrics* 19 (1982):287–299.

Azzam, A., and E. Pagoulatos. "Testing for Oligopolistic and Oligopsonistic Behavior: An Application to the U.S. Meat-Packing Industry." *Journal of Agricultural Economics* 41 (1990):362–370.

Azzam, A., and T. Park. "Testing for Switching Market Conduct." *Applied Economics* 25 (1993):795–800.

Buschena, D.E., and J.M. Perloff. "The Creation of Dominant Firm Market Power in the Coconut Oil Export Market." *American Journal of Agricultural Economics* 73 (1991):1000–1008.

Cornick, J., and T.L. Cox. "Endogenous Switching Systems: Issues, Options, and Application to the U.S. Dairy Sector." *Journal of Agricultural Economics Research* 44 (1994):28–38.

Durham, C.A., and R.J. Sexton. "Oligopsony Potential in Agriculture: Residual Supply Estimation in California's Processing Tomato Market." *American Journal of Agricultural Economics* 74 (1992):962–972.

Kaiser, H.M., D.H. Streeter, and D.J. Liu. "Welfare Comparisons of U.S. Dairy Policies with and without Mandatory Supply Control." *American Journal of Agricultural Economics* 70 (1988):848–858.

LaFrance, J.T., and H. de Gorter. "Regulation in a Dynamic Market: The U.S. Dairy Industry." *American Journal of Agricultural Economics* 67 (1985):821–832.

Leading National Advertisers, Inc. *Ad $ Summary*. New York NY, selected issues.

Lee, L.F. "Simultaneous Equations Models with Discrete and Censored Dependent Variables." *Structural Analysis of Discrete Data with Econometric Applications*. C. Manski and D. McFadden, eds., pp. 346–364. Cambridge MA: MIT Press, 1981.

Liu, D.J., H.M. Kaiser, T.D. Mount, and O.D. Forker. "Modeling the U.S. Dairy Sector with Government Intervention." *Western Journal of Agricultural Economics* 16 (1991):360–373.

Maddala, G.S. *Limited-Dependent and Qualitative Variables in Econometrics*. Cambridge England: Cambridge University Press, 1983.

Rotemberg, J.J., and G. Saloner. "A Supergame-Theoretic Model of Price Wars during Booms." *American Economic Review* 76 (1986):390–407.

Schroeter, J.R. "Estimating the Degree of Market Power in the Beef Packing Industry." *Review of Economics and Statistics* 70 (1988):158–162.

Schroeter, J.R., and A. Azzam. "Measuring Market Power in Multi-Product Oligopolies: The U.S. Meat Industry." *Applied Economics* 22 (1990):1365–1376.

———. "Marketing Margins, Market Power, and Price Uncertainty." *American Journal of Agricultural Economics* 73 (1991):990–999.

Shonkwiler, J.S., and G.S. Maddala. "Modeling Expectations with Bounded Prices: An Application to the Market of Corn." *Review of Economic Statistics* 67 (1985):697–702.

Snedecor, G.W., and W.G. Cochran. *Statistical Methods*. Ames IA: Iowa State University Press, 1980.

Suzuki, N., H.M. Kaiser, J.E. Lenz, and O.D. Forker. "An Analysis of U.S. Dairy Policy Deregulation Using an Imperfect Competition Model." *Agricultural and Resource Economics Review* 23 (1994):84–93.

Wann, J.J., and R.J. Sexton. "Imperfect Competition in Multiproduct Food Industries with Application to Pear Processing." *American Journal of Agricultural Economics* 74 (1992):980–990.

U.S. Department of Agriculture, Agricultural Marketing Service. *Federal Milk Order Market Statistics*. Washington DC, selected issues, 1975–1990.

U.S. Department of Agriculture, Economic Research Service. *Dairy Situation and Outlook*. Washington DC, selected issues, 1975–1990.

U.S. Department of Labor, Bureau of the Census. *Concentration Ratios in Manufacturing*. Washington DC, selected issues.

U.S. Department of Labor, Bureau of Labor Statistics. *Consumer Price Index*. Washington DC, selected issues.

———. *Employment and Earnings*. Washington DC, selected issues, 1975–1990.

5

Oligopoly Power and Allocative Efficiency in U.S. Food and Tobacco Industries

Sanjib Bhuyan and Rigoberto A. Lopez

Most countries throughout the world have competition policies against restraints of trade to prevent significant social welfare and consumer losses from market power (Boner and Krueger 1991).[1] Since Harberger's (1954) seminal work, a considerable volume of research has been conducted to measure allocative efficiency losses from oligopoly power, and the literature continues to grow, showing evidence of continued interest in estimating such losses (e.g., Cowling and Mueller 1978; Willner 1989; Bhuyan and Lopez 1995; Peterson and Connor 1995). Part of this interest is due to the continued reliance by authorities enforcing competition policies on estimates of actual or potential welfare losses as a measure of industry performance (Preston and Connor 1992). Given their economic importance and previous evidence of anticompetitive behavior (Mueller 1983; Connor et al. 1985; Bhuyan and Lopez 1997), the U.S. food and tobacco manufacturing sectors offer a reasonable context in which to analyze allocative efficiency losses from oligopoly power.

This chapter originally appeared as an article titled "Oligopoly Power and Allocative Efficiency in U.S. Food and Tobacco Industries" by Bhuyan and Lopez in the *Journal of Agricultural Economics* in 1998. This material is reprinted by permission of the *Journal of Agricultural Economics*.

1 In the United States, the principal policy instrument against monopoly and monopolization is Section Two of the Sherman Act, enacted in 1890 and largely based on British common-law court decisions against restraint of trade (Scherer and Ross 1990). Two agencies, the Federal Trade Commission (FTC) and the Antitrust Division of the Department of Justice, have responsibilities to enforce competitive policies (more commonly known as antitrust enforcement) in the United States. These agencies, particularly the FTC, are similar to the United Kingdom's Monopolies Commission or the Federal Cartel Office of Germany. For a survey of major aspects of antitrust policies in the United States, the United Kingdom, Germany, Japan, and France, among others, see Boner and Krueger (1991).

As shown by Peterson and Connor (1995) and Bhuyan and Lopez (1995), the welfare losses are critically determined by assumptions about the modes of conduct, marginal costs, and demand elasticities. One serious limitation of previous studies that measured welfare losses in these sectors is that they used industry profitability rates such as price-cost margins to measure market power in order to compute allocative efficiency losses (e.g., Gisser 1982; Willner 1989; Peterson and Connor 1995; Bhuyan and Lopez 1995).[2] Another limitation of past studies is that they assumed constant marginal costs, which likely biased their loss estimates for industries that exhibit either increasing or decreasing returns to scale.[3] Finally, the food-industry demand elasticities used in previous studies (e.g., the elasticities provided by Pagoulatos and Sorensen (1986) based on 1957–1972 data) were external to the data used in the computation of the losses or were simply assumed; Parker and Connor (1979), for example, assumed demand elasticities to be 0.5 for the entire sector and Willner (1989) assumed these elasticities to be 1.0 for all four-digit SIC (standard industrial classification) food industries.

An alternative is to measure oligopoly power using the new empirical industrial organization (NEIO) approach since this approach does not rely on accounting definitions of oligopoly power (such as profitability rates) and it explicitly parameterizes industry conduct. The NEIO approach yields oligopoly power tests and measures that are well defined theoretically and are less subject to the specificity of a particular year chosen for analysis (Bresnahan 1989). Being in essence an econometric model of market power, the NEIO approach can allow for nonconstant returns to scale and provide demand elasticity estimates within an integrated framework.

The main objective of this study is to provide, for the first time, an upgraded and cross-sectional measure of allocative efficiency losses for

---

2 The shortcomings of using profit rates as indices of market power are outlined by Schmalensee (1989), Bresnahan (1989), and Fisher and McGowen (1983), among others.

3 The constant marginal cost assumption implies that industries are operating at constant returns to scale which, as the productivity literature argues (e.g., Morrison 1990), might be inappropriate in most cases.

U.S. food and tobacco industries using demand, marginal cost, and conduct parameter estimates from an NEIO approach, thereby avoiding to a degree biases introduced by assumptions of demand, cost, or mode of conduct that were subject to the researcher's discretion in former studies. Some comparisons also are drawn between the set of NEIO-based estimates of welfare losses and those from previous studies based on other models.

The Model

The NEIO oligopoly model presented by Appelbaum (1982) is adopted as the analytical framework. Let

$$(1) \qquad \pi_i\,(P, W) = P\,(Y)\,y_i - c_i\,(W, y_i)$$

be the profit function of firm i producing a homogenous product where P is the product price, W is a nonnegative input price vector, y_i is the output produced by firm i, $P\,(Y)$ is the output demand function, c_i is firm i's cost function, and $Y\,(= \sum_i y_i;\, i = 1, 2, \ldots k)$ is the total industry output with k firms.

The profit-maximization condition yields

$$(2) \qquad S_{y_i}\,(1 - \theta_i/\eta) = \partial \ln c_i \,/\, \partial \ln y_i$$

where $S_{y_i} = Py_i \,/\, c_i$ is the revenue-to-cost ratio, $\theta_i = \partial \ln Y \,/\, \partial \ln y_i$ is the conjectural-variation elasticity of firm i, and $\eta = -\partial \ln Y \,/\, \partial \ln P$ is the price elasticity of industry output demand (in absolute value).

From (2), the Lerner Index of firm i's market power is obtained as

$$(3) \qquad L_i = \frac{P - MC_i}{P} = \frac{\theta_i}{\eta}$$

where MC_i denotes the marginal cost of firm i. Using market shares ($S_i = y_i \,/\, Y$) as weights and summing over i yields the industry aggregate Lerner Index of oligopoly power as

$$(4) \qquad L = \frac{P - MC}{P} = \frac{\Phi}{\eta}$$

where MC is the industry-level (weighted) marginal cost and Φ is the aggregated conjectural-variation elasticity.[4] Following Ferguson (1979), the elasticity of marginal cost for each industry is defined as

$$(5) \qquad \rho = \psi + \varepsilon_C - 1$$

where $\psi = \partial \ln \varepsilon_C / \partial \ln Y$ and $\varepsilon_C = \partial \ln C / \partial \ln Y$ and where C is the industry total cost.

Following Dickson and Yu (1989) and Bhuyan and Lopez (1995), let the competitive price and output be indexed to one. Assuming constant elasticity of demand, the industry demand curve for a homogeneous product is represented by $Y = 1 / P^{\eta}$, $\eta > 0$, where Y is industry demand, η is the absolute value of the price elasticity of market demand, and the other notations are the same as defined earlier. Similarly, the inverse of the industry marginal cost curve is then denoted by $Y = MC^{\gamma}$ where γ is the inverse of marginal cost elasticity ($\gamma = 1 / \rho$).

Using the oligopoly power index in (4), the oligopoly price (P_o) for the oligopoly output (Y_o) is given by

$$(6) \qquad P_o = \left(\frac{\eta}{\eta - \Phi}\right) Y_o^{1/\gamma}.$$

Substituting (6) for P_o in the market demand curve yields the oligopoly output:

$$(7) \qquad Y_o = P_o^{-\eta} = \left(\frac{\eta - \Phi}{\eta}\right)^{\frac{\eta\gamma}{\eta+\gamma}}.$$

The allocative efficiency loss (AEL) from oligopoly power is given by

$$(8) \qquad AEL = \int_{Y_o}^{1} \left((1 / Y)^{1/\eta} - Y^{1/\gamma}\right) dY.$$

The allocative efficiency loss in dollar value can be depicted by $AEL \times S$ where S is the actual dollar value of industry sales. The loss as a percent of indexed sales can be expressed as $AEL / (P_o Y_o)$, which is equivalent to the dollar value welfare loss as a percent of actual sales in that industry.

---

4 Note that $\Phi = (0, 1)$ for (competition, pure monopoly). Cowling and Waterson (1976) interpret the aggregate Φ as industry-average conduct and L as the industry-average markup.

Empirical Procedures

The first step of the empirical procedures was to estimate Φ, η, and γ from market-equilibrium conditions for each industry separately as in Appelbaum (1982), in which the demand function for each industry is assumed to take the following form:[5]

$$\ln Y = a + \eta \ln (P / d) + \zeta_q \ln (q / d) + \zeta_Z \ln (Z / d) + \zeta_T \ln T + u_y \quad (9)$$

where d is a price deflator, q is income, Z is the price of substitute products from related industries, T is a time variable to capture trends in taste and preferences, and u_y is an error term.[6]

Following the aggregation cost models used by Ball and Chambers (1982) and Seldon and Bullard (1992), a translog cost function was specified.[7] Following Bhuyan and Lopez (1997), let the industry-level cost function be given by

$$\begin{aligned} \ln C_t = {} & \alpha 0 + \alpha_y \ln Y_t + \Sigma_i \alpha_i \ln w_{i,t} + 0.5\, \beta_{yy} (\ln Y_t)^2 + 0.5\, \beta_{yy} (\ln Y_t)^2 \\ & + 0.5\, \Sigma_i \Sigma_j \beta_{ij} \ln w_{i,t} \ln w_{j,t} + \Sigma_i \beta_{yi} \ln Y_t \ln w_{i,t} + r_t T + 0.5\, r_{tt} T^2 \\ & + r_{ty} T \ln Y_t + \Sigma_i r_{ti} T \ln w_{i,t} \end{aligned} \quad (10)$$

where t is a time subscript, C_t is the industry total cost, $w_{i,t}$ or $w_{j,t}$ is an input price, and T is a time variable. Other notation is as defined previously. Applying Shephard's lemma to (10) yields the industry demand for inputs represented by the factor share equations given by

$$\begin{aligned} \partial \ln C / \partial \ln w_i = w_i x_i / C = S_i = {} & \alpha_i + \Sigma_j \beta_{ij} \ln w_j \\ & + \beta_{yi} \ln Y + r_{ti} T + u_i \end{aligned} \quad (11)$$

---

5 The specification of Equations (9) through (12) is taken verbatim from Bhuyan and Lopez (1997).

6 Note that the price elasticity of substitutes was included in the demand function in an attempt to incorporate product substitutability *among* industries, *not within* an industry.

7 While the necessity to allow nonconstant returns to scale was critical in the decision to assume a translog cost function, the necessity to obtain constant elasticities drove the log-linear specification of output demand.

where x_i is the quantity of the ith input and u_i is a random error that includes errors in optimization.

Using (7), the empirical equation for industry-level profit-maximizing behavior (based on Equation (2)) is given by

$$(12) \qquad S_Y \left(1 - \Phi/\eta\right) = \alpha_y + \beta_{yy} \ln Y + \sum \beta_{yi} \ln w_i + \gamma_{ty} T + u_y$$

where S_Y is the ratio of aggregate reserve to total cost for a product and u_y denotes random disturbances.

Accordingly, following Berndt (1991), a nonlinear three-stage least squares technique was employed to estimate the parameters of equations (9), (11), and (12). The sample consisted of annual data for the 1972–1987 period for thirty-eight U.S. food and tobacco industries at the four-digit SIC level. Details on data sources and management are in Bhuyan (1995). The estimated demand, inverse marginal cost elasticity, and conjectural-variation elasticity for each industry are presented in Table 5.1.

Equation (8) was used to compute an index of welfare losses due to oligopoly power for each industry in the U.S. food and tobacco manufacturing sector. Then the allocative efficiency losses for 1987, both as percent of sales and in terms of actual dollar value, were computed as previously indicated. All estimations were carried out using SHAZAM software.

For comparative purposes, five alternative welfare loss estimates were selected from the literature, all of which used price-cost margins as a measure of oligopoly power and external assumptions or data on demand and marginal cost elasticities. Following Peterson and Connor (1995), three of these scenarios were categorized as traditional structure-conduct-performance (SCP) since they did not model oligopoly pricing behavior, while the other two were categorized as "newer oligopoly models" since they used explicit models of oligopoly pricing. None of the estimates used the NEIO approach to estimate market and industry parameters econometrically. Because most of the estimates did not include the tobacco industries, those industries were excluded from the comparison.

Table 5.1. Allocative Efficiency Loss and Parameter Estimates for U.S. Food and Tobacco Industries

		Allocative Efficiency Loss			Estimated Parameters						
SIC	Industry	Rank	Percent of 1987 Sales	Million Dollars	$\hat{\Phi}$	t_{Φ}	$\hat{\eta}$	t_{η}	$\hat{\rho}$	t_{ρ}	1987 Sales in Million Dollars
2011	Meat Packing	12	4.96	2,232.95	0.219	9.090[b]	0.528	10.371[b]	–0.240	–28.706[b]	44,991
2013	Sausage & Prepared Meats	23	1.70	282.47	0.146	4.746[b]	0.694	13.952[b]	–0.068	–5.440[b]	16,623
2016	Poultry & Egg Processing	6	7.94	1,183.39	0.289	15.152[b]	0.738	24.581[b]	–0.293	–17.037[b]	14,912
2022	Cheese	21	1.79	231.28	0.128	7.119[b]	0.503	22.783[b]	–0.228	–14.800[b]	12,948
2023	Cond. & Evap. Milk	13	4.45	260.46	0.192	4.928[b]	0.324	7.503[b]	–0.242	–15.506[b]	5,857
2024	Ice Cream	20	1.80	70.55	0.112	5.011[b]	0.337	6.506[b]	–0.296	–5.701[b]	3,915
2026	Fluid Milk	16	2.51	517.29	0.182	10.305[b]	0.770	10.810[b]	–0.086	–6.779[b]	20,591
2032	Canned Specialties	34	0.23	12.46	0.063	1.202	0.543	38.204[b]	1.151	35.115[b]	5,350
2033	Canned Fruit & Vegetables	24	1.52	180.85	0.176	6.351[b]	0.728	23.974[b]	0.784	17.157[b]	11,890
2034	Dried Fruit & Vegetables	37	0.12	2.22	0.035	2.455[a]	0.430	3.915[b]	0.391	3.124[b]	1,820
2035	Pickled, Sauces, etc.	4	10.36	523.27	0.348	11.007[b]	0.656	13.518[b]	0.015	1.053	5,050
2041	Flour & Grain Mill	3	26.17	540.90	0.482	6.252[b]	0.710	6.401[b]	–0.931	–39.683[b]	2,067
2043	Cereal, Breakfast Prepared	1	33.43	2,191.84	0.550	22.307[b]	0.767	25.597[b]	–0.365	–5.858[b]	6,557
2044	Rice Milling	35	0.16	2.00	0.032	1.271	0.294	10.664[b]	0.155	5.845[b]	1,235
2047	Pet Food	38	0.07	4.02	0.014	1.130	0.122	9.534[b]	0.565	1.782	5,639
2048	Prepared Feeds	14	3.60	392.01	0.173	11.521[b]	0.386	21.298[b]	–0.187	–44.419[b]	10,899
2051	Bread & Bakery	25	1.12	251.63	0.145	2.377[a]	0.661	13.915[b]	0.841	8.438[b]	22,511
2061	Refined Sugar	17	2.45	135.72	0.146	3.213[b]	0.443	10.255[b]	–0.188	–4.951[b]	5,531

Continued on following page

Table 5.1 (continued)

		Allocative Efficiency Loss			Estimated Parameters						1987 Sales in Million Dollars
SIC	Industry	Rank	Percent of 1987 Sales	Million Dollars	$\hat{\Phi}$	t_{Φ}	$\hat{\eta}$	t_{η}	$\hat{\rho}$	t_{ρ}	
2065	Candy & Confectionary	30	0.46	42.14	0.078	4.869[b]	0.489	36.052[b]	0.801	14.554[b]	9,158
2066	Chocolate & Cocoa Products	27	0.96	28.53	0.110	3.689[b]	0.521	10.271[b]	0.470	7.945[b]	2,960
2067	Chewing Gum	36	0.14	1.53	0.024	0.599	0.163	51.250[b]	1.247	12.658[b]	1,090
2074	Cottonseed Oil Mills	32	0.46	2.14	0.063	8.675[b]	0.430	14.708[b]	–0.009	–0.514	471
2075	Soybean Oil Mills	2	31.79	2,884.58	0.484	19.790[b]	0.938	8.465[b]	–0.475	–85.439[b]	9,074
2076	Vegetable Oil Mills	22	1.77	7.62	0.131	4.667[b]	0.471	8.470[b]	0.033	0.347	432
2077	Animal & Marine Fats & Oils	19	2.12	37.31	0.139	5.833[b]	0.469	17.589[b]	–0.136	–4.567[b]	1,763
2079	Edible Fats & Oils	15	3.20	132.80	0.167	9.163[b]	0.430	19.366[b]	–0.173	–14.688[b]	4,151
2082	Malt Beverages	7	7.22	1,021.64	0.252	9.289[b]	0.515	9.649[b]	–0.588	–40.994[b]	14,150
2084	Wine & Brandy Spirits	28	0.66	21.09	0.065	5.670[b]	0.285	9.226[b]	0.155	8.042[b]	3,179
2085	Distilled Liquor	11	5.19	176.95	0.225	13.186[b]	0.394	14.131[b]	0.376	59.206[b]	3,411
2086	Soft Drinks	5	10.18	2,240.49	0.327	12.206[b]	0.550	11.645[b]	–1.208	–18.719[b]	22,006
2087	Flavor Extracts & Syrups	29	0.56	25.99	0.107	2.298[a]	0.583	11.901[b]	1.524	6.074[b]	4,646
2092	Fresh or Frozen Prep. Fish	33	0.23	13.42	0.061	2.599[a]	0.660	25.274[b]	0.369	27.143[b]	5,752
2095	Roasted Coffee	8	6.87	439.81	0.267	14.907[b]	0.527	25.625[b]	–0.073	–5.062[b]	6,401
2098	Macaroni & Spaghetti	31	0.46	5.99	0.074	3.320[b]	0.436	22.513[b]	0.940	12.601[b]	1,315
2111	Cigarettes	9	5.94	1,032.06	0.336	6.516[b]	0.789	10.460[b]	0.694	1.881	17,372
2121	Cigars	10	5.51	10.56	0.225	12.166[b]	0.533	29.783[b]	–0.391	–7.597[b]	192

Continued on following page

Table 5.1 (continued)

		Allocative Efficiency Loss			Estimated Parameters						
SIC	Industry	Rank	Percent of 1987 Sales	Million Dollars	$\hat{\Phi}$	t_{Φ}	$\hat{\eta}$	t_{η}	$\hat{\rho}$	t_{ρ}	1987 Sales in Million Dollars
2131	Chewing & Smoking Tobacco	18	2.23	24.83	0.207	2.773^{b}	0.662	7.926^{b}	1.014	1.633	1,114
2141	Tobacco Stemm. & Redry.	26	0.96	20.05	0.076	4.031^{b}	0.240	9.250^{b}	0.561	131.163^{b}	2,079
20	All Food Products	–	5.19	14,946.87	–	–	–	–	–	–	288,341
21	All Tobacco Products	–	3.66	759.71	–	–	–	–	–	–	20,752
20–21	All Food & Tobacco	–	5.03	15,706.60	–	–	–	–	–	–	309,099

Notes: Rank= rank of *AEL* where *AEL* equals the baseline estimates of allocative efficiency losses. Percent allocative efficiency losses at the two-digit level are simple averages of respective two-digit-level values. Data at the two-digit level are simple averages of the sample industries. Φ = conjectural-variation elasticity. η = absolute value of elasticity of output demand. ρ = elasticity of marginal cost for 1987. t_{Φ} is the t-statistic for estimated conjectural-variation elasticity ($\hat{\Phi}$).

[a] = significant at 5 percent level.

[b] = significant at 1 percent level.

Results and Discussion

The NEIO-based allocative efficiency loss estimates for each of the thirty-eight industries are presented in Table 5.1, both in dollar terms and as percentages of 1987 sales. Oligopoly-induced allocative efficiency losses in the U.S. food and tobacco manufacturing industries were 5.03 percent of sales in 1987, which amounted to more than $15 billion (Table 5.1). Average welfare losses in the food industries (SIC 20) were 5.19 percent of sales or almost $14.95 billion, while similar losses in the tobacco industries (SIC 21) amounted to more than 3.6 percent of sales or more than $759 million.

In terms of their rankings, the cereal breakfast preparation industry (SIC 2043) ranked at the top with the highest percent of allocative efficiency losses, while the pet food industry (SIC 2047) showed the smallest percent loss. Other food industries with relatively large percent welfare losses were poultry and egg processing (SIC 2016), pickles and sauces (SIC 2035), flour and grain mills (SIC 2041), soybean oil mills (SIC 2075), malt beverages (SIC 2082), soft drinks (SIC 2086), and roasted coffee (SIC 2095)—all of which ranked among the top ten. Among tobacco industries, cigarette (SIC 2111) and cigar (SIC 2121) industries also ranked among the top ten for the entire sector. Previous studies with similar objectives (e.g., Willner 1989; Bhuyan and Lopez 1995) also found evidence of higher than average welfare losses in the same industries. It is noteworthy that some of these industries (e.g., SIC 2043, 2082, 2095, and 2111) have been the focus of antitrust investigations in the past.

To put the allocative efficiency loss estimates in Table 5.1 in perspective, Table 5.2 presents a comparison of the baseline results (i.e., loss estimates of Table 5.1) and five selected past estimates of allocative efficiency losses in the U.S. food-processing industry. The first three scenarios represent losses based on the traditional SCP models while the other two estimates are drawn from "newer" oligopoly models.[8]

For comparison of the NEIO estimates to the five alternative model estimates, the null hypothesis tested is H_o: $\mu_i - \mu_j = 0$ where μ_i is the

---

8 To assess the difference in magnitudes across the six scenarios shown in Table 5.2, t-tests of difference of means of the welfare losses for these six scenarios were conducted following the procedure outlined by Neter, Wasserman, and Whitmore (1993, chapter 14). The analysis was conducted for the thirty-four industries that could be matched across studies. The four tobacco industries reported in Table 5.1 were dropped because they were not common to the six scenarios.

Table 5.2. Comparison of the NEIO-Based Oligopoly-Induced Allocative Efficiency Loss Estimates in U.S. Food Manufacturing Industries with Earlier Estimates

Estimate Type	Mean Losses (percent of sales)	Mean Difference t-Statistics Estimate No. 1	2	3	4	5	6
1. Price-cost margins, monopoly pricing, workable competition	0.436						
2. Price-cost margins, monopoly pricing	1.607	3.317[b]					
3. Price-price margins, monopoly pricing	0.639	2.111[a]	– 2.536[b]				
4. Price-cost margins, Stackelberg price leadership	5.925	3.289[b]	2.851[b]	3.177[b]			
5. Price-cost margins, industrywide oligopoly pricing	2.146	4.376[b]	0.983	3.435[b]	– 2.240[a]		
6. NEIO-based estimates (from Table 5.1)	5.195	3.298[b]	2.290[a]	3.162[b]	– 0.345	2.302[a]	

Note: All computations are based on thirty-four food-processing industries. The allocative efficiency loss estimates for scenarios 1 through 4 are from Connor and Peterson (1994) and for scenario 5 are from column 1 of table 1 of Bhuyan and Lopez (1995).

[a] Mean difference at the 5 percent level of significance.

[b] Mean differences at the 1 percent level of significance.

mean welfare loss in the *i*th row and *j*th column ($i, j = 1, 2, 3, \ldots, 6$). Of the fifteen resultant t-statistics in Table 5.2, nine are statistically significant at the 1 percent level and four more are statistically significant at the 5 percent level. That is, in thirteen of fifteen cases, the mean allocative efficiency loss estimate scenarios are different from each other. Furthermore, with the exception of scenario 4, the NEIO-based loss estimates are significantly different (at either a 1 percent or 5 percent level of significance) from previous estimates. More specifically, the NEIO-based loss estimates are significantly higher than those based on the traditional SCP approach.

It is obvious from Equation (8) that the magnitude of allocative efficiency losses depends on the elasticity of demand (η), the marginal cost elasticity (γ), and the market-power parameters (Φ).

Willner (1989) and Lopez and Bhuyan (1996) raised the issue of the impact of pricing behavior, demand, and marginal cost assumptions on estimated welfare losses. Thus, differences in allocative efficiency loss estimates among the models presented in Table 5.2 arise in part from differences in these parameter values in their respective models and in the different time periods. To illustrate, consider models 5 and 6 in Table 5.2. While both models used Equation (8) to compute allocative efficiency loss, model 5 used demand elasticities from Pagoulatos and Sorensen (1986) and a constant marginal cost assumption. A mean difference test between demand elasticities in Table 5.1 and those of Pagoulatos and Sorensen (1986) shows that these two sets of values are significantly different ($t = 2.325$, $N = 38$) and have a correlation coefficient of 0.135. Similarly, models 1 through 3 also assumed constant marginal costs for each industry. On average, the NEIO-based estimates are statistically indistinguishable from those under the Stackelberg price leadership (model 4).

Conclusions

This study used an NEIO-based approach to estimate allocative efficiency losses in thirty-eight U.S. food and tobacco manufacturing industries. All the necessary parameters, such as elasticities of cost and demand, were first estimated independently for each industry using the same empirical model and then allocative efficiency losses were computed and compared across industries as well as with five alternative estimates available in the literature.

The empirical findings show that welfare losses due to oligopoly power in the U.S. food and tobacco manufacturing industries amounted to 5.03 percent of sales, or slightly more than $15 billion, in 1987. The estimated losses range from 33.43 percent of 1987 industry sales in the cereal breakfast preparation industry (SIC 2043) to 0.07 percent of 1987 sales in the pet food industry (SIC 2047). In terms of food industries alone, the estimated welfare losses amounted to approximately 5.2 percent (or $14.95 billion), while such losses in the tobacco industries were more than 3.6 percent of sales ($759 million) in 1987. These results are consistent with the findings of previous studies as well as with the fact that most of the industries with the highest estimated welfare losses relative to sales have been the focus of public scrutiny for their alleged anticompetitive behavior in the United States. On average, the NEIO-based allocative efficiency loss estimates are significantly larger than previous estimates of allocative efficiency losses in the food sector.

Even though this study has incorporated some novel features, including the use of NEIO-based parameters to compute allocative efficiency losses, the NEIO model used has some important limitations. First, welfare loss estimates do not include endogenous productivity growth, which must be based on dynamic models and is extremely complex and often intractable or inestimable (Geroski 1988). Second, possible price leadership behavior as well as product differentiation within an industry were excluded, although both may be strongly exhibited in certain industries. Another important limitation is the preclusion of possible market failures. For instance, losses from market power in the tobacco industries may be counterbalanced by health-cost savings from higher prices. Incorporation of these dimensions could improve this study and provide fruitful avenues for future research.

References

Appelbaum, E. "The Estimation of the Degree of Oligopoly Power." *Journal of Econometrics* 19 (1982):287–299.

Ball, V.E., and R.G. Chambers. "An Economic Analysis of Technology in the Meat Products Industry." *American Journal of Agricultural Economics* 64(4) (1982):699–709.

Berndt, E.R. *The Practice of Econometrics: Classic and Contemporary*. Edison MA: Addison-Wesley, 1991.

Bhuyan, S. "Oligopoly Power and Allocative Efficiency in Food and Tobacco Industries." Doctoral dissertation, Department of Agricultural and Resource Economics, University of Connecticut, Storrs, 1995.

Bhuyan, S., and R.A. Lopez. "Welfare Losses under Alternative Oligopoly Regimes: The U.S. Food and Tobacco Manufacturing Industries." *Journal of Agricultural and Applied Economics* 27 (1995):577–587.

———. "Oligopoly Power in the Food and Tobacco Industries." *American Journal of Agricultural Economics* 79 (1997):1035–1043.

Bresnahan, T.F. "Empirical Studies of Industries with Market Power." *Handbook of Industrial Organization*, vol. 2. R. Schmalensee and R.D. Willig, eds., pp. 1011–1057. Amsterdam The Netherlands: Elsevier, 1989.

Boner, R.A., and R. Krueger. "The Basics of Antitrust Policy: A Review of Ten Nations and the European Communities." Washington DC: The World Bank Technical Paper 160, 1991.

Connor, J.M., and E.B. Peterson. "New Estimates of Welfare and Consumer Losses in U.S. Food Manufacturing." West Lafayette IN / Blacksburg VA: Purdue University / Virginia Polytechnic Institute and State University, 1994.

Connor, J.M., R.T. Rogers, B.W. Marion, and W.F. Mueller. *The Food Manufacturing Industries: Structure, Strategies, Performance and Policies*. Lexington MA: Lexington Books, 1985.

Cowling, K., and D.C. Mueller. "The Social Costs of Monopoly Power." *Economic Journal* 88 (1978):727–748.

Cowling, K., and M. Waterson. "Price-Cost Margins and Market Structure." *Economica* 43 (1976):267–274.

Dickson, V.A., and W. Yu. "Welfare Losses in Canadian Manufacturing under Alternative Oligopoly Regimes." *International Journal of Industrial Organization* 7 (1989):257–267.

Ferguson, C.E. *The Neoclassical Theory of Production and Distribution*. Cambridge England: Cambridge University Press, 1979.

Fisher, F.M., and J.J. McGowan. "On the Misuse of Accounting Rates of Return to Infer Monopoly Profits." *American Economic Review* 73 (1983):82–97.

Geroski, P.A. "In Pursuit of Monopoly Power: Recent Quantitative Work in Industrial Economics." *Journal of Applied Econometrics* 3 (1988):107–123.

Gisser, M. "Welfare Implications of Oligopoly in U.S. Food Manufacturing." *American Journal of Agricultural Economics* 63 (1982):616–624.

Harberger, A.C. "Monopoly and Resource Allocation." *American Economic Review* 2 (1954):77–87.

Lopez, R.A., and S. Bhuyan. "A Comparison of Oligopoly Welfare Loss Estimates for U.S. Food Manufacturing: Comment." *American Journal of Agricultural Economics* 78 (1996):706–709.

Morrison, C.J. "Market Power, Economic Profitability and Productivity Growth Measurement: An Integrated Structural Approach." Cambridge MA: National Bureau of Economic Research Working Paper 3355, 1990.

Mueller, W.F. "Market Power and Its Control in the Food System." *American Journal of Agricultural Economics* 65(5) (1983):855–863.

Neter, J., W. Wasserman, and G.A. Whitmore. *Applied Statistics*. Boston MA: Allyn and Bacon, 1993.

Pagoulatos, E., and R. Sorensen. "What Determines the Elasticity of Industry Demand?" *International Journal of Industrial Organization* 3(4) (1986):237–250.

Parker, R.C., and J.M. Connor. "Estimates of Consumer Loss due to Monopoly in the U.S. Food Manufacturing Industries." *American Journal of Agricultural Economics* 6(4) (1979):626–639.

Peterson, E.B., and J.M. Connor. "A Comparison of Oligopoly Welfare Loss Estimates for U.S. Food Manufacturing." *American Journal of Agricultural Economics* 77 (1995):300–308.

Preston, W.P., and J.M. Connor. "An Economic Evaluation of Federal Antitrust Activity in the Manufacturing Industries." *Antitrust Bulletin* Winter (1992):969–996.

Scherer, F.M., and D. Ross. *Industrial Market Structure and Economic Performance.* Boston MA: Houghton Mifflin Co., 1990.

Schmalensee, R. "Inter-industry Studies of Structure and Performance." *Handbook of Industrial Organization*, vol. 2. R. Schmalensee and R.D. Willig. eds., pp. 951–1010. Amsterdam The Netherlands: Elsevier, 1989.

Seldon, B.J., and S.H. Bullard. "Input Substitution, Economies of Scale and Productivity Growth in the U.S. Upholstered Furniture Industry." *Applied Economics* 24 (1992):1017–1025.

Willner, J. "Price Leadership and Welfare Losses in U.S. Manufacturing: Comment." *American Economic Review* 79(3) (1989):604–613.

Part IV
Oligopoly and Oligopsony

6

Testing Oligopolistic and Oligopsonistic Behavior: An Application to the U.S. Meat-Packing Industry

Azzeddine M. Azzam and Emilio Pagoulatos

Recent empirical work in industrial organization has popularized the use of conjectural-elasticity models to test price-taking behavior. In these models, firms are hypothesized to simultaneously and independently choose their output levels given their beliefs about rivals' reactions to their output choice, and these beliefs are called conjectural variations. A firm's conjectural elasticity is the conjectural variation of the firm multiplied by its market share. Several studies have estimated such models (see Geroski (1988) and Bresnahan (1987) for surveys), but virtually all applications consider price-taking behavior only in the output market. Only one (Schroeter 1988) considers the possibility of imperfections in output and factor markets. From the standpoint of econometric estimation, inferring pricing distortions from an oligopoly/oligopsony model would involve a simple adaptation of an oligopoly model (such as that of Appelbaum (1982) for example) where, in addition to the parameter estimates of a cost function, an output demand function, and conduct as parameterized by the conjectural elasticity, one also needs information on the parameter estimates of the supply function(s) of the oligopsonized input(s). However, the problem is that such adaptation is possible only if certain restrictions are imposed on the firm's cost function.

Since the cost function has as arguments the price(s) of the input(s), deriving an expression for the conjectural elasticity in the factor market

---

This chapter originally appeared as an article titled "Testing Oligopolistic and Oligopsonistic Behavior: An Application to the U.S. Meat-Packing Industry," by Azzam and Pagoulatos in the *Journal of Agricultural Economics* in 1990. This material is reprinted by permission of the *Journal of Agricultural Economics*.

is not straightforward unless the production technology is restricted to be of fixed proportions between the output and the oligopsonistically purchased input. Consequently, the conjectural elasticities in the imperfect output and input markets turn out to be identical since the oligopsonized input and output can be represented by the same variable in the profit function (see Schroeter (1988)). The problem with this formulation is that identical conduct in the two markets is not an implication of oligopoly/oligopsony theory but a result of the imposed technology.

The purpose of this chapter is to propose an alternative empirical model to test price-taking behavior in output and factor markets without restricting the conjectural elasticities in the respective markets to being identical. To do so, we use a production-function approach that (1) allows all inputs to be used in variable proportions, and (2) allows the derivation of market-specific conjectural elasticities. This enables us to provide a parametric test for identical conduct and to measure the degree of market power in output and factor markets.

The framework is applied to a time series of aggregate data for the U.S. meat-packing industry. This particular industry has often been at the center of controversy over conditions of competition in both its livestock procurement and wholesale meat markets (Nicholls 1941; Yeager 1981). Concern about the potential exercise of oligopsony and oligopoly power by meat packers has grown out of a reversal that began in the 1970s of an earlier trend toward deconcentration in the market. The four largest beef packers increased their share of the national market from 29 percent in 1972 to 45 percent in 1982. The four largest firms in hog slaughtering and processing increased their national share from 32 percent in 1972 to 36 percent in 1982 (U.S. House of Representatives 1980).

In the next section, the theoretical model is presented. The estimating model and empirical results are presented in the third section. Based on published testimonies to various congressional committees and previous research, our prior notion was that the U.S. meat-packing industry exerts some degree of market power. Moreover, we suspected that market power exercised by the industry is different in the two markets and that it is likely to be greater in the input (livestock) markets. In contrast to the meat market, which is national in scope, the market for slaughter animals generally extends a few hundred miles (*Monfort of Colorado v. Cargill, Inc. and Excel Corp.*) and has over time involved fewer and fewer outlets to which producers can market their live animals.

Theoretical Framework

Consider an industry in which N firms produce a homogeneous output using M inputs. Let the jth firm's technology be defined by the production process f:

$$(1) \qquad q_j = f(x_{1j}, X_{2j}, \ldots, X_{Mj})$$

where q_j is output produced (meat for our intended purpose), x_{1j} is the farm input (livestock), and x_{kj} (indexed $k = 2, \ldots, M$) represents nonfarm inputs. Furthermore, assume that each firm exercises some market power in purchasing the farm input x_{1j} and in selling its output q_j but is a price-taker in the market for other inputs.

Let the inverse (wholesale) market demand curve facing the industry in its output market be given by

$$(2) \qquad P = g\,(Q)$$

where P and $Q = \Sigma_{j=1}^{N} q_j$ are the market meat price and total industry output, respectively. The inverse market supply function for the agricultural input is given by

$$(3) \qquad w_1 = h\,(X_1)$$

where w_1 and $X_1 = \Sigma_{j=1}^{N} x_{1j}$ are the market livestock price and total industry livestock input, respectively. Denoting the price of nonfarm inputs by $w_2, \ldots, w_m$ and assuming that each firm is a profit maximizer, the problem for the jth firm is to choose X_{kj} (indexed $k = 1, 2, \ldots, M$) so as to maximize profits, II_j:

$$(4) \qquad \text{II}_j = Pq_j - \sum_{k=1}^{M} w_k x_{kj} \quad \text{for } j=1, 2, \ldots, N$$

subject to (2) and (3). The first-order conditions corresponding to this profit-maximization problem are given by:

$$(5) \qquad \partial \text{II}_j / \partial x_{1j} = P\,(1 - \theta_j/\eta)\, f_{x1j} - w_1\,(1 + \varphi_j/\in) = 0$$

$$(6) \qquad \partial \text{II}_j / \partial x_{kj} = P\,(1 - \theta_j/\eta)\, f_{xkj} - w_k = 0 \qquad j = 1, \ldots, N;\, k = 2, \ldots, M$$

or

(7) $w_1 / P = (1 - \theta_j/\eta) f_{x1j} - w_1/P \mid \varphi_j / \in \mid = 0$

(8) $w_k / P = (1 - \theta_j/\eta) f_{xkj}$ $\quad j = 1, \ldots, N; \quad k = 2, \ldots, M$

where $\eta = \partial Q/\partial P \times P/Q$ is the absolute value of the price elasticity of output demand;

$\in = \partial X_1/\partial w_1 \times w_1/X_1$ is the market price elasticity of the farm input supply;

$\theta_j = \partial Q/\partial q_j \times q_j/Q$ is the jth firm's conjectural elasticity in the output market;

$\varphi_j = \partial X_1/\partial x_{1j} \times x_{1j}/X_1$ is the jth firm's conjectural elasticity in the farm input market; and

$f_{xkj} = \partial q_j/\partial x_{kj}$ is the marginal product of the kth input used by firm j.

In theory, the conjectural elasticities, θ_j and φ_j, as shown in (7) and (8), provide useful benchmarks for testing for price-taking behavior or degree of competitiveness (Appelbaum 1982). Assuming positive marginal products, if both θ_j and φ_j are equal to zero, we have the perfectly competitive case where each firm equates the marginal product of each input to its real price. In the extreme case where both θ_j and φ_j are equal to one, we obtain the monopoly and monopsony case where the firm equates the marginal revenue product to the marginal expense of the farm input. Other combinations of market structures can be identified: monopsony in the farm input market and perfect competition in the output market ($\theta_j = 0$ and $\varphi_j = 1$) or vice versa. Alternatively, one can identify the location of the firm on the continuum between the two poles of market structure as θ_j and φ_j can take on individual values between zero and one. The ratios of the conjectural elasticities to the demand and supply elasticities measure the degree of market power in the output and input market, respectively.

In practice, an absence of panel data on firm-level output and employment levels of factors of production means that the system embodied in (7) and (8) cannot be readily estimated. This limitation leads us to consider the problem at the industry level. In doing so, however, an additional assumption must be maintained to make the preceding analysis applicable to the behavior of the industry as whole. The assumption is that, in equilibrium, the conjectural elasticities are invariant across firms (Appelbaum 1982); i.e., $\theta_1 = \theta_2 = \ldots = \theta_N = \Theta$ and $\varphi_1 = \varphi_2 = \ldots = \varphi_N = \Phi$. Maintaining the invariance of the conjectural elasticities across firms allows us to drop the subscript *j* on the marginal products of the *M* inputs and think of the result as a weighted industry marginal product where the weights are the respective firms' shares in the input market; i.e., $f_{xk} = \sum_{j=1}^{N} f_{xkj}$ for $k = 1, \ldots, M$.[1] The aggregate analog of the optimality conditions, (7) and (8), may now be written as

(9) $$w_1 / P = (1 - \Theta/\eta)\, fx_1 - w_1/P \times \Phi/\in \text{ and}$$

(10) $$w_k / P = (1 - \Theta/\eta)\, fx_k \quad \text{for } k = 2, \ldots, M.$$

The ratios Θ/η and $\Phi/\in$ represent industrywide indices of market power in the output and input markets, respectively.

Empirical Model and Results

To estimate the model of industry oligopoly/oligopsony behavior developed in the previous section (Equations 9 and 10), test the hypothesis of price-taking behavior, and measure the indices of market power in the meat-packing industry, we need to select a functional form for the production function. The preferred form would not impose severe *a priori* constraints on the production characteristics in the industry. One function suitable for our purposes is the transcendental logarithmic (translog) production function (Christensen, Jorgenson, and Lau 1971).

---

1 Alternatively, one could assume that the marginal products are identical across firms and have the industry-level conjectural elasticities be a weighted average of the conjectural elasticities of the individual firms in which the weights are the individual firms' output shares. However, as a reviewer pointed out, using output market shares as weights in a time-series model means that the conjectural elasticities are no longer constant since the changing structure of the industry over time will change the aggregate values of the conjectural elasticities over time as well.

(11) $$\ln Q = \beta_0 + \sum_{k=1}^{4} \beta_k \ln X_k + 1/2 \sum_{k=1}^{4} \sum_{j=1}^{4} \beta_{kj} \ln X_k \ln X_j$$

where Q is red meat production. We assume that, in addition to the livestock input (X_1), there are three competitively priced inputs utilized by the meat-packing industry: labor (X_2), capital (X_3), and nonlivestock material (X_4).

From Equation (11), the marginal product for the kth input is

(12) $$fx_k = [\beta_k + \sum_{j=1}^{4} \beta_{kj} \ln X_j]\ Q/x_k \text{ for } k = 1, 2, 3, 4.$$

Substituting (12) into (9) and (10) and rearranging lead to the following system of $k + 1$ equations:

$$\ln Q = \beta_0 + \sum_{k=1}^{4} \beta_k \ln X_k + 1/2 \sum_{k=1}^{4} \sum_{k=1}^{4} \beta_{kj} \ln X_k \ln X_j$$

$$S_1 = \{\frac{1-(\Theta/\eta)}{1+(\Phi/\in)}\} \{\beta_1 + \sum_{j=1}^{4} \beta_{1j} \ln X_j\}$$

(13) $$S_2 = \{1-(\Theta/\eta)\} \{\beta_2 + \sum_{j=1}^{4} \beta_{2j} \ln X_j\}$$

$$S_3 = \{1-(\Theta/\eta)\} \{\beta_3 + \sum_{j=1}^{4} \beta_{3j} \ln X_j\}$$

$$S_4 = \{1-(\Theta/\eta)\} \{\beta_4 + \sum_{j=1}^{4} \beta_{4j} \ln X_j\}$$

where $S_k = w_k X_k / PQ$ (for $k = 1, 2, 3, 4$) is the share equation for the kth input.

The data used in the estimation of the system of equations (13) are annual aggregate time series for the U.S. meat-packing industry from 1959 through 1982. The sample period was dictated by data availability. The year 1982 is the most recent year for which census data were available. The definition of variables and sources are as follows.

$Q =$ Total U.S. commercial red meat (beef, pork, sheep, and lamb) production (million pounds carcass weight) obtained from various issues of *Livestock and Meat Statistics* by the U.S. Department of Agriculture's (USDA's) Economic Research Service.

$S_1 = w_1 X_1 / PQ$ Cost of livestock input relative to the value of shipment (U.S. Department of Commerce, Bureau of the Census, *Census of Manufacturers*, various issues).

$S_2 = w_2 X_2 / PQ$ Total labor wages in millions of dollars relative to the value of shipments (U.S. Department of Commerce, Bureau of the Census, *Census of Manufacturers;* U.S. Department of Labor, Bureau of Labor Statistics, *Employment and Earnings*). Labor wages were calculated as the product of the total hours worked by production workers (X_2) and the average hourly earnings (w_2) of production workers for SIC2011 (meat-packing plants).

$S_3 = w_3 X_3 / PQ$ The value of capital input relative to the value of shipments obtained from the U.S. Department of the Treasury's *Corporation Source Book of Statistics of Income* (1958–1982). The values for the capital input and capital cost were calculated following the procedure outlined in Gollop and Roberts (1979). The capital input (X_3) is the sum of net depreciable and depletable assets, land, and inventories. The cost of capital (w_3) was computed by dividing the sum of interest, depreciation, and depletion expenses by x_3.

$S_4 = w_4 X_4 / PQ$ Cost of nonlivestock material input relative to the value of shipments. The nonlivestock material input (X_4) was computed as the quotient formed by dividing the cost of nonlivestock material by the price index of intermediate inputs in manufacturing (w_4) published by the U.S. Department of Commerce, Bureau of the Census, in *Statistical Abstracts* (various issues).

For empirical implementation, we assume that the production function and the share equations are stochastic due to technical errors and errors in optimization, respectively. The errors are assumed to be additive and jointly normally distributed with zero mean and a constant variance-covariance matrix. Since output and all inputs in the model are endogenous, we use the method of instrumental variables: each regressor in (13) is regressed on a set of variables considered exogenous to the meat-packing industry. The instruments include the labor wage, the price of capital, the price of materials, the price of poultry, the number of animal units as of January 1, per capita disposable income, and a time trend. For estimation, we use the iterative nonlinear three-stage least squares (IT3SLS) technique (Gallant 1987). Finally, as exogenous point estimates of market price elasticities (Gelfand and Spiller 1987; Gollop and Roberts 1979; Roberts 1984), we use the following values: 0.49 (demand, η) and 0.16 (supply, $\in$).[2]

In Table 6.1 we present the results. The R^2 figures, computed as one minus the ratio of the sum of squared errors and the corrected sum of the squares of the actual variables, are 0.56 for S_1, 0.52 for S_2, 0.57 for S_3, 0.57 for S_4, and 0.95 for Q. The asymptotic t-ratios (in parentheses) indicate that all but three of the production-function parameters are statistically significant at least at the 5 percent level. All estimates are after correction for first-order serial correlation.

The hypothesis of price-taking conduct in both the input and the output markets, $Ho: \Theta = \Phi = 0$, was soundly rejected as the observed likelihood-ratio test of 24.15 is above the 1 percent value of 9.21 for an X^2 distribution with two degrees of freedom. The hypothesis of identical conduct or identical degree of competition in the input and output markets could not be rejected. The likelihood-ratio test of 1.26 was below the 1 percent value of 6.635. Thus, we conclude that, while the industry may be characterized as noncompetitive in both the factor and the output markets, the degree of noncompetitiveness is similar in the two markets.

Having obtained estimates of the conjectural elasticities, we next examine the degree of oligopoly and oligopsony power. As was shown

---

2 The demand and supply elasticities were estimated from the following simultaneous demand-supply system for red meat. Per capita demand for output (Q) was specified as a function of the price index of total red meat and poultry (USDA, *Agricultural Statistics*, various issues) and of per capita disposable income (Council of Economic Advisors, *Economic Report to the President*, various issues). Supply of livestock (in millions of pounds) was specified as a function of the index of prices received by farmers for meat animals, the price index of prices received for feed grains, and an index of the inventory of meat animal units (USDA, *Agricultural Statistics*, various issues).

in the text, these measures are respectively given by Θ / η and $\Phi / \in$. Thus, the lower the price elasticity of demand (supply) is, the higher is the degree of oligopoly (oligopsony) power. An examination of Table 6.1 shows that the average degrees of oligopoly and oligopsony power over the sample period are 0.46 with a standard error of 0.04 and 1.1 with a standard error of 0.13, respectively. Indeed, as suspected, the degree of market power in the factor market is higher, and the value of 1.1 is not statistically different from a collusive oligopsony.

Table 6.1. Estimated Parameters of the Model

Translog Production-Function Parameters	**Estimate**	**t-Ratio**
β_0	–0.005	(–1.0)
β_1	0.932	(15.9)
β_2	0.009	(6.2)
β_3	0.026	(6.4)
β_4	0.085	(6.3)
β_{11}	0.212	(3.7)
β_{22}	0.010	(1.1)
β_{33}	0.023	(4.9)
β_{44}	0.102	(4.2)
β_{12}	0.030	(3.2)
β_{13}	0.018	(2.11)
β_{14}	–0.116	(–5.3)
β_{23}	–0.0006	(–0.39)
β_{24}	–0.014	(–3.6)
β_{34}	–0.032	(–4.0)
Conjectural Elasticity Parameters		
Output Market: Θ	0.223	(2.3)
Input Market: Φ	0.178	(3.0)
Indices of Market Power		
Output Market: Θ / η	0.460	(11.5)
Input Market: $\Phi / \in$	1.100	(8.5)

Conclusions and Limitations

The purpose of this chapter was to extend the traditional conjectural approach to the analysis of imperfections in output and factor markets simultaneously. Starting from the specification of a production function, the econometric analysis is based on the formulation and estimation of a simultaneous-equation model consisting of a production function, first-order conditions associated with factor employment, and two conjectural elasticities to parameterize the industry's oligopoly and oligopsony equilibria. Our results suggest that, for the sample period considered, while the U.S. meat-packing industry may be characterized as noncompetitive in both the factor and output markets, the degree of noncompetitiveness is similar in the two markets. However, the degree of market power in the input (livestock) market was found to be significantly higher than that in output (meat) market.

As with any study of any industry, our results should be interpreted in light of data constraints and model construction. A more appropriate data set would contain information on output and factor employment by type of meat. Unfortunately, the data from the census are reported for aggregate meat output only. More importantly, the relevant unit of observation in a model of oligopolistic interdependence is the firm. Until such data are available, little can be known about how the presence or absence of market power is obscured by too much or too little aggregation. Finally, as is well known, the choice of a flexible cost or production function also can affect the final results of any behavioral model.

References

Appelbaum, E. "The Estimation of the Degree of Oligopoly Power." *Journal of Econometrics* 19 (1982):287–299.

Bresnahan, T.F. "Empirical Studies of Industries with Market Power." Stanford CA: Stanford University Center for Economic Policy Research Publication 95, 1987.

Christensen, L.R., D.W. Jorgenson, and L.J. Lau. "Conjugate Duality and the Transcendental Logarithmic Production Function." *Econometrica* 39 (1971):255–256.

Council of Economic Advisors. *Economic Report to the President*. Washington DC: U.S. Government Printing Office, various issues.

Gallant, R.A. *Nonlinear Statistical Models*. New York NY: Wiley, 1987.

Gelfand, M.D., and P.T. Spiller. "Entry Barriers and Multiproduct Oligopolies: Do They Forbear or Spoil?" *International Journal of Industrial Organization* 5 (1987):101–113.

Geroski, P.A. "In Pursuit of Monopoly Power: Recent Quantitative Work in Industrial Economics." *Journal of Applied Econometrics* 3 (1988):107–124.

Gollop, F., and M. Roberts. "Firm Interdependence in Oligopolistic Markets." *Journal of Econometrics* 10 (1979):313–331.

Monfort of Colorado v. Cargill, Inc. and Excel Corp. Civil Action 83-F-1318, Opinion and Order (U.S. District Court, Colorado).

Nicholls, W.H. *Imperfect Competition within Agricultural Industries.* Ames IA: Iowa State College Press, 1941.

Roberts, M.J. "Testing Oligopolistic Behavior." *International Journal of Industrial Organization* 2 (1984):367–383.

Schroeter, J.R. "Estimating the Degree of Market Power in the Beef Packing Industry." *Review of Economics and Statistics* 70 (1988):158–162.

U.S. Department of Agriculture, Economic Research Service. *Livestock and Meat Statistics.* Washington DC, various issues.

U.S. Department of Agriculture. *Agricultural Statistics.* Washington DC, various issues.

U.S. Department of Commerce, Bureau of the Census. *Census of Manufacturers.* Washington DC, various issues.

———. *Statistical Abstracts.* Washington DC, various issues.

U.S. Department of Labor, Bureau of Labor Statistics. *Employment and Earnings.* Washington DC, various issues.

U.S. Department of the Treasury, Internal Revenue Service. *Corporation Source Book of Statistics of Income.* Washington DC, annual issues, 1958–1982.

U.S. House of Representatives Committee on Small Business. *Small Business Problems in the Marketing of Meat and Other Commodities.* Washington DC, U.S. 96th Congress, 2nd Session, 1980.

Yeager, M. *Competition and Regulation: The Development of Oligopoly in the Meat Packing Industry.* Greenwich CT: JAI Press, 1981.

7

An Assessment of the Impact of Food-Industry Market Power on U.S. Consumers

Richard J. Sexton and Mingxia Zhang

Marketing activities account for the majority share of costs for most foods, and the share of costs due to marketing is rising over time. Consider that the farm share of the U.S. Department of Agriculture's (USDA's) "market basket" of food products remained stable at about 40 percent from 1960 through 1980 but has declined rapidly since then to 30 percent in 1990 and 22 percent in 1998. In 1996, U.S. consumers spent $547 billion on food, excluding imports and seafood. A second measure of the farm-retail price spread, the marketing bill, was $424 billion, leaving $123 billion in farm value or 23 percent. This measure of the farm share also has been declining steadily over time, falling from 41 percent in 1950 to 31 percent in 1980 and then to 24 percent in 1990.

What explains this growth in the marketing sector's share of the U.S. food dollar? Certainly, changes in consumers' buying habits are one explanation. Increased demand for convenience, such as reduced time for meal preparation, and consumption of more meals away from home are examples of trends that naturally cause more marketing services to be consumed per food dollar expended. Another consideration is the rapid increase in concentration in all stages of the food marketing sector. High concentration may be associated with the exercise of market power. Fundamental to the exercise of market power is restriction of sales relative to the competitive level so as to influence prices and increase profits. Because

---

A shorter version of this chapter originally appeared as an article titled "An Assessment of the Impact of Food Industry Market Power on U.S. Consumers" by Sexton and Zhang in *Agribusiness: An International Journal* in 2001. This material is reprinted with permission of Wiley-Liss Inc., a subsidiary of John Wiley and Sons, Inc.

consumers' welfare from food consumption is, in general, monotonic in the quantity of food brought to the marketplace, exercise of market power *anywhere* in the market chain will reduce consumers' welfare.

This chapter focuses specifically on the possible impacts on consumers of concentration and market power in the food chain. We first present an overview of key trends in food-market concentration. Then we develop a simple and convenient, but relatively flexible, model of a food-market channel to study the potential impacts of market power at alternative stages of the market chain on the magnitude and distribution of economic welfare. We then turn to the impact of market power in the food sector on economic welfare and the distribution of returns among producers, consumers, and marketers. This topic is of longstanding concern among agricultural economists and has featured some relatively recent contributions. One limitation is that work to date has emphasized food manufacturers' market power as sellers, yet concerns also have been expressed about manufacturers' oligopsony power as buyers of raw agricultural products and about the emerging power of retailers in the food chain. What are the potential welfare implications of both buyer and seller market power and of market power at successive stages in a food market? To help answer these questions, we utilize our analytical model to conduct a simulation analysis of potential welfare impacts for plausible ranges of market parameters, including processor and retailer market power. We conclude with policy implications of the analysis.

To maintain focus on the consequences of market power, we necessarily abstract away from several factors that contribute to determining welfare from food consumption. In particular, we focus on individual food commodities and utilize a partial equilibrium framework. Issues of product differentiation, variety, and interactions among food products are ignored. We also ignore for the most part the possible, even likely, linkages between concentration and efficiency in food marketing.

A Brief Overview of Trends and Issues in Food-Market Concentration

Concentration in food manufacturing in the U.S. has increased due to rapid growth of the largest manufacturers, fueled by mergers and consolidations. Rogers (1997) estimates that the top one hundred U.S. food manufacturers accounted for 80 percent of sales in 1995, doubling the cumulative share they held in 1954. Rogers reports increasing concentration in most of the fifty-three food and tobacco industries classified by four-digit standard industrial classification (SIC) codes; the average four-firm concentration ratio (CR4) in 1992 was 53.3 percent compared to 43.9 percent in 1967.

These reported concentration ratios apply only to food manufacturers in their role as sellers, although the power of food handlers as buyers from farmers also has long been an issue in agriculture. The relevant geographic and product markets for the purchase of raw agricultural commodities are usually narrower and, thus, concentration is normally higher than in the relevant markets for associated finished products (Rogers and Sexton 1994). Geographic markets are narrower because farm products are often bulky and/or perishable, causing shipping costs to be high, restricting the products' geographic mobility, and limiting farmers' access to buyers located close to the production site.[1] In the product market dimension, processors' needs for agricultural products are highly specialized. Other inputs normally cannot be substituted for a given farm product, nor can a given farm product substitute for other agricultural product inputs in alternative production processes. For example, although pork, beef, and poultry may substitute for one another in consumers' budgets, they do not substitute at all as inputs into a particular processing plant.

On the farm-product supply side, farmers are often specialized in the production of particular commodities through extensive investments in sunk assets, which represent exit barriers for these farmers and cause the farm-product supply to be inelastic. High buyer concentration in relevant farm-product markets and an inelastic farm supply represent structural conditions that are conducive to the exercise of oligopsony power by processors and handlers. Although such power acts directly on farm producers, its consequence is to reduce production and sales at the farm level, an impact that reverberates through the marketing chain, causing higher prices and reduced welfare to food consumers.

Food retailing in the U.S. is dominated increasingly by large chains of grocery supermarkets and supercenters. Retail markets are inherently local in nature, making national or regional measures of concentration meaningless from the perspective of retailer oligopoly power over consumers. In terms of local markets, the average CR4 in grocery retailing

---

1 In the U.S., maximum hauling distances are about fifty miles for broilers (Rogers 1998), 150 miles for processing tomatoes (Durham, Sexton, and Song 1996), and 650 miles for beef (Hayenga, Koontz, and Schroeder 1996).

for U.S. metropolitan areas has increased due to consolidations and mergers among grocery chains, reaching 68 percent in 1987, the most recent data available (Franklin and Cotterill 1993). However, because the large chains normally use consolidated purchasing, concentration at the national level has implications for retailers' interactions with food manufacturers and wholesalers.

A Simple Model of an Agricultural Market with Both Processing and Retailing Sectors

Consider a market where a primary agricultural product is procured from farmers by a processing sector, which then transfers the finished product to an independent retailing sector, where it is sold to consumers. Consumers' inverse demand for the retail product is

$$(1) \qquad P^r = D\,(Q^r \mid \boldsymbol{X})$$

where Q^r is the market quantity of processed product, P^r is the market price, and $\boldsymbol{X}$ denotes unspecified demand shifters.

Farmers are assumed to be price-takers in their output market. Inverse farm supply of the raw commodity is expressed as

$$(2) \qquad P^f = S\,(Q^f \mid \boldsymbol{Y})$$

where P^f is the price received at the farm, Q^f is the total volume of farm shipments, and $\boldsymbol{Y}$ represents unspecified supply shifters.

To sharpen the focus of the model on the implications of possible market power in the industry, we make a number of simplifying assumptions about technologies for the processing and retailing sectors. Processors are assumed to utilize a fixed-proportion, constant-returns technology to convert the raw farm product into a finished product and, similarly, food retailers' technology also exhibits both fixed proportions and constant returns and is separable across the various products sold at retail. For convenience and without further loss of generality, we choose to measure units so that $Q^f = Q^w = Q^r = Q$ where the superscripts f, w, and r denote the farm, wholesale, and retail sectors, respectively. Given these assumptions, changes in market concentration have no cost-side effects, enabling the analysis to focus solely on competitive impacts.

Denote a representative processing firm's volume of raw product purchases by q^f. Given our assumptions, the representative processor's variable cost function can be written as

(3) $$C^w = c^w\,(\boldsymbol{V}^w)\,q^f + P^f q^f$$

where $c^w\,(\boldsymbol{V}^w)$ represents the constant processing costs per unit of raw product processed and $\boldsymbol{V}^w$ is the vector of prices for variable processing inputs.

Denote a representative retailer's volume of wholesale purchases by q^w. A representative retailer's variable cost function for selling the product is

(4) $$C^r = c^r\,(\boldsymbol{V}^r)\,qw + P^w q^w$$

where $c^r\,(\boldsymbol{V}^r)$ represents the constant retailing costs per unit of wholesale product sold and $\boldsymbol{V}^r$ is the vector of prices for variable retailing inputs. To simplify notation, we drop further reference to the exogenous variables $\boldsymbol{X}$, $\boldsymbol{Y}$, $\boldsymbol{V}^w$, and $\boldsymbol{V}^r$.

We now derive the implications of various combinations of oligopoly and oligopsony power in the processing and/or the retailing sector on total market surplus and the distribution of surplus among consumers, producers, and marketers. Throughout the analysis, we assume that farm producers and consumers act as competitive price-takers.

Case I: Manufacturers or Retailers May Have Both Oligopsony and Oligopoly Power

In this case, we assume that *either* retailers or manufacturers are price-takers; in other words, given market power in one of the marketing sectors, the other sector is competitive. Given the model structure, the output, farm price, consumer price, and aggregate welfare effects are identical for a given magnitude of market power regardless of whether the power is held by food processors or by food retailers. To simplify the exposition, we develop the case where food manufacturers may exercise market power and retailers are competitive. In this case, the retail price is $P^r = P^w + c^r$.

A representative processing firm's profit function can be expressed as

(5) $$\pi^W = (D(Q) - c^r)\, q - S(Q)\, q - c^W q$$

where q is the representative firm's level of output and volume of farm product purchases. The first-order necessary condition for maximizing Equation (5) is

(6) $$\partial\pi^W / \partial q = P^W + \frac{\partial D(Q^W)}{\partial Q^W}\frac{\partial Q^W}{\partial q}\, q - (P^f + c^W) - \frac{\partial S(Q^f)}{\partial Q^f}\frac{\partial Q^f}{\partial q}\, \mathrm{q} = 0.$$

Equation (6) can be written in elasticity form as

(6') $$P^W\left(1 - \xi^W/\eta^W\right) = P^f\left(1 + \theta^f/\varepsilon^f\right) + c^W$$

where

$$\varepsilon^f = \frac{\partial Q^f}{\partial P^f}\frac{P^f}{Q^f}$$

is the market price elasticity of supply of the raw product,

$$\eta^W = \frac{\partial Q^W}{\partial P^W}\frac{P^W}{Q^W}$$

is the absolute value of the market price elasticity of derived demand for the processed product, and

$$\theta^f = \frac{\partial Q^f}{\partial q}\frac{q}{Q^f} \text{ and } \xi^W = \frac{\partial Q^W}{\partial q}\frac{q}{Q^W}$$

are familiar conjectural elasticities in which $\theta^f \in (0,1)$ measures the processing firm's oligopsony market power in procuring the farm product and $\xi^W \in (0,1)$ measures the firm's oligopoly power in selling the product to retailers. In either instance, a value of zero denotes perfect competition and a value of one denotes pure monopsony or monopoly. Forms of oligopoly and/or oligopsony competition, such as Cournot competition, are represented by intermediate values of ξ and/or θ with higher values denoting progressively greater departures from competition.

Various authors have noted that θ^f and ξ^W can be utilized independent of any specific interpretation as conjectural variations and, rather, can be viewed simply as indexes of market power that are the outcome of an unknown oligopsony and/or oligopoly game (Karp and Perloff 1996).[2] Aggregation from the firm to the industry is accomplished readily within this model framework. Because firms produce a homogeneous product and have identical technologies, optimizing behavior compels that all firms' conjectures *ex post* are identical (Wann and Sexton 1992). Equation (6') thus represents an equilibrium condition that, in conjunction with the retail-demand and farm-supply functions specified in (1) and (2), respectively, and the retailer cost function, yields equilibrium values for P^r, P^w, P^f, and Q.

For the purpose of conducting simulations, it is necessary to assign specific functional forms for the retail-demand and farm-supply functions. We choose linear models for this purpose:

(1') $Q^r = a - \alpha P^r$, retail demand, and

(2') $P^f = b + \beta Q^f$, inverse farm supply.

In addition, we invoke the normalizations that are available without loss of generality by choosing units so that the quantity and retail price in competitive equilibrium, (Q_c, P_c^r), are both unity: $Q_c = 1$ and $P_c^r = 1$, in which case $P_c^w = 1 - c^r$ and $P_c^f = 1 - c^r - c^w = f$ where f is farmers' revenue share under perfect competition. Given the normalizations, the following relationships among the model's parameters are readily derived:

(7) $\alpha = \eta_c^r$, $\beta = f / \varepsilon_c^f$, $a = 1 + \alpha$, and $b = f - \beta$

where η_c^r is the absolute value of the retail price elasticity of demand and ε_c^f is the price elasticity of farm supply, both evaluated at the competitive equilibrium. Economic welfare under perfect competition is apportioned among the market participants as

(8) $CS_c = \int_{P_c^r}^{a/\alpha} (a - \alpha P)\, dP = 1 / 2\alpha = 1 / 2\eta_c^r$, consumer surplus;

2 The converse of this practice is the approach of Gasmi and Vuong (1991) and Gasmi, Laffont, and Vuong (1992) to econometrically impose the parameter restrictions implied by specific oligopoly/oligopsony models and use nonnested methods of hypothesis testing to distinguish among the various models.

(9) $PS_c = \int_b^{P_c^f} (P - b / \beta)\, dP = \beta / 2 = f / 2\varepsilon_c^f$, producer surplus; and

(10) $\Pi_c = 0$, processor + retailer profit.

Total social benefit is maximized under perfect competition and is computed as

$$SB_c = PS_c + CS_c = 1 + \alpha\beta / 2\alpha.$$

Given that $P^r = P^w + c^r$ under competitive retailer behavior, we can solve (1'), (2'), and (6') simultaneously to obtain the following solutions for the linear model when the processing sector is imperfectly competitive:

(11) $Q_1 = 1 + \alpha\beta / \Omega_1$,

$P_1^w = a - Q_1/\alpha - c^r$,

$P_1^r = P_1^w + c^r$, and

$P_1^f = b + \beta Q_1$

where $\Omega_1 = (1 + \xi^w) + (1 + \theta^f)\,\alpha\beta = (1 + \xi^w) + (1 + \theta^f)\, f\phi_c$ and $\phi_c = \eta_c^r / \varepsilon_c^f$ is the ratio of the retail-demand elasticity to the farm-supply elasticity. The expression Ω_1 measures the total distortion due to oligopoly and oligopsony power, and $Q_1 < 1 = Q_c$ whenever either θ^f or ξ^w is positive. Figure 7.1 illustrates the model.[3]

---

3 Figures 7.1, 7.2, and 7.3 rely on the observation by Melnick and Shalit (1985) that an industry with oligopoly power acts as if it faces a perceived marginal revenue curve that consists of a linear combination of the marginal revenue curve, $\partial[D(Q)Q] / \partial Q$, and the market demand curve with ξ representing the weight attributed to the marginal revenue curve and $(1 - \xi)$ representing the weight attributed to the demand curve. Similarly, for an industry with oligopsony power, the perceived marginal-factor cost curve is $\theta MFC\,(Q) + (1 - \theta)\, S\,(Q)$ where $MFC = \partial[S(Q)Q] / \partial Q$ denotes the marginal-factor cost curve. Figure 7.1 illustrates the situation where $\theta = \xi = 0.5$.

Figure 7.1. Market Equilibrium under Processor Oligopoly and Oligopsony Power

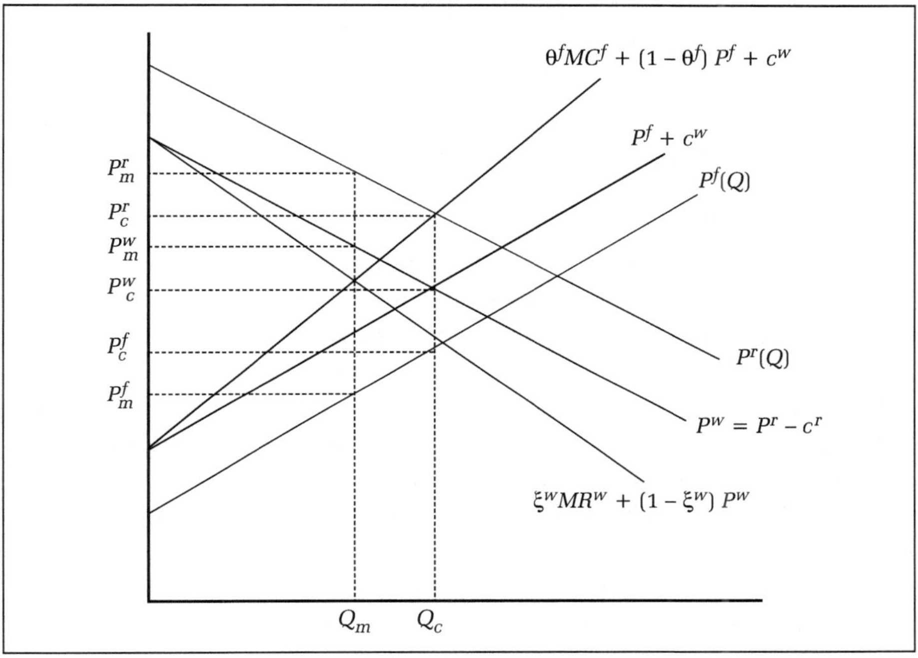

Note: Figure 7.1 illustrates the case where $\xi^w = \theta^w = 0.5$ (see footnote 3).

Economic surplus under processor *or* retailer market power is distributed as follows:

(12) $$CS_1 = \int_{P_1^r}^{a/\alpha} (a - \alpha P)\, dP = (a - \alpha P_1^r)^2 / 2\alpha,$$

(13) $$PS_c = \int_{b}^{P_1^f} (P - b / \beta)\, dP = (P_1^f - b)^2 / 2\beta, \text{ and}$$

(14) $$\Pi_1 = \Pi_1^w + \Pi_1^r = (P_1^r - P_1^f - 1 + f)\, Q_1.$$

In this model, the market equilibrium and distribution of economic welfare are determined solely by four market parameters: ξ^w (seller oligopoly power), θ^f (buyer oligopsony power), ϕ (ratio of the elasticity of retail demand to farm supply), and f (farm revenue share under perfect competition). The equilibrium output, farm price, retail price, and distribution of welfare among producers, marketers, and consumers are identical if the same magnitudes of market power are exercised by the retail sector instead of the manufacturing sector. These same results also hold for the rather plausible case in which processors exercise oligopsony power over farmers and retailers exercise oligopoly power over consumers, but the interactions between processors and retailers are conducted under conditions of perfect competition.

Case II: Market Power at Successive Market Stages

Here we consider scenarios in which retailers may exercise oligopoly power over consumers and processors may exercise oligopsony power over farmers and, in addition, in which processor-retailer interactions may be characterized by imperfect competition. We consider two alternative subcases for the processor-retailer interactions. The first involves processor oligopoly power and retailer price-taking in the processor-retailer interactions and the second involves retailer oligopsony power and processor price-taking in the processor-retailer interactions. The case in which both retailers and processors attempt to exercise market power in their mutual interactions would need to be studied within a bargaining environment, which is beyond the scope of the present study.[4]

A representative retailer's profit function can be expressed as

$$\pi^r = D(Q)q - P^w q - c^r q. \tag{15}$$

The first-order condition for maximizing Equation (15) is

$$P^r \left(1 - \xi^r/\eta^r\right) = P^w + c^r. \tag{16}$$

Using the retail-demand function $P^r = D(Q)$ to substitute for P^r in (16), we can solve Equation (16) for the retailer's inverse derived demand function for processed product $P^w = D^w(Q \mid \xi^r, c^r)$. A representative processing firm's profit function can then be expressed as

$$\pi^w = D^w(Q)q - S(Q)q - c^w q. \tag{17}$$

---

4 One plausible outcome of processor-retailer bargaining is that they agree on the volume of trade that maximizes their mutual benefit with bargaining restricted to determining the division of surplus between the bargainers. This outcome is identical in terms of output, retail price, farm price, and welfare distribution to our Case I equilibria.

The first-order condition for maximizing Equation (17) is

$$(18) \qquad P^w\left(1 - \xi^w/\eta^w\right) = P^f\left(1 + \theta^f/\varepsilon^f\right) + c^w$$

where ξ^w denotes processors' oligopoly power and $\eta^w = -\partial Q/\partial P^w \times P^w/Q$ is the price elasticity of derived demand for the processed product. For the linear version of the model, the market equilibrium is defined by Equations (1'), (2'), (16), and (18):

$$Q_2 = (1 + \alpha\beta) / \Omega_2,$$

$$P_2^W = b + \beta Q_2 + c^w,$$

$$P_2^r = (a - Q_2) / \alpha, \text{ and}$$

$$P_2^f = b + \beta Q_2$$

where

$$\Omega_2 = (1 + \xi^r)(1 + \xi^w) + (1 + \theta^f)\,\alpha\beta = (1 + \xi^r)(1 + \xi^w) + (1 + \theta^f)\,f\phi.$$

In this case, the market equilibrium and welfare distribution are determined by five parameters: ξ^r, ξ^w, θ^f, ϕ, and f. In addition to the parameters contained in Case I, we now have a second ξ parameter reflecting the degree of seller market power at successive stages of the market chain. Figure 7.2 illustrates this scenario.

For the subcase of successive oligopsony power, a representative processor's profit function can be expressed as

$$(19) \qquad \pi^w = P^w q - S(Q)q - c^w q.$$

The first-order condition for maximizing Equation (19) is

$$(20) \qquad P^w = P^f\left(1 + \theta^f/\varepsilon^f\right) + c^w.$$

Equation (20) can be used in conjunction with the inverse farm-supply curve $S(Q)$ to yield the inverse derived supply curve of processed product $P^w = S^w(Q \mid \theta^f, c^w)$. A representative retailer's profit function can be expressed as

Figure 7.2. Successive Oligopoly Power with Processor Oligopsony Power

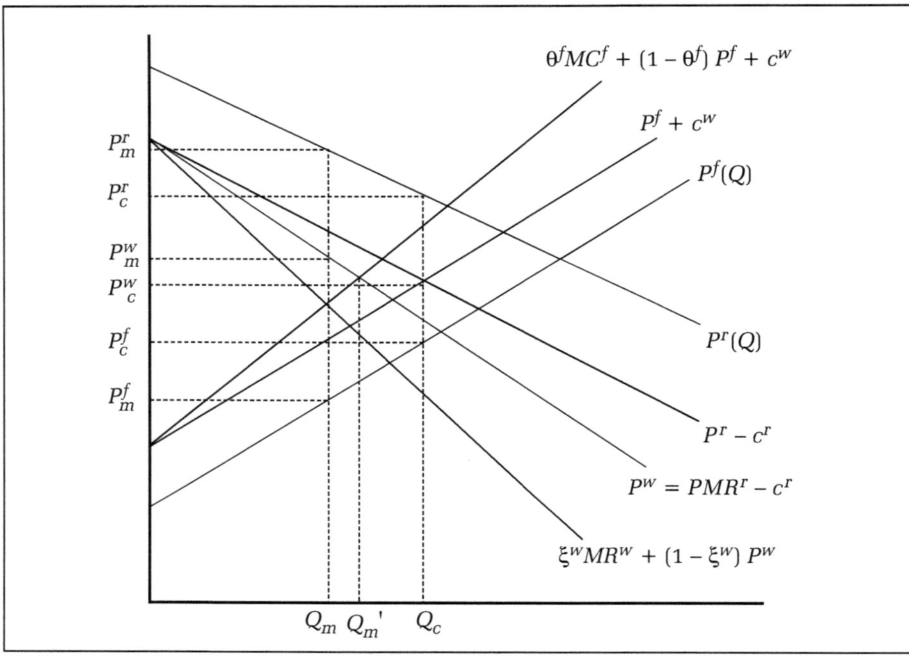

Note: Figure 7.2 illustrates the case where $\xi^w = \xi^r = \theta^f = 0.5$ (see footnote 3).

(21) $$\pi^r = P^r(Q)q - S^w(Q)q - c^r q.$$

The first-order condition for maximizing Equation (21) is

(22) $$P^r \left(1 - \xi^r/\eta^r\right) = P^w \left(1 + \theta^w/\varepsilon^w\right) + c^r.$$

For the linear version of the model, the market equilibrium with successive oligopsony power is defined by equations (1'), (2'), (20), and (22). Solving the system yields the following solutions for the endogenous variables:

$$Q_3 = (1 + \alpha\beta) / \Omega_3,$$

$$P^w_3 = b + \beta Q_3 + c^w,$$

$$P^r_3 = (a - Q_3) / \alpha, \text{ and}$$

$$P^f_3 = b + \beta Q_3$$

where $\Omega_3 = (1 + \xi^r) + (1 + \theta^f)(1 + \theta^w)\alpha\beta = (1 + \xi^r) + (1 + \theta^f)(1 + \theta^w) f\phi$. The market equilibrium and welfare distribution are determined by five parameters: ξ^r, θ^w, θ^f, ϕ, and f, with the fifth parameter in this subcase reflecting the possibility of oligopsony power at successive stages. The successive oligopsony case is illustrated in Figure 7.3.

In summary, given linear retail-demand and farm-supply curves, this model expresses the market equilibrium and welfare distribution in terms of four parameters when market power resides at most in one stage of the market chain (or the equivalent case of processor oligopsony power and retailer oligopoly power) and five parameters when market power is possible at successive stages. By focusing directly on the end product of oligopoly/oligopsony power, as measured by the ξ and θ parameters, we need not be concerned with particular market structures or oligopoly/oligopsony games. This makes the model a very convenient tool for conducting simulations.

Figure 7.3. Successive Oligopsony Power with Retailer Oligopoly Power

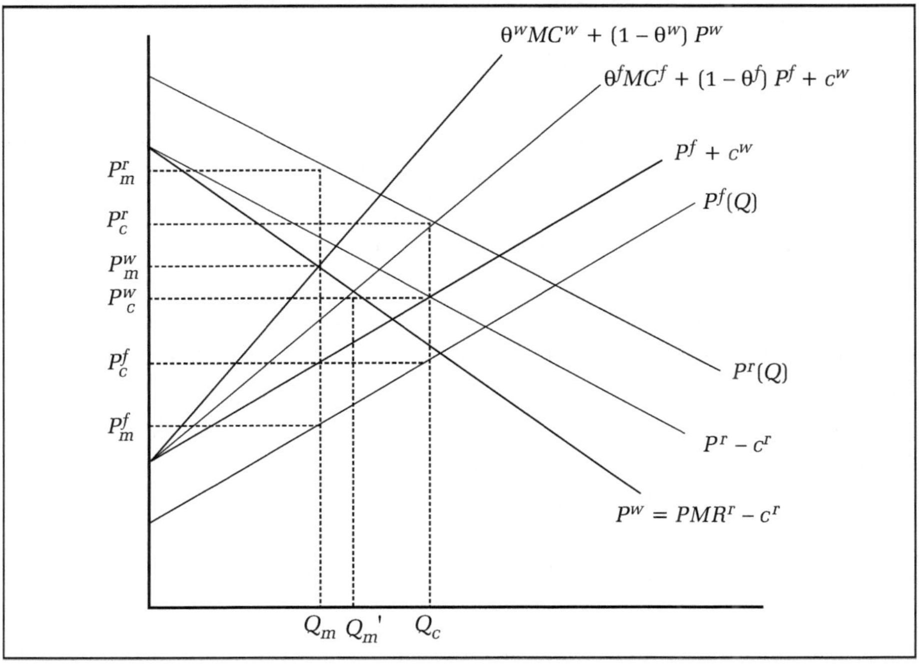

Note: Figure 7.3 illustrates the case where $\theta^f = \theta^w = \xi^r = 0.5$ (see footnote 3).

Empirical Evidence of Market Power in the U.S. Food System

We do not attempt to provide an exhaustive survey of empirical analyses of market power in food processing and distribution. Rather, we focus on some key industries or sectors that have been the objects of several studies and on key issues pertaining to this body of work.

Beef Packing. The U.S. beef sector has been the object of particularly frequent attention due to the dramatic rise in seller and buyer concentration. Congress in 1992 commissioned USDA to investigate the effects of concentration in the industry. This study alone resulted in seven technical reports that are summarized by USDA (1996).[5] The rise in concentration in beef packing was fueled at least partially by technological change. During the 1960s, boxed-beef technology was introduced and carcasses were processed into individual cuts, packed, and shipped from the same plant where slaughter took place. This capital-intensive technology resulted in expanded economies of size in the industry. In addition, declining consumption of red meat led to excess capacity, triggering a wave of mergers and acquisitions during the 1970s and 1980s. In 1977, the four leading packers were estimated to hold 30 percent of total slaughter capacity. By 1992, the CR4 was estimated at 82 percent (Purcell 1990).

Studies of the beef-packing sector conducted within the structure-conduct-performance (SCP) framework include Menkhaus, St. Clair, and Ahmaddaud (1981); Quail et al. (1986); and Marion and Geithman (1995). These studies found a negative relationship between concentration (generally measured as the four-firm concentration ratio, CR4) and the price paid to ranchers and a positive correlation between feedlot size and the price received, suggesting possible countervailing power. Ward (1981, 1982, 1988, 1992) pioneered the use of transaction price data to examine the determinants of fed-cattle prices. In Ward (1981, 1992), feedlot prices are positively correlated with the number of buyers bidding for the purchase. The Texas A&M University Agricultural Market Research Center (1996), however, found only a weak effect on the fed-cattle price for a regional Herfindahl Index variable.

Using a structural model in the evolving tradition of new empirical industrial organization (NEIO) and a sample period of 1951–1983,

5 Among the technical reports is a detailed survey of research on the competitiveness of the U.S. meat-packing industry by Azzam and Anderson (1996).

Schroeter (1988) rejected price-taking behavior but found that distortions from competitive pricing were modest in magnitude—on the order of 3 percent in output sales and 1 percent in input purchases. Several extensions of this work followed soon after. Azzam and Pagoulatos (1990) studied meat packing as an aggregate industry. Through their production-function formulation, they were able to obtain point estimates of conjectural elasticities of both oligopoly power ($\xi = 0.223$) and oligopsony power ($\theta = 0.178$). Schroeter and Azzam (1990) developed a multiproduct model of the meat-packing industry, treating pork and beef as separate products but not allowing oligopoly and oligopsony conjectures to differ. This study also rejected price-taking behavior, although the estimated $\theta = \xi$ parameters were small in magnitude. Azzam (1992) rejected price-taking behavior by U.S. beef packers in farm-product purchases but not in processed-product sales.

More recent tests of beef-packer market power include Muth (1996), who analyzed oligopoly power, and Kambhampaty et al. (1996) and Muth and Wohlgenant (1999), who analyzed oligopsony power. Muth and Muth and Wohlgenant failed to find any evidence of market power, in contrast to the majority of prior studies. They attributed their results to the prior authors' assumption of a fixed-proportion and constant-return processing technology. Kambhampaty et al. treated the conjectural elasticity/farm-product supply elasticity ratio (θ / ε) as a single parameter. This parameter was positive and significant in the estimation, but other estimated parameters did not comport with economic theory.

Fruits and Vegetables. Just and Chern's (1980) analysis of oligopsony power in the California tomato-processing industry was the first application of the comparative statics approach, exposited more generally by Bresnahan (1982), to testing for market power. Just and Chern argued that substitution of a fixed capital input (the mechanical harvester) in place of a variable input (farm labor) made farm supply less elastic, offering a natural experiment in which a competitive market's response could be distinguished from the response of a market with oligopsony power. Whereas the competitive model predicts unambiguously that the supply shift will lead to greater production of the farm product, the effect is ambiguous in the oligopsony model. Empirical results supported the oligopsony model's prediction. The industry was later analyzed by Durham and Sexton (1992), who estimated residual supply elasticities for six production and processing regions in California. The estimated elasticities were large, ranging from 8.6 to essentially infinity, which caused the authors to conclude that oligopsony potential in the industry is limited. Structural changes in the industry since the time of the Just and Chern analysis may account for the different conclusions about competitiveness.[6]

Applications of conjectural-variations models include Wann and Sexton (1992) to U.S. pear processing, Taylor and Kilmer (1988) to Florida celery, and several studies of behavior under U.S. marketing orders. Estimates from Wann and Sexton's multiple-product framework suggested limited oligopoly power in pear processing ($\xi = 0.08$) but greater power in the sale of fruit cocktail ($\xi = 0.48$). Hypotheses of perfect competition and pure monopsony in pear procurement were both rejected. Florida celery is marketed through a single cooperative supported by a federal marketing order. Taylor and Kilmer investigated whether this organization was able to exercise market power. Estimation results found modest ($\xi \: \varepsilon \: (0.03, 0.15)$) and insignificant levels of oligopoly power.[7]

Dairy. Masson and Eisenstat (1980) estimated that U.S. dairy cooperatives succeeded in raising retail fluid-milk prices by $0.07 to $0.10 per gallon (3.78 liters) with an annual loss to consumers of $71 million from 1967 to 1975 before their alleged anticompetitive behavior was halted by antitrust action. A subsequent study by Madhavan, Masson, and Lesser (1994) focused on the monopoly power of a single cooperative, Associated Milk Producers, Inc., and concluded that it was able to raise margins by $0.026 per gallon prior to the 1975 consent decree.[8] Suzuki et al. (1994) exploited observed price differentials between fluid milk and manufacturing milk to derive nonparametric estimates of the

---

6 A processor that was dominant during the time of the Just and Chern study had since seen its share erode, a grower bargaining association had risen to prominence, and longer hauls may have stimulated greater interregional competition.

7 This result is consistent with the observation that Florida is a relatively minor player in celery sales in the U.S. In most metropolitan areas and in most months, most sales are from California, where marketing is not coordinated (Sexton, Kling, and Carman 1991).

8 Baumer, Masson, and Masson (1986) argued that the monopoly power needed in U.S. milk markets to exercise price discrimination over and above that mandated through U.S. marketing orders was caused by a wave of mergers among milk marketing cooperatives during the 1960s. The mergers were not challenged because of a belief that they were protected under the Capper-Volstead Act.

U.S. dairy cooperatives' market power. The manufacturing-milk price plays the role of a competitive benchmark framework in this study. The imputed ξ values in Suzuki et al. ranged from 0.06 to 0.08. Liu, Sun, and Kaiser (1995) estimated mean values of θ for U.S. dairy processors equal to 0.10 for manufacturing milk and 0.18 for fluid milk.

Grocery Retailing. Rising concentration and consolidation of sales among large supermarket chains and supercenters in the U.S. have made retailer market power in the food industry a topical issue. Food retailing is not amenable to the application of NEIO methods because it involves the production and sale of a vast number of different products—an average of 30,000 different items for U.S. supermarkets.[9] SCP methods are useful, however, because prices can be observed readily and aggregated into indices. Data often are also available on the explanatory variables that might be utilized within the loose theoretical structure of the SCP paradigm to account for demand- and cost-side determinants of price. Studies such as Hall, Schmitz, and Cothern (1979); Lamm (1981); Newmark (1990); Marion, Heimforth, and Bailey (1993); and Binkley and Connor (1998) have examined average retail price relationships using cities as the unit of observation.

Cotterill (1986), Kaufman and Handy (1989), and Cotterill and Harper (1995) focused on the behavior of individual stores, giving them an opportunity for increased precision and relevance in construction and use of explanatory variables relative to earlier studies. Cotterill studied food retailer monopoly power in Vermont, a sparsely populated state that provided an almost ideal setting in which to delineate relevant geographic markets for identifying concentration.[10] Concentration variables —CR4 (four-firm), CR1 (one firm), and the Herfindahl Index—were positively associated with price and statistically significant. A parallel study

---

9 Hyde and Perloff's (1998) study of market power in the Australian meat sector is an attempt to apply NEIO methods to measure grocer market power. They avoided issues of multiproduct technology by using simple linear marginal cost specifications for each meat product studied, implicitly assuming that the unspecified cost function is strongly separable.

10 This geographic isolation has, ironically, been the primary basis for criticizing the study on the grounds that the very high levels of concentration observed are atypical.

of Arkansas supermarkets by Cotterill and Harper (1995) reached similar conclusions as to the impacts of retailer concentration on food prices.[11]

However, not all studies of grocery retailing have found a positive association between concentration and price. Kaufman and Handy (1989) studied 616 supermarkets chosen from twenty-eight cities selected at random. Both firm market share and a four-firm Herfindahl Index were negatively but insignificantly correlated with price. Newmark (1990) also obtained a negative and insignificant coefficient on CR4 in a study of the price of a market basket of goods for twenty-seven cities. However, Newmark's conclusions have been questioned in recent work by Yu and Connor (2002). Part five of Cotterill (1993) contains a debate on the issue of market power in grocery retailing, including a critique of the Kaufman-Handy study and a response by the authors.[12]

Evaluation of the Empirical Work

What have we learned from the various empirical studies of market power in agriculture? First, in highly concentrated industries, a positive (negative) correlation between concentration and selling (purchasing) price exists. This correlation has been found rather consistently across many SCP studies of food processor oligopoly and oligopsony power and food retailer oligopoly power.

NEIO studies of processor behavior generally have found some statistical evidence of market power, although the measured departures from competition have mostly been small, with point estimates of θ or ξ often being less than 0.2 (the market-power equivalent of that produced in a five-firm symmetric Cournot equilibrium). Because these studies have naturally been conducted in industries where structural conditions suggest the possible presence of market power, these results on the whole suggest that in the past market power has not been a very important factor in the food-processing sector.

---

11 Studies conducted at the city level that have found a positive structure-price relationship include Hall, Schmitz, and Cothern (1979); Lamm (1981); and Marion, Heimforth, and Bailey (1993).

12 Binkley and Connor (1998) suggested one explanation for the conflicting results in terms of product coverage in the price variable. They found a positive and significant concentration-price correlation for dry groceries and a negative and insignificant correlation for fresh and chilled food items.

Various criticisms can be leveled against both bodies of work. The most well known are the critiques of the SCP studies. These include the Demsetz (1973) critique of the interpretation of profit-structure studies and Fisher and McGowan's (1983) critique of the use of accounting data to infer market power. Although responses to these critiques can and have been made, the drift of the SCP literature away from studies of profits and toward the analysis of price-structure relationships within a single industry is appropriate.[13] In our view, studies of the price-structure relationship such as Cotterill (1986) that focus on a single industry and that control effectively for factors other than structure that may influence price provide some of the most convincing evidence of market power in the food chain. Nonetheless, a modified version of the Demsetz critique has been leveled against SCP studies of price based on a quality argument. According to this reasoning, the most successful firms provide the best quality products and related service, thereby receiving price premiums or paying discounted prices and attaining large market shares. For example, in the context of the beef-sector studies, it might be argued that large buyers offer better service than small buyers (e.g., prompt and reliable payment, secure market outlets, technical assistance), thereby enabling them to pay lower prices. A similar argument can be constructed for why large sellers may earn price premiums.

Perhaps due to their more recent vintage, there has been less formal criticism of NEIO studies, although some criticisms of SCP studies (e.g., Fisher and McGowan (1983)) apply equally to NEIO studies (Cotterill 1993). One aspect that has been discussed critically is the conceptual underpinning or lack thereof provided by the conjectural-variations framework. These criticisms are rather unimportant as they pertain to the empirical research. Conjectural variations attempt to model a dynamic phenomenon (i.e., action and reaction) within a static framework. An example of this problem is the tendency of conjectured behavior to fail to coincide with rivals' actual optimal response. As it pertains to the theory, the profession has moved to address dynamic interactions with explicitly dynamic models. As to empirical work, recent practice has been to specify first-order conditions such as (6') without any reference to θ or ξ representing conjectural variations. They are simply empirical

13 A good source on the debate between SCP analysts and their critics is part five of Cotterill (1993).

indices that measure the departure of a given market from competitive outcomes

Two general criticisms can be leveled against empirical work on the food and beverage industries conducted within the NEIO framework. The first is that essentially all of the studies have relied on maintained functional forms for market demand and/or supply and processor technology. The researcher is, thus, always testing a joint hypothesis—whatever is intended to be tested plus the maintained hypothesis of functional form. This criticism applies, of course, not just to studies of market power and is the launching point for nonparametric analyses of demand, production, and market power. The problem is mitigated partially when the researcher utilizes flexible functional forms. In NEIO studies, the processing technology is often represented by such functions, but retail or wholesale demand and/or farm supply are either usually represented by simple linear or double log functions or the needed elasticities are obtained from extraneous estimates.

Even more vexing is the issue of technical change in food processing. Most applications of NEIO models proceed with annual data at the industry level. To obtain sufficient observations, these applications may study thirty or more years of industry data, during which time significant technical change will have almost inevitably occurred. NEIO studies that have addressed technological change have done so through very simple means such as time trends. Incorporating more sophisticated methods is not necessarily straightforward because of data limitations and convergence problems in the highly nonlinear empirical models.

Another dimension of the functional-form argument that arises mainly in agricultural applications concerns the elasticity of substitution, σ, between the farm input and other inputs in producing a finished product. Most authors have assumed that no such substitution exists, whereas others, such as Gardner (1975), Wohlgenant (1989), and Holloway (1991), have considered substitution possibilities to be an integral part of agricultural market models. Muth (1996) and Muth and Wohlgenant (1999) have attributed several authors' empirical findings of market power in the beef industry in part to their failure to allow for substitution in the model frameworks.

A second criticism of the NEIO studies generally has been their architects' failure in many cases to think carefully about the markets they intend to analyze before conducting the actual analysis. Antitrust actions that evolve around market power begin with definitions of the relevant market, both in geographic and in product-form dimensions. Only when these issues are settled does the action proceed to assess the actual exercise of market power. This sequence is fundamental to studying market power. Of what relevance is it to ask whether a group of firms exercises market power without having first determined the relevant geographic and product markets within which the firms operate? Most NEIO applications in agriculture have treated market-definition issues superficially at best. The work on meat provides examples of the problem. Some applications have studied meat packing in aggregate. Others have focused on specific meats. Despite evidence that cattle are seldom shipped as far as 300 miles, various studies have investigated packer oligopsony power using data aggregated to the national level without questioning whether the relevant geographic markets are regional in scope.

A factor contributing to poorly defined product or geographic markets is that many NEIO and aggregate SCP studies use data collected for reasons other than economic analysis, such as U.S. SIC data. Data at the four-digit level of aggregation often cover a variety of products that are mostly linked through a common agricultural input, not by end use. For example, SIC 2015 is poultry and egg processing and Bhuyan and Lopez (1997) found moderate oligopoly power ($\theta = 0.289$) in this category for the period 1972–1987. Five-digit SIC categories within SIC 2015 include young chickens, turkeys, other poultry, and liquid, dried, and frozen eggs. Other examples of empirical analyses of market power that clearly have failed to define relevant markets include Holloway (1991) and, more recently, Reed and Clark (2000), in which processed fruits and vegetables, fresh fruit, and fresh vegetables were among the "markets" studied.

A related concern is the tendency of NEIO studies to investigate processor oligopoly or oligopsony power on one side of the market while maintaining an assumption of perfect competition on the other side of the market.[14] Although there may be good economic rationales for this

14 An alternative approach has been to assume that market power is identical in both the farm-product and finished-product markets—i.e., that $\theta = \xi$ (Schroeter 1988; Schroeter and Azzam 1990). Because of likely differences in the relevant geographic and product markets for the farm product versus the finished product, this practice also has limitations. Azzam and Pagoulatos (1990) and Wann and Sexton (1992) presented two alternative approaches that allowed θ and ξ to differ within a vertical market.

decision, more often the structural bases for concern about oligopoly power imply parallel concerns about oligopsony power and vice versa. How does an erroneous assumption about competition on one side of the market affect inferences about market power on the other side of the market?

More research is needed on the implications of the various maintained hypotheses for our ability to draw inferences about market power. Additional Monte Carlo studies along the lines of Hyde and Perloff (1994) and Raper, Love, and Shumway (2000) could help to answer these questions. Also helpful will be more studies patterned after Genesove and Mullin (1998), who utilized cost and price data and a demand elasticity estimate for the U.S. cane sugar industry to compute a direct estimate of ξ using (8'). Estimates of ξ derived from alternative NEIO approaches were then compared to the "true" value and were found to comport quite closely and to be insensitive to the functional form of demand.

Welfare Losses from Market Power in the U.S. Food System

How do estimates of market power translate into measures of social loss and loss to consumers? Measuring the static welfare loss from seller market power is not difficult conceptually. Given the volume of output, Q^*, in an industry, one needs to obtain an estimate of the markup of price over cost and an estimate of the elasticity of demand in the market in order to compute the deadweight loss or Harberger triangle. In Figure 7.4, this is the area B given retail demand $D_1(Q)$ and the area $B + C$ given demand $D_0(Q)$. An immediate issue is whether to also include the rectangle A of profits in Figure 7.3 transferred to the monopolist/oligopolists (the so-called Posner (1975) rectangle). Arguments in support of including the rectangle are that monopoly power encourages resource waste or X-inefficiency and wasteful expenditures such as advertising and lobbying to achieve and maintain the monopoly position and that these costs can be approximated by the profit rectangle.

Parker and Connor (1979) estimated welfare losses in each U.S. food manufacturing industry by first estimating SCP models of the price-cost margin and then using fitted values from the estimation to calculate the monopoly overcharge in each industry. Parker and Connor produced loss estimates including the Posner rectangle of about 5.7 percent of U.S. food expenditures in 1975. In contrast to the empirical approach of Parker and Connor, Gisser (1982) generated welfare loss estimates from a conceptual

Figure 7.4. Impact of the Choice of Demand Elasticity on the Calculation of Welfare Loss due to Imperfect Competition

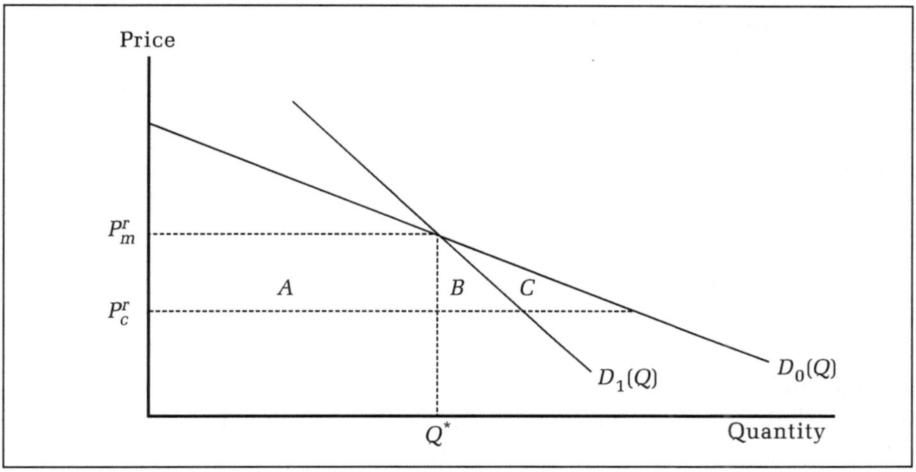

oligopoly model that assumed that 36 percent of firms in an industry act as collusive price leaders and the remaining 64 percent operate as a competitive price-taking fringe.[15] He adduced gross losses including the Posner rectangle that were similar to those of Parker and Connor. However, Gisser also analyzed the determinants of factor productivity in food manufacturing and found it to be positively associated with increases in food-industry concentration. Incorporating these cost savings into the welfare analysis caused the welfare losses due to oligopoly power to disappear. Young (1997) provides a detailed critique of the use of the dominant-firm price leadership model to estimate welfare loss.

Recent welfare loss studies for the U.S. food sector include Bhuyan and Lopez (1995) and Peterson and Connor (1996). Peterson and Connor utilized results from a study of brand-name versus private-label price differences in food manufacturing by Connor and Peterson (1992) to estimate price-cost margins in each industry, from which the Harberger triangle and the Posner rectangle were computed.[16] Losses on average were 6.9

15 The justification for this assumption was that 36 percent was the weighted CR4 in food manufacturing for 1972.

16 The premise underlying this approach is that the private-label market is competitive so that the private-label price equals marginal cost. If this premise is correct and if private-label production costs are the same as brand products' production costs, then the price difference between a branded product and its private-label counterpart will represent the oligopoly markup.

percent of the value of 1979/80 food shipments.[17] Bhuyan and Lopez (1995) inferred a value of ξ for each U.S. food manufacturing industry from published data on industry sales and costs. Given additional data on elasticity of demand in each industry from a study by Pagoulatos and Sorensen (1986) and the Herfindahl Index and an assumption of constant marginal production costs, they then estimated the resulting welfare loss in terms of the Harberger triangle only. They pegged the average loss at 3 percent of total sales for 1987 with nearly 40 percent of that loss coming from the tobacco industries alone.

Given that no consensus has been reached on the importance of market power in the food system, it is not surprising that little agreement exists on the welfare costs of market power. An important difference among analysts concerns market power's effect on costs. Some believe that market power breeds inefficiency and waste and, thus, justifies inclusion of Posner's rectangle as a welfare loss. Others of a Chicago-School persuasion believe market power is mostly efficiency driven so the benefits must be measured and weighed against the costs of any supracompetitive pricing.

The demand elasticity also plays a crucial role in studies such as Gisser (1982) and Bhuyan and Lopez (1995) that derive the price-cost margin from a structural oligopoly model rather than by estimating it from an SCP regression. Under the latter methodology, the elasticity plays only a minor role in calculating the Harberger triangle, *given* the estimated price-cost margin, as the area B versus the area $B+C$ in Figure 7.3. In the structural models, the imputed elasticity value is crucial to *deriving* the price-cost markup given the assumed mode of oligopoly behavior—see, for example, Equation (6'). Notably absent from the literature are studies that attempt to measure welfare losses due to the buying and selling power of food retailers or to the power of food processors as buyers of farm products. Our subsequent simulation analysis gives a sense of possible welfare losses when market power is present at multiple stages of an industry.

Overall, this body of work is too ambitious in attempting to apply the same set of methods, whether an SCP paradigm or a structural

17 The key measure of concentration in Connor and Peterson (1992), the Herfindahl Index divided by the industry elasticity of demand, was subsequently found to have the erroneous sign (Hinloopen and Martin 1997). The loss estimate reported here is based on the correct sign.

one, across many diverse industries. Studies that focus on the welfare implications of market power in individual industries are likely to yield more accurate indications of the welfare costs of market power. Recent examples of such work include Azzam and Schroeter (1995) and Azzam (1997) on the trade-off between efficiency gains and losses from oligopsony power from consolidation in the U.S. beef industry and Tiffin and Dawson (1997) on the losses from oligopoly in frozen potato products in the United Kingdom.[18]

Simulation Analysis

Because no empirical studies have evaluated the welfare consequences of both oligopoly and oligopsony power or of successive oligopoly/oligopsony in a food market, simulation analysis will play a useful role in indicating the potential welfare implications of such market behavior. The linear version of the conceptual model set forth in this chapter shows that market equilibrium can be expressed in terms of four parameters for markets featuring possible oligopoly and oligopsony power but not successive oligopoly/oligopsony; the presence of successive oligopoly or oligopsony requires addition of a fifth parameter.

In structuring the simulation, the parameter f, the farm share of revenue under perfect competition, was fixed at either $f = 0.5$ (i.e., a 50 percent farm share) or $f = 0.2$. A 50 percent farm share is roughly reflective of the situation for some meat products and eggs, while a 20 percent share is reflective of both fresh and processed fruit and of vegetable markets. The primary effect of f in the model is to moderate the importance of oligopsony power on output and consumer welfare in the market. When f is small, the farm input does not contribute importantly to the final product and, thus, oligopsony power in the farm sector has only a minor impact on total market output and consumer welfare.

The farm-supply and retail-demand elasticities always affect the market equilibrium and the distribution of welfare in the relative form $\phi = \eta / \varepsilon$, enabling the simulation to focus solely on alternative values for ϕ. In considering values of ϕ for the simulation, note that in many cases both retail demand and farm supply for a food product are rather inelastic, suggesting that the ratio of the elasticities will often be near unity.

18 Azzam (1997) estimated that oligopsony power increased with rising concentration in beef packing but that the costs of market power were more than offset by the reduction in packing costs due to economies of size.

We treat $\phi = 1$ as a base value and also consider values of $\phi = 0.5$ (i.e., demand is less elastic than supply) and $\phi = 2$ (i.e., demand is more elastic than supply). In all cases, the elasticities are set at the competitive equilibrium, which is always set via normalization to be $P^r_c = 1$, $Q_c = 1$. Of course, given the linear formulations for retail supply and farm demand, the elasticities change as one moves along the curves to identify alternative forms of imperfectly competitive equilibria. However, the relative magnitudes of alternative elasticity specifications are the same across the various equilibria to be explored in the simulations.

The most important parameters for the purposes of the simulation analysis are the market-power parameters, θ and ξ, which both range in the unit interval. As noted, most point estimates of θ and/or ξ from prior empirical studies have been 0.2 or less. Bhuyan and Lopez's (1997) ambitious attempt to estimate ξ values for all four-digit SIC food and beverage industries, however, yielded somewhat higher point estimates, with some categories (2043 cereal preparation, 2041 flour and grain mills, 2075 soybean oil mills) yielding estimates of ξ of about 0.5. Given the rapid increase in consolidation of both food manufacturing and retailing, past studies may understate current levels of market power.

Given a base value of $f = 0.5$, Figure 7.5 illustrates the percent loss in consumer surplus from processor or retailer oligopoly power only while Figure 7.6 provides comparable information for the case of processor or retailer oligopsony power only. Each figure illustrates results

Figure 7.5. The Effect of Oligopoly Power on Consumer Welfare

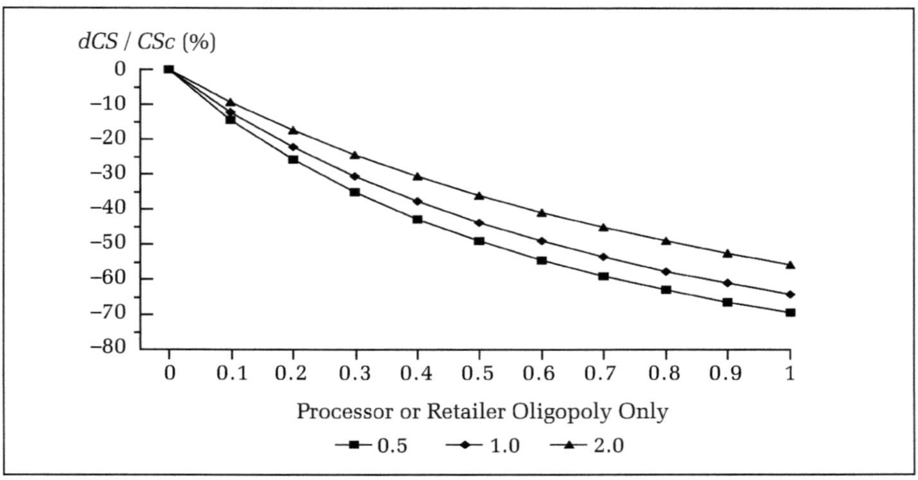

for $\phi_C \in \{0.5, 1.0, 2.0\}$. Recall that $\varphi_C = \eta_C^r / \varepsilon_C^f$. Thus, higher values of ϕ indicate a retail demand that is elastic relative to the farm supply. Figure 7.5 shows that even the modest exercise of oligopoly power as manifest by $\xi = 0.2$ causes a loss in consumer surplus relative to the competitive outcome that ranges from 17 percent ($\phi = 2$) to 26 percent ($\phi = 0.5$), with the loss being larger when consumer demand relative to the farm supply is more elastic. Market power in the magnitude of $\xi = 0.5$ (e.g., a Cournot duopoly) results in consumer surplus losses ranging from 36 percent ($\phi = 2$) to 49 percent ($\phi = 0.5$).

The impact of oligopsony power on consumers (Figure 7.6) is less severe because, as noted, the farm product comprises only part of the final product's value. Figure 7.6 illustrates the case of $f = 0.5$ (i.e., a 50 percent farm revenue share under perfect competition), in which case modest oligopsony power as manifest by $\theta = 0.2$ results in a loss of consumer surplus ranging from 817 percent ($\phi = 2$) to 26 percent ($\phi = 0.5$). With oligopsony, the loss to consumers from a given θ is magnified by increasing price inelasticity in the farm supply relative to retail demand because the distortion to output is greater when supply is more inelastic. For situations where $\theta = 0.50$ (e.g., a Cournot duopsony), the loss to consumers ranges from 17 percent ($\phi = 2$) to 36 percent ($\phi = 0.5$).

These losses are roughly reflective of meat and egg markets where the farm revenue share is rather large. The impact declines as f declines. For

Figure 7.6. The Effect of Oligopsony Power on Consumer Welfare

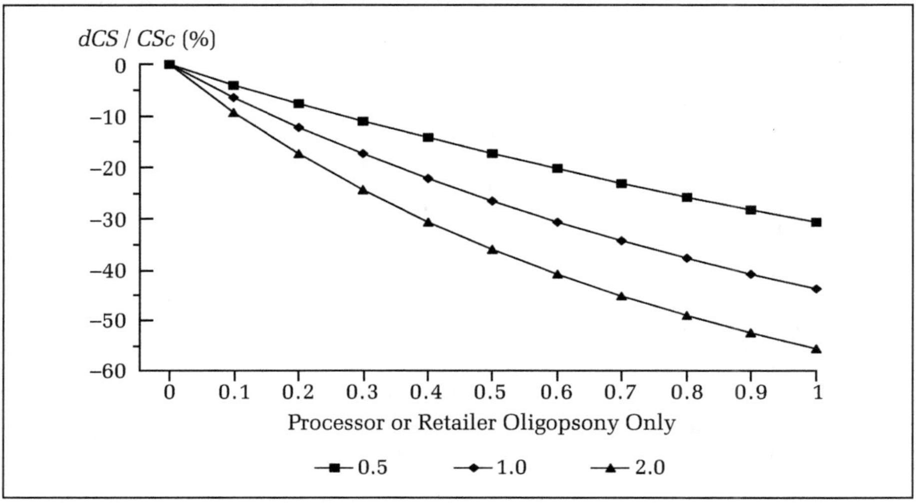

the case where $f = 0.2$, which is reflective of many fruit and vegetable markets, the consumer loss from $\theta = 0.5$ ranges from 8.5 percent to 23 percent.

Further perspective on the impact of food-sector market power on consumers is provided in Figure 7.7. Focusing on the case where $\phi = 1$, Figure 7.7 compares the percentage reduction in consumer surplus relative to the competitive outcome from oligopoly and oligopsony power only and from joint oligopoly and oligopsony power. For reasons already noted, the impact of a given level of oligopoly power is more severe than the equivalent exercise of oligopsony power. When both types of market power are present, losses to consumers are compounded. Modest oligopoly and oligopsony power on the magnitude of $\theta = \xi = 0.2$ causes consumer welfare to fall by 31 percent relative to the competitive outcome, while $\theta = \xi = 0.5$ results in consumer welfare that is 56 percent less than under perfect competition.

Figure 7.8 illustrates deadweight losses and Figure 7.9 indicates the distribution of welfare among farmers, marketers, and consumers for oligopoly only, oligopsony only, and both oligopoly and oligopsony scenarios shown in Figure 7.7. Jointly, Figures 7.8 and 7.9 communicate an important piece of information concerning modest levels of market power: the impacts are primarily distributional. For $f = 0.5$ and $\phi = 1$, the percentage reduction in total welfare relative to a competitive solution

Figure 7.7. The Effect of Market Power on Consumer Welfare: $\phi = 1.0$

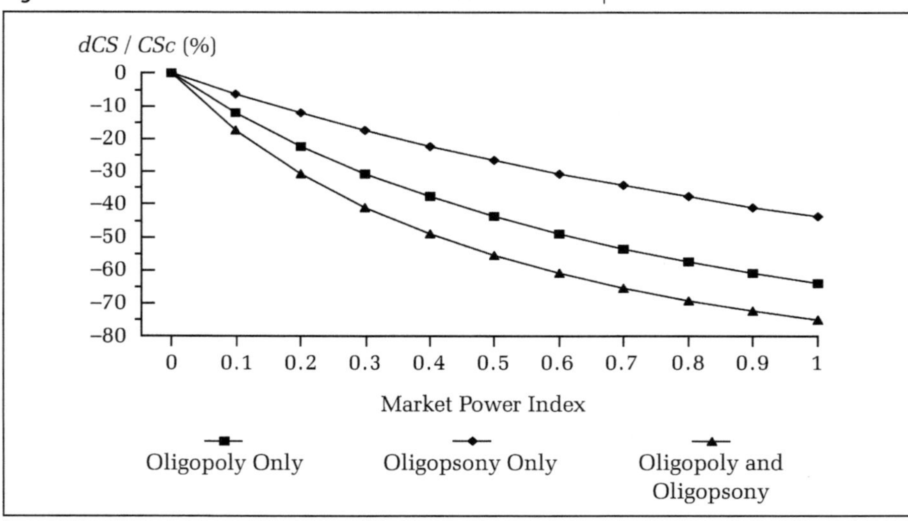

Figure 7.8. The Effect of Market Power on Total Welfare: $\phi = 1.0$

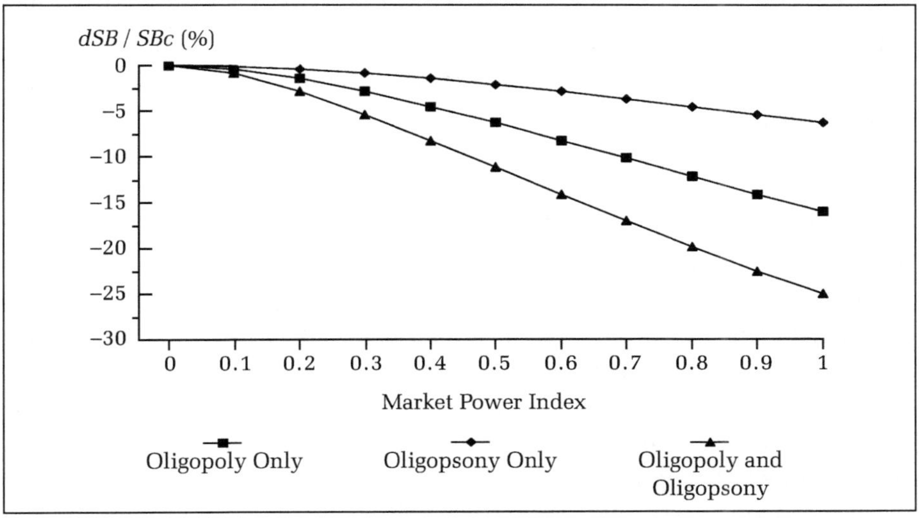

is only 0.4 percent for $\theta = 0.2$ and $\xi = 0$, 1.4 percent for $\theta = 0$ and $\xi = 0.2$, and 2.8 percent for $\xi = \theta = 0.2$. Given this same scenario and perfect competition, consumers capture two-thirds of the total welfare in the market and producers capture the other one-third. Under oligopoly power only ($\xi = 0.2$), the distribution of surplus from the market becomes 53 percent for consumers, 26 percent for producers, and 21 percent for

Figure 7.9. The Effect of Market Power on the Distribution of Welfare: $\phi = 1.0$

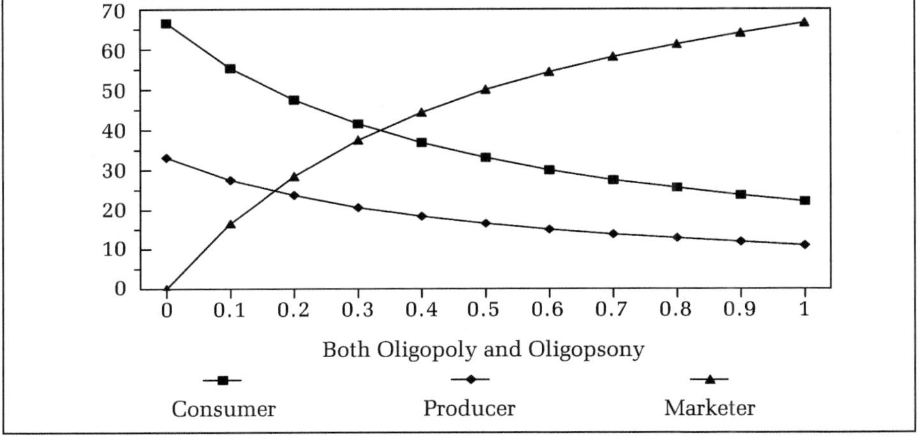

marketers. Under oligopsony only ($\theta = 0.2$), the distribution is consumers 59 percent, producers 29 percent, and marketers 12 percent. Under both oligopoly and oligopsony ($\theta = \xi = 0.2$), marketers capture 29 percent of the total surplus compared to 48 percent for consumers and 24 percent for producers.

As Figure 7.8 illustrates, the percent deadweight loss grows at an increasing rate as a function of θ and ξ. For $\xi = 0.5$, the deadweight loss is 6.25 percent, while for $\theta = 0.5$, it is 2.0 percent. For $\theta = \xi = 0.5$, the deadweight loss escalates to 11.1 percent. The distributional effects of market power of this order of magnitude are even more pronounced. For $\xi = 0.5$, the distribution is 40 percent for consumers, 20 percent for producers, and 40 percent for marketers. For $\theta = 0.5$, the comparable distribution is 50 percent for consumers, 25 percent for producers, and 25 percent for marketers. When marketers hold both oligopoly and oligopsony power on this order of magnitude, the distribution of benefits is one-third for consumers, one-sixth for producers, and one-half for marketers. As we have shown in other work (Alston, Sexton, and Zhang 1997; Sexton and Zhang 2000), this same pattern of distribution effects applies to benefits generated from industry initiatives for research, promotion, etc., when markets are imperfectly competitive.

Successive Oligopoly/Oligopsony

Figures 7.10 through 7.13 summarize behavior for the food industry that is characterized by successive oligopoly or oligopsony power. Figure 7.10 examines losses to consumers for the base case where $f = 0.5$ and $\phi = 1$. A modest exercise of successive oligopoly power in conjunction with processor oligopsony power, as characterized by $\xi^w = \xi^r = \theta^f = 0.2$, reduces consumer surplus by nearly half (46 percent) relative to the competitive outcome. The impact of the comparable case of successive oligopsony power in conjunction with retailer oligopoly power, $\theta^f = \theta^w = \xi^r = 0.2$, is somewhat less severe, generating a consumer surplus loss of 39 percent. If successive oligopoly power rises to the level of Cournot duopoly (i.e., $\xi^w = \xi^r = \theta^f = 0.5$), consumer surplus is reduced by fully 75 percent from its level at the competitive benchmark, with the comparable successive oligopsony case, $\theta^f = \theta^w = \xi^r = 0.5$, reducing consumer surplus by two-thirds relative to perfect competition.

Figure 7.10. The Effect of Successive Market Power on Consumer Welfare: $\phi = 1.0$

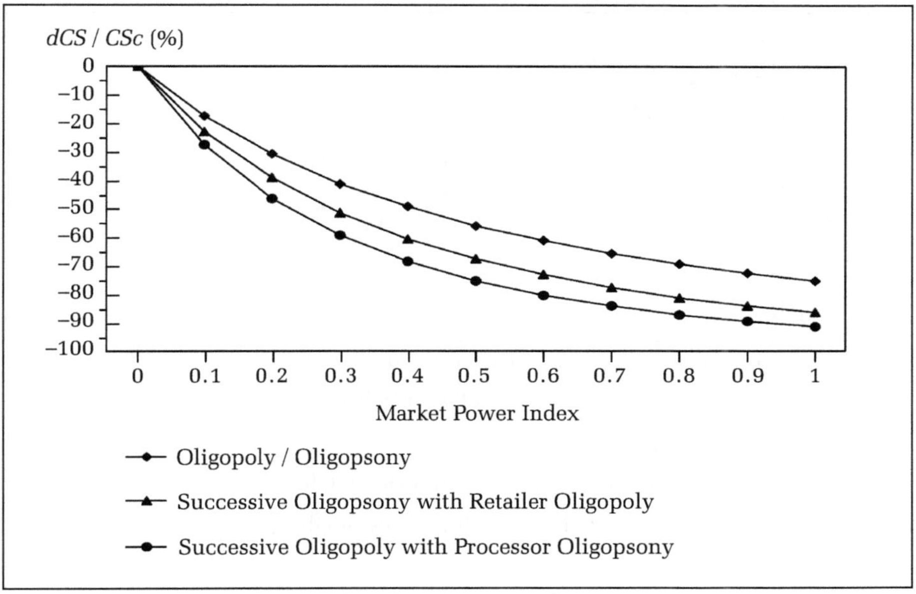

Figure 7.11 indicates the magnitude of deadweight losses under successive oligopoly or oligopsony. The modest exercise of successive oligopoly power, $\xi^w = \xi^r = \theta^f = 0.2$, induces a deadweight loss of 7 percent. When $\xi^w = \xi^r = \theta^f = 0.5$, the deadweight loss rises to 25 percent, again reflecting that deadweight loss rises at an increasing rate as a function of the degree of market power exercised. Deadweight losses are more modest under successive oligopsony—4.8 percent for $\theta^f = \theta^w = \xi^r = 0.2$ and 18.4 percent for $\theta^f = \theta^w = \xi^r = 0.5$.

Figure 7.12 provides information on the distribution of benefits in the market under successive oligopoly power for the base case where $f = 0.5$ and $\phi = 1$, while Figure 7.13 provides the same information for the successive oligopsony case. Modest market power, when it is distributed throughout the marketing sector as manifest by the case of $\xi^w = \xi^r = \theta^f = 0.2$, enables the marketing sector to capture the plurality of benefits from the production and sale of the product: 42 percent versus 39 percent for consumers and 19 percent for producers. (Recall that, given the setup of the model, marketers capture no surplus under perfect

Figure 7.11. The Effect of Successive Market Power on Total Welfare: $\phi = 1.0$

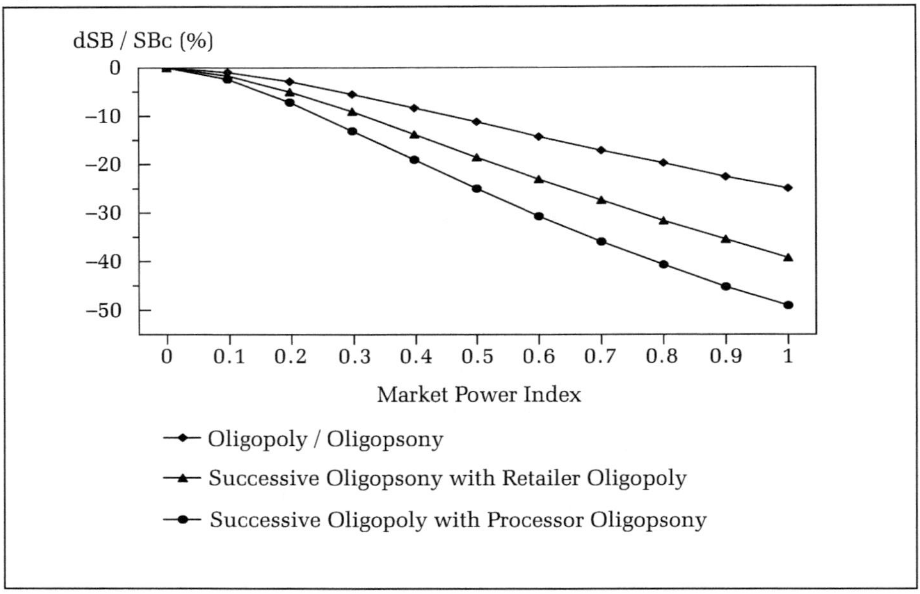

Figure 7.12. The Effect of Successive Oligopoly Power on the Distribution of Welfare: $\phi = 1.0$

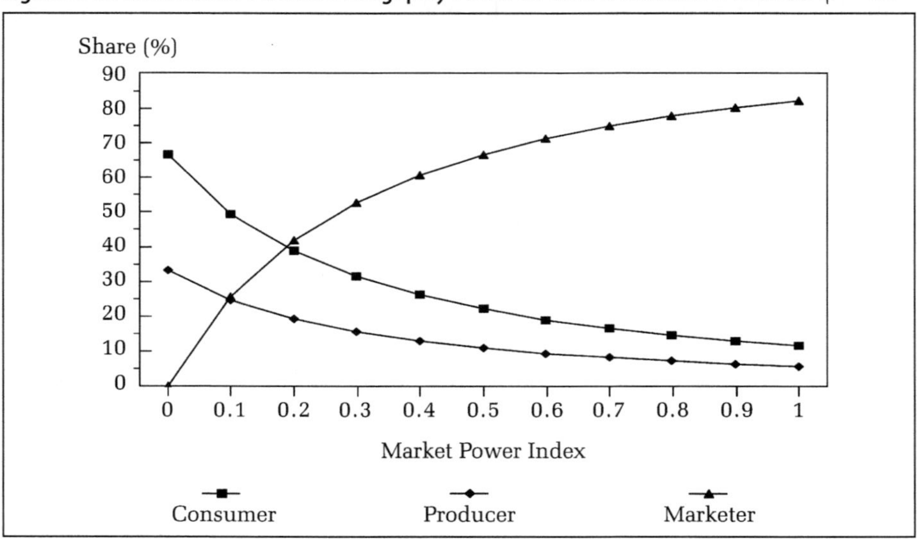

Figure 7.13. The Effect of Successive Oligopsony Power on the Distribution of Welfare: $\phi = 1.0$

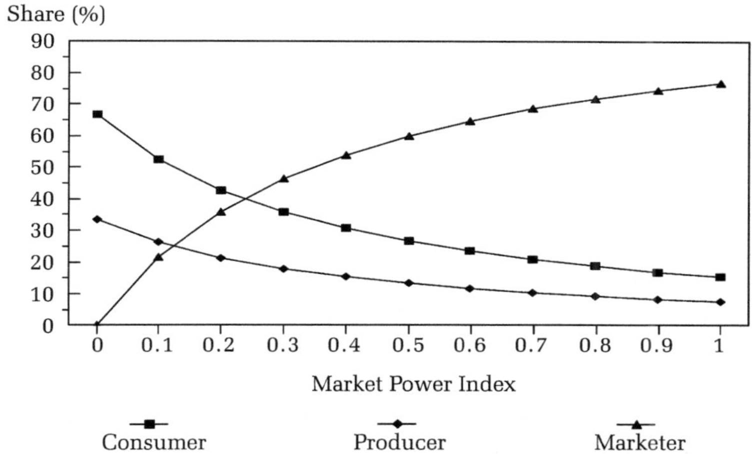

competition.) If successive oligopoly power should rise to the level represented by $\xi^w = \xi^r = \theta^f = 0.5$, marketers would capture fully two-thirds of the surplus in the market, relegating consumers and farmers to shares of two-ninths and one-ninth respectively. The distributional impacts of successive oligopsony power are somewhat less severe. The case of $\theta^f = \theta^w = \xi^r = 0.2$ results in a distribution of benefits of 43 percent for consumers, 36 percent for marketers, and 21 percent for farmers. The marketers' share rises to 60 percent for $\theta^f = \theta^w = \xi^r = 0.5$, with 27 percent going to consumers and 13 percent to producers.

Policy Implications

The policy implications of market power in the food chain are not clear despite the significant amount of work done to estimate the departures from competition and the attendant welfare costs. Key unresolved issues include the following.

- How often are departures from competitive pricing observed and how quantitatively important (in the sense of generating large welfare losses and/or redistributions of income) are those cases of imperfect competition?

- What are the efficiency implications of increasing concentration? Is increasing concentration primarily efficiency driven or does monopoly power breed inefficiencies and wasteful competition?

- How important is quality and variety as a determinant of consumer welfare from food consumption and how are quality and variety dimensions of food markets influenced by concentration and market power?

- How important is countervailing power as an endogenous market response to monopoly or monopsony power?

- Even if monopoly and/or monopsony power is significant in the food marketing sector, how effective can antitrust actions or regulations be in curtailing it?

To date, most of the resources among agricultural economists have been directed, with limited success, to the first issue. Although the economic bases for economies of size—X-inefficiencies, excessive advertising, and product differentiation—are well established, there is little empirical evidence of their actual effects in reinforcing or offsetting welfare costs from supracompetitive pricing. Countervailing power is only now re-emerging as an important subject of economic inquiry following years of neglect after the famous exchange between Galbraith (1952) and Stigler (1954). Key issues are the extent to which farmers can countervail buyer power through collective action and the extent to which the power of food manufacturers and of food retailers offsets in their mutual dealings.

Finally, as to the subject of antitrust, the current practice of enforcement in the U.S. is far from the activist policy advocated by the pioneers of industrial organization analysis in agriculture such as Hoffman (1940). Compare, for example, the subdued tone and modest recommendations contained in the recent report by the USDA Advisory Committee on Agricultural Concentration (USDA 1996) with the activist recommendations issued thirty years earlier in a similar report by the National Commission on Food Marketing (1966). Although markets are generally much more concentrated now than thirty years ago, the advisory committee's main recommendation was for enhanced disclosure and improved reporting of information, and the committee distanced itself from recommendations that would "ultimately stunt opportunities for growth within the industry" or "slow or prevent the industry's need to adapt to a changing market place" (p. 15).[19]

References

Alston, J.M., R.J. Sexton, and M. Zhang. "The Effects of Imperfect Competition on the Size and Distribution of Research Benefits." *American Journal of Agricultural Economics* 79 (1997):1252–1265.

Azzam, A. "Testing the Competitiveness of Food Price Spreads." *Journal of Agricultural Economics* 43 (1992):248–256.

———. "Measuring Market Power and Cost-Efficiency Effects of Industrial Concentration." *Journal of Industrial Economics* 45 (1997):377–386.

Azzam, A., and D. Anderson. *Assessing Competition in Meatpacking: Economic History, Theory, and Evidence.* Washington DC: U.S. Department of Agriculture (USDA) Grain Inspection, Packers and Stockyards Administration Report GIPSA-RR 96-6, 1996.

Azzam, A., and E. Pagoulatos. "Testing Oligopolistic and Oligopsonistic Behavior: An Application to the U.S. Meat Packing Industry." *Journal of Agricultural Economics* 41 (1990):362–370.

Azzam, A., and J.R. Schroeter. "The Tradeoff between Oligopsony Power and Cost Efficiency in Horizontal Consolidation: An Example from Beef Packing." *American Journal of Agricultural Economics* 77 (1995):825–836.

Baumer, D.L., R.T. Masson, and R.A. Masson. "Curdling the Competition: An Economic and Legal Analysis of the Antitrust Exemption for Agriculture." *Villanova Legal Review* 31 (1986):183-252.

Bhuyan, S., and R.A. Lopez. "Welfare Losses under Alternative Oligopoly Regimes: The U.S. Food and Tobacco Manufacturing Industries." *Journal of Agricultural and Applied Economics* 27 (1995):577–587.

———. "Oligopoly Power in the Food and Tobacco Industries." *American Journal of Agricultural Economics* 79 (1997)1035–1043.

Binkley, J.K., and J.M. Connor. "Market Competition and Metropolitan-Area Grocery Prices." *Journal of Retailing* 74 (1998):273–293.

Bresnahan, T.F. "The Oligopoly Solution Concept Is Identified." *Economics Letters* 10 (1982):87–92.

Connor, J.M., and E.B. Peterson. "Market-Structure Determinants of National Brand – Private Label Price Differences of Manufactured Food Products." *Journal of Industrial Economics* 40 (1992):157–171.

---

19 Ironically, mandatory disclosure of terms of trade such as price can have anticompetitive consequences, presumably not what the committee intended. In markets such as those for live cattle where prices are set through buyer bidding, mandatory disclosure of bids is an excellent device by which to reinforce bidder cartels (McAfee and McMillan 1987).

Cotterill, R.W. "Market Power in the Retail Food Industry: Evidence from Vermont." *Review of Economics and Statistics* 68 (1986):379–386.

Cotterill, R.W., ed. *Competitive Strategy Analysis in the Food System*. Boulder CO: Westview Press, 1993.

Cotterill, R.W., and C.D. Harper. *Market Power and the Demsetz Quality Critique: An Evaluation for Food Retailing*. Storrs CT: University of Connecticut, Food Marketing Policy Research Center Research Report 29, 1995.

Demsetz, H. "Industry Structure, Market Rivalry, and Public Policy." *Journal of Law and Economics* 16 (1973):1–9.

Durham, C.A., and R.J. Sexton. "Oligopsony Potential in Agriculture: Residual Supply Estimation in California's Processing Tomato Market." *American Journal of Agricultural Economics* 74 (1992):962–972.

Durham, C.A., R.J. Sexton, and J.H. Song. "Spatial Competition, Uniform Pricing, and Transportation Efficiency in the California Processing Tomato Industry." *American Journal of Agricultural Economics* 78 (1996):115–125.

Fisher, F.M., and J.J. McGowan. "On the Misuse of Accounting Rates of Returns to Infer Monopoly Profits." *American Economic Review* 73 (1983):82–87.

Franklin, A.W., and R.W. Cotterill. *An Analysis of Local Market Concentration Levels and Trends in the U.S. Grocery Retailing Industry*. Storrs CT: University of Connecticut, Food Marketing Policy Research Center Research Report 19, 1993.

Galbraith, J.K. *The Concept of Countervailing Power*. Boston MA: Houghton-Mifflin, 1952.

Gardner, B. "The Farm Retail Price Spread in a Competitive Industry." *American Journal of Agricultural Economics* 57 (1975):399–409.

Gasmi, F., J.J. Laffont, and Q. Vuong. "Econometric Analysis of Collusive Behavior in a Soft-Drink Market." *Journal of Economics and Management Strategy* 1 (1992):277–311.

Gasmi, F., and Q. Vuong. "An Econometric Analysis of Some Duopolistic Games in Prices and Advertising." *Advances in Econometrics: Econometric Methods and Models for Industrial Organization*. G. Rhodes and T. Fomby, eds., pp. 225–254. Greenwich CT: JAI Press, 1991.

Genesove, D., and W.P. Mullin. "Testing Static Oligopoly Models: Conduct and Cost in the Sugar Industry, 1890–1914." *Rand Journal of Economics* 29 (1998):355–377.

Gisser, M. "Welfare Implications of Oligopoly in U.S. Food Manufacturing." *American Journal of Agricultural Economics* 64 (1982):616–624.

Hall, L., A. Schmitz, and J. Cothern. "Beef Wholesale-Retail Marketing Margins and Concentration." *Economica* 46 (1979):295–300.

Hayenga, M.L., S.R. Koontz, and T.C. Schroeder. *Definition of Regional Cattle Procurement Markets*. Washington DC: U.S. Department of Agriculture Grain Inspection, Packers and Stockyards Administration Report GIPSA-RR 96-1, 1996.

Hinloopen, J., and S. Martin. "Market-Structure Determinants of National Brand – Private Label Price Differences of Manufactured Food Products: Comment." *Journal of Industrial Economics* 45 (1997):219–223.

Hoffman, A.C. *Large-Scale Organization in the Food Industry*. Washington DC: U.S. Government Printing Office, 1940.

Holloway, G. "The Farm-Retail Price Spread in an Imperfectly Competitive Food Industry." *American Journal of Agricultural Economics* 73 (1991):979–987.

Hyde, C.E., and J.M. Perloff. "Can Monopsony Power Be Estimated?" *American Journal of Agricultural Economics* 76 (1994):1151–1155.

———. "Multimarket Market Power Estimation: The Australian Retail Meat Sector." *Applied Economics* 30 (1998):1169–1176.

Just, R.E., and W.S. Chern. "Tomatoes, Technology, and Oligopsony." *Bell Journal of Economics* 11 (1980):584–602.

Kambhampaty, S.M., P.J. Driscoll, W.D. Purcell, and E.B. Peterson. *Effects of Concentration on Prices Paid for Cattle*. Washington DC: U.S. Department of Agriculture Grain Inspection, Packers and Stockyards Administration Report GIPSA-RR 96-4, 1996.

Karp, L.S., and J.M. Perloff. "Dynamic Models of Oligopoly in Rice and Coffee Export Markets." *Agricultural Markets: Mechanisms, Failures and Regulations*. D. Martimort, ed., pp. 171–204. Amsterdam Netherlands: Elsevier, 1996.

Kaufman, P., and C. Handy. "Supermarket Prices and Price Differences." Washington DC: U.S. Department of Agriculture, Economic Research Service Technical Bulletin 1176, 1989.

Lamm, R.M. "Prices and Concentration in the Food Retailing Industry." *Journal of Industrial Economics* 30 (1981):67–78.

Liu, D.J., C. Sun, and H.M. Kaiser. "Market Conduct under Government Price Intervention in the U.S. Dairy Industry." *Journal of Agricultural and Resource Economics* 20 (1995):301–315.

Madhavan, A.N., R.T. Masson, and W.H. Lesser. "Cooperation for Monopolization?: An Empirical Analysis of Cartelization." *Review of Economics and Statistics* 76 (1994):161–175.

Marion, B.W., and F.E. Geithman. "Concentration-Price Relations in Regional Fed Cattle Markets." *Review of Industrial Organization* 10 (1995):1–19.

Marion, B.W., K. Heimforth, and W. Bailey. "Strategic Groups, Competition, and Retail Food Prices." *Competitive Strategy Analysis in the Food System*. R.W. Cotterill, ed., pp. 179–199. Boulder CO: Westview Press, 1993.

Masson, R.T., and P.M. Eisenstat. "Welfare Impacts of Milk Orders and the Antitrust Immunities for Cooperatives." *American Journal of Agricultural Economics* 62 (1980):270–278.

McAfee, R.P., and J. McMillan. "Auction and Bidding." *Journal of Economic Literature* 25 (1987):699–738.

Melnick, R., and H. Shalit. "Estimating the Market for Tomatoes." *American Journal of Agricultural Economics* 6 (1985):147–153.

Menkhaus, D.J., J.S. St. Clair, and A.Z. Ahmaddaud. "The Effects of Industry Structure on Price: A Case in the Beef Industry." *Western Journal of Agricultural Economics* 6 (1981):147–153.

Muth, M. "Identifying the Degree of Oligopoly Power in the Beef Packing Industry." Unpublished, North Carolina State University, 1996.

Muth, M., and M.K. Wohlgenant. "Measuring the Degree of Oligopsony Power in the Beef Packing Industry in the Absence of Marketing Input Quantity Data." *Journal of Agricultural and Resource Economics* 24 (1999):299–312.

National Commission on Food Marketing. *Food from Farmer to Consumer*. Washington DC: U.S. Government Printing Office, 1966.

Newmark, C.M. "A New Test of the Price-Concentration Relationship in Grocery Retailing." *Economics Letters* 33 (1990):369–373.

Pagoulatos, E., and R. Sorensen. "What Determines the Elasticity of Industry Demand?" *International Journal of Industrial Organization* 4 (1986):237–250.

Parker, R.C., and J.M. Connor. "Estimates of Consumer Loss due to Monopoly in the U.S. Food-Manufacturing Industries." *American Journal of Agricultural Economics* 61 (1979):628–639.

Peterson, E.B., and J.M. Connor. "Consumer Welfare Loss Estimates for Differentiated Food Product Markets." *Review of Agricultural Economics* 18 (1996):233–246.

Posner, R.A. "The Social Costs of Monopoly." *Journal of Political Economy* 83 (1975):807–827.

Purcell, W.D. "Structural Change in the Livestock Sector: Causes, Implications, Continuing Issues." *Structural Change in Livestock: Causes, Implications, Alternatives*. W.D. Purcell, ed., pp. 2–57. Blacksburg VA: Research Institute on Livestock Pricing, 1990.

Quail, G., B. Marion, F. Geithman, and J. Marquardt. "The Impact of Packer Buyer Concentration on Live Cattle Prices." Madison WI: North Central Agricultural Experimental Stations North Central Project Working Paper 117, 1986.

Raper, K.C., H.A. Love, and C.R. Shumway. "Determining Market Power Exertion between Buyers and Sellers." *Journal of Applied Econometrics* 15 (2000):225–252

Reed, A.J., and J.S. Clark. "Structural Change and Competition in Seven U.S. Food Markets." Washington DC: U.S. Department of Agriculture, Economic Research Service Technical Bulletin 1881, 2000.

Rogers, R.T. "The Role of Marketing Cooperatives in Increasingly Concentrated Agricultural Markets: Reaction." *Cooperatives: Their Importance in the Future Food and Agricultural System*. M. Cook, R. Torgerson, T. Sporleder, and D. Padberg, eds., pp. 49–65. Washington DC: National Council of Farmer Cooperatives, 1997.

———. "Broilers: Differentiating a Commodity." *Industry Studies*. L.L. Duetsch, ed., pp. 59–95. Armonk NY: M.E. Sharpe, 1998.

Rogers, R.T., and R.J. Sexton. "Assessing the Importance of Oligopsony Power in Agricultural Markets." *American Journal of Agricultural Economics* 76 (1994):1143–1150.

Schroeter, J.R. "Estimating the Degree of Market Power in the Beef Packing Industry." *Review of Economics and Statistics* 70 (1988):158–162.

Schroeter, J.R., and A. Azzam. "Measuring Market Power in Multi-Product Oligopolies: The U.S. Meat Industry." *Applied Economics* 22 (1990):1365–1376.

Sexton, R.J., C.L. Kling, and H.F. Carman. "Market Integration, Efficiency of Arbitrage, and Imperfect Competition: Methodology and Application to U.S. Celery." *American Journal of Agricultural Economics* 73 (1991):568–580.

Sexton, R.J., and M. Zhang. "Optimal Commodity Promotion in Imperfectly Competitive Markets." University of California, Davis, Department of Agricultural and Resource Economics, working paper, 2000.

Stigler, G. "The Economist Plays with Blocs." *American Economic Review* 44 (1954):7–14.

Suzuki, N., H.M. Kaiser, J.E. Lenz, and O.D. Forker. "An Analysis of U.S. Dairy Policy Deregulation Using an Imperfect Competition Model." *Agricultural and Resource Economics Review* 23 (1994):84–93.

Taylor, T.G., and R.L Kilmer. "An Analysis of Market Structure and Pricing in the Florida Celery Industry." *Southern Journal of Agricultural Economics* 20 (1988):35–43.

Texas A&M University, Agricultural Market Research Center. *Price Determination in Slaughter Cattle Procurement.* Washington DC: U.S. Department of Agriculture (USDA) Grain Inspection, Packers and Stockyards Administration Report GIPSA-RR 96-2, 1996.

Tiffin, A.L., and P.J. Dawson. "Measuring Oligopolistic Distortion in the UK Frozen Potato Product Sector: A Calibration Modeling Approach." *Journal of Agricultural Economics* 48 (1997):300–312.

U.S. Department of Agriculture, Agricultural Marketing Service. *Concentration in Agriculture: A Report of the USDA Advisory Committee on Agricultural Concentration.* Washington DC: USDA Agricultural Marketing Service, 1996.

Wann, J.J., and R.J. Sexton. "Imperfect Competition in Multiproduct Food Industries with Application to Pear Processing." *American Journal of Agricultural Economics* 74 (1992):980–990.

Ward, C.E. "Short-Period Pricing Models for Fed Cattle and Impacts of Wholesale Carcass Beef and Live Cattle Futures Market Prices." *Southern Journal of Agricultural Economics* 13 (1981):125–132.

———. "Relationship between Fed Cattle Market Shares and Prices Paid by Beefpackers in Localized Markets." *Western Journal of Agricultural Economics* 7 (1982):79–86.

———. *Meatpacking Competition and Pricing.* Blacksburg VA: Research Institute on Livestock Pricing, 1988.

———. "Inter-Firm Differences in Fed Cattle Prices in the Southern Plains." *American Journal of Agricultural Economics* 74 (1992):480–485.

Wohlgenant, M.K. "Demand for Farm Output in a Complete System of Demand Functions." *American Journal of Agricultural Economics* 71 (1989):241–252.

Young, D.P.T. "Dominant Firms, Price Leadership and the Measurement of Monopoly Welfare Losses." *International Journal of Industrial Organization* 15 (1997):533–547.

Yu, C-C., and J. Connor. "The Price-Concentration Relationship in Grocery Retailing: Retesting Newmark." *Agribusiness* 18 (2002):413–426.

8

Measuring Market Power in Bilateral Oligopoly: The Wholesale Market for Beef

John R. Schroeter, Azzeddine M. Azzam, and Mingxia Zhang

The last two decades have witnessed the development of a variety of approaches to the measurement of market power.[1] In general, these studies base their inferences of market power on econometric models of firm conduct in which monopoly (or monopsony) and price-taking behavior have empirically distinguishable comparative static properties. One often cited and particularly illuminating example of this body of work is Bresnahan's 1982 paper. In it, he develops a simple two-equation empirical model of price and quantity determination in a homogeneous product oligopoly. The model consists of a linear market demand equation and a generalized supply relation incorporating a "conduct parameter" that indexes the degree of market power and nests some conventional oligopoly solution concepts: With λ (the conduct parameter) equal to zero, the model's equations reduce to perfectly competitive supply and demand. A value of one for λ implies the monopoly or perfect cartel solution. Intermediate values of λ correspond to outcomes "between" the polar cases of perfect competition and pure monopoly. Within this context, Bresnahan's fundamental insight is that the comparative statics of price and quantity with respect to exogenous parallel *shifts* of demand or marginal cost are, by themselves, insufficient to reveal the degree of competition. For λ to

This chapter originally appeared as an article titled "Measuring Market Power in Bilateral Oligopoly: The Wholesale Market for Beef" by Schroeter, Azzam, and Zhang in *Southern Economic Journal* in 2000. This material is reprinted by permission of the Southern Economic Association.

1 A short list of examples of these works includes Appelbaum (1982), Bresnahan (1987), Just and Chern (1980), Porter (1983), and Roberts (1984).

be econometrically identified, it is necessary that the model admit exogenous *rotations* of the demand curve as well.

Bresnahan's (1982) use of a conduct parameter as a means of parameterizing a range of potential solution concepts is a modeling device that has been applied in a wide variety of approaches using several different means of achieving econometric identification of the degree of market power.[2] This conduct-parameter method can be readily adapted, moreover, to the measurement of *oligopsony* power.[3] To the best of our knowledge, however, all previous applications of these methods assumed that the participants on a particular side of the market take price as given while the agents on the other side may or may not exercise market power. For example, in studies of oligopoly, buyers are assumed to be passive price-takers while sellers may exploit their influence on price. In studies of oligopsony, sellers are viewed as price-takers while buyers may or may not take advantage of an upward-sloping supply relation.

In this chapter, we address the problem of measuring market power without a maintained hypothesis of price-taking behavior on one side of the market or the other. This issue arises naturally in the context of bilateral oligopoly, a market structure in which both the buyers' and the sellers' sides are relatively concentrated.[4] In such settings, there are at least three stylized equilibrium concepts that suggest themselves: bilateral

---

2 See Bresnahan (1989) for a survey of these methods. This methodology is not immune from criticism, however. Corts (1999) argued that the conduct parameter estimates obtained in these studies provide accurate gauges of the average level of market power only if the static optimization model underlying the estimating equations is an accurate representation of the data-generating process. If the data were actually generated by some other process (for example, a supergame model of tacit collusion through repeated interaction), the estimated conduct parameters can be quite misleading. Genesove and Mullin (1998) performed a "test" of the conduct parameter method for a case in which reliable measures of market power are available from accounting data.

3 Schroeter (1988) is one example of this extension.

4 Related theoretical work includes Waterson (1980) and Blair, Kaserman, and Romano (BKR) (1989). Waterson deals with the problem of sellers exercising monopoly power in successive stages of a marketing channel but does not treat the possibility of monopsony power in the intermediate good market. BKR focus primarily on a bilateral monopoly solution concept that we do not consider: joint profit maximization with the intermediate good price determined through profit-share bargaining. They do not address the possibility of variable degrees of market power exercised by buyers or sellers over passive counterparts on the market's opposite side. Related empirical studies include Azzam (1996) and Azzam and Zhang (1992).

price-taking (leading to the competitive outcome; both buyers and sellers treat price as given), seller price-taking (sellers behave passively while buyers may or may not exercise some degree of monopsony power), and buyer price-taking (buyers behave passively while sellers may or may not exercise some degree of monopoly power). Given the structural relations that specify the model's "primitives" (demand and marginal costs), a set of equations jointly determining price and quantity can be derived for each of the equilibrium concepts. For the seller price-taking and buyer price-taking equilibrium concepts, moreover, a conduct parameter can be used, just as in Bresnahan's model, to index the degree of market power. But it is at this point that the identification issue arises. For certain formulations of the model's structural relations, the equilibrium concepts are not empirically distinguishable from one another. We will see that the easiest way to insure that the alternative equilibrium concepts for the bilateral oligopoly model *do* possess empirically distinguishable comparative static properties and that the degrees of market power are identified within each is to employ a device very closely parallel to the Bresnahan (1982) solution to the identification problem in the oligopoly case: Allow the sellers' marginal cost to *rotate*—not merely *shift*—with changes in exogenous variables. In formulations of the model that possess this feature, the task of choosing among the equilibrium concepts comes down to a test of nonnested, nonlinear, simultaneous-equation models.[5] We will show how to apply existing econometric procedures to carry out such tests. So, to summarize, this chapter first makes a modest contribution to the methodology of market-power measurement—it shifts the assumption of price-taking behavior on a particular side of the market from the category of maintained hypothesis to the category of testable hypothesis.

The larger contribution of this chapter stems from its findings pertaining to a specific bilateral oligopoly application. It is very rare for final consumers of a product to exhibit any appreciable degree of concentration, so the standard paradigm for discussions of bilateral oligopoly is that of a wholesale market in which a concentrated manufacturing industry produces a product and sells it to a concentrated retailing industry. In the U.S., the wholesale market for fresh beef in particular is a good

5 Bresnahan's (1987) and Gasmi, Laffont, and Vuong's (1992) studies of differentiated product oligopolies (with price-taking buyers) also involve nonnested hypothesis testing.

fit for this paradigm. National concentration levels are high among both sellers (beef packers) and buyers (primarily retail grocery chains) and concerns about market power have supplied the motivation for a number of previous economic studies of both of these industries. So, ideally, an econometric analysis of price and quantity determination in the wholesale beef market should leave open the question of whether buyers or sellers (or both) behave competitively while allowing for the possible exercise of market power on either side. Such an analysis is carried out in this chapter.

Our starting point in the next section is a brief outline of the alternative equilibrium concepts that we will consider for bilateral oligopoly. We then address the identification issue in a simple illustrative way and provide an example of a structural model in which the solution concept is not identified. Thereafter, we show how a slight modeling alteration as in Bresnahan (1982) yields an example in which the solution concept and the degree of market power are identified. Estimation and testing issues are then discussed before the final summary and conclusions.

Equilibrium Concepts for Bilateral Oligopoly

Consider a wholesale market in which a concentrated manufacturing industry produces a product and sells it to a concentrated retailing industry. Retailers, in turn, sell the product to final consumers. We assume that the product is homogeneous at both the wholesale and retail levels. Retail firms may effect some transformation of the wholesale version of the product through packaging, transportation, storage, bundling with customer services, or possibly some actual physical alteration of the product itself. But it is assumed that any transformation of the product from wholesale to retail versions is one of fixed proportions. So, with appropriately chosen units, wholesale and retail quantities can be measured with the same variable. Final consumers of the product are assumed to be price-takers but retailers may exercise some degree of monopoly power.

Our interest focuses primarily on the determination of price and quantity in the wholesale market. Because both sides of the market (that is, both manufacturing and retailing industries) are relatively concentrated, the outcome need not be competitive. In fact, at least three stylized equilibrium concepts suggest themselves as plausible candidates.

The wholesale market may be competitive; that is, it may be characterized by bilateral price-taking (BPT). In this scenario, illustrated in

Figure 8.1a, the derived demand curve of the retail industry is obtained by the following construction: Start with the industry's perceived marginal revenue curve (PMR_r). If retailers exercise no monopoly power over final consumers, PMR_r is coincident with retail demand (D_r). On the other hand, if the retail market is perfectly cartelized, PMR_r will be the curve marginal to D_r. With some intermediate degree of oligopoly power, PMR_r will assume a position between these two extremes. From PMR_r subtract retailers' marginal cost (MC_r) to get perceived marginal revenue net of marginal cost ($PNMR_r$), which, in the bilateral price-taking case, is the retail industry's derived demand curve (DD_r). The manufacturing industry's supply curve is simply its marginal cost curve ($MC_m = S_m$). Under bilateral price-taking, wholesale (and retail) quantity and wholesale price are determined as the coordinates of the point of intersection of DD_r and S_m (Q_{BPT} and $P_{w,BPT}$). Retail price is the final demand price at quantity Q_{BPT} ($P_{r,BPT}$).

We refer to the second candidate solution concept as manufacturer price-taking (MPT): Manufacturers take wholesale price as given while retailers exercise some degree of oligopsony power. In this scenario,

Figure 8.1a. The Bilateral Price-Taking Solution

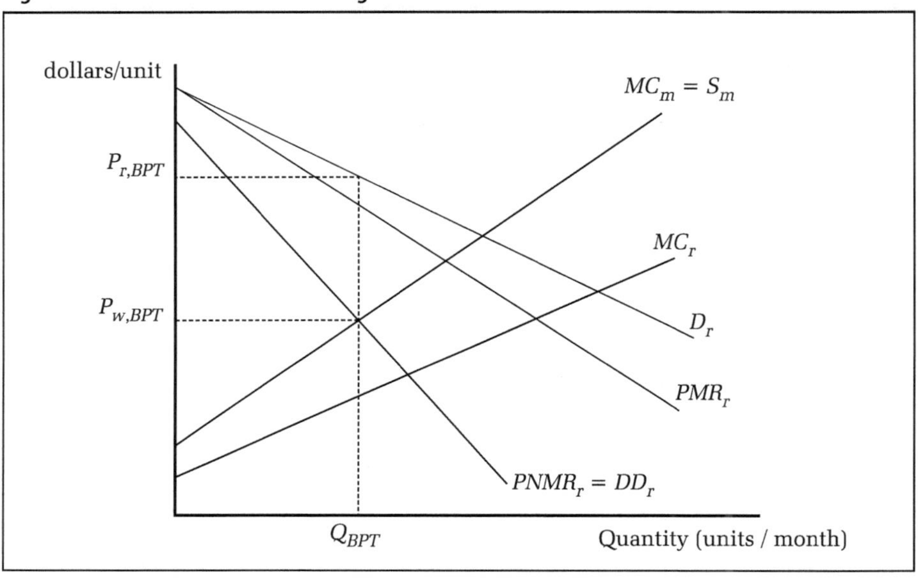

illustrated in Figure 8.1b, manufacturers will sell output up to the point at which $P_w = MC_m$. Hence retailers face a supply curve given by MC_m. The position of the retail industry's perceived marginal expenditure curve (PME_r) will then reflect the industry's degree of oligopsony power over price-taking manufacturers. If retailers collude in their input purchase decisions so as to operate as a pure monopsony, PME_r will be the curve marginal to MC_m; if they exercise no oligopsony power, PME_r will be coincident with MC_m. Retailers will then purchase the product up to the point at which PME_r equals $PNMR_r$. Wholesale (and retail) output quantity will be Q_{MPT}. Retail and wholesale prices ($P_{r,MPT}$ and $P_{w,MPT}$) will be, respectively, the final demand price and the manufacturers' supply price at this quantity.

The third stylized equilibrium concept is referred to here as retailer price-taking (RPT): Retailers take the manufacturers' asking price as given while manufacturers exercise some degree of monopoly power over retailers. Under these circumstances, retailers will purchase the product up to the point at which $P_w = PNMR_r$. So manufacturers face a demand curve for their product given by the $PNMR_r$ curve. The position of the manufacturing industry's perceived marginal revenue curve (PMR_m) is a reflection of its degree of monopoly power over retailers. A perfectly

Figure 8.1b. The Manufacturer Price-Taking Solution

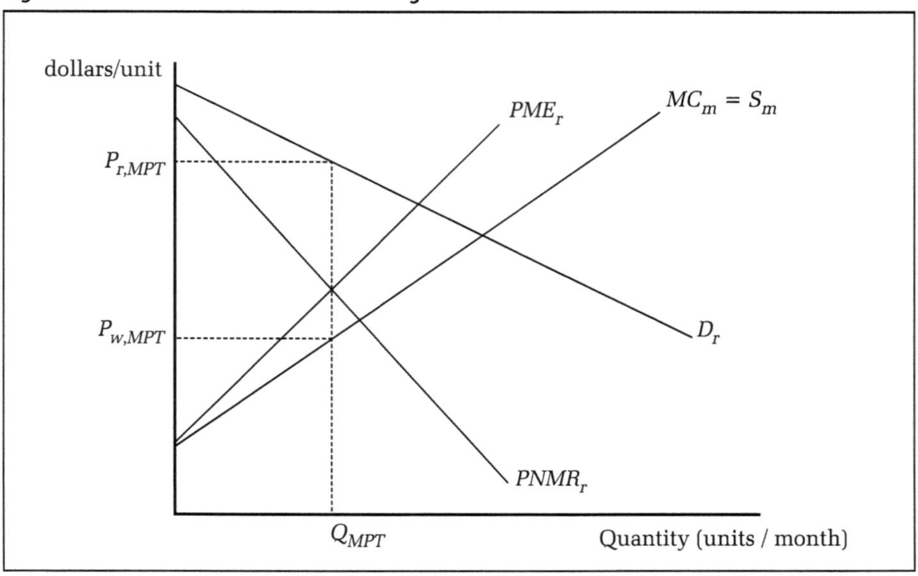

cartelized manufacturing industry would perceive its marginal revenue to be represented by the curve marginal to $PNMR_r$. If the manufacturing industry behaved competitively in its output market, PMR_m would be coincident with $PNMR_r$. Again, an intermediate degree of market power would result in a perceived marginal revenue curve lying between these two extremes. In any event, the solution, illustrated in Figure 8.1c, is determined by the intersection of PMR_m and MC_m. The manufacturing industry would offer a quantity Q_{RPT} for sale at a wholesale price ($P_{w,RPT}$) given by the height of $PNMR_r$ at this quantity. The retail industry would, in turn, sell output to consumers at the corresponding final demand price ($P_{r,RPT}$).

The primitives of this model are the final demand, retailer marginal cost, and manufacturer marginal cost curves. The shapes and positions of these three curves are, of course, influenced by a set of exogenous variables: typically income and the prices of related goods in the case of the demand curve and most likely factor prices in the cases of the two marginal cost curves. Each of the three proposed solution concepts explains the determination of three endogenous variables: output quantity, wholesale prices, and retail prices. Is it possible, given sufficiently many and varied observations on quantity, retail prices, and wholesale

Figure 8.1c. The Retailer Price-Taking Solution

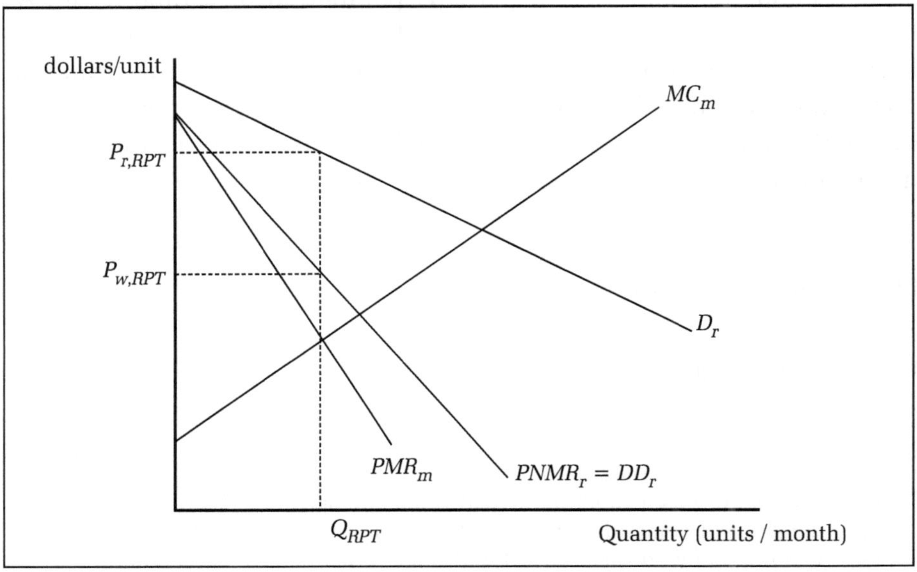

prices and the exogenous demand and marginal cost determinants, to distinguish among the three proposed solution concepts? To put it differently, are the solution concepts identified by the comparative statics of quantity and price with respect to changes in the position of demand and/or marginal cost?

The next two sections address this question through reference to formalized versions of the model we have described. Two examples serve to demonstrate that identification of the solution concept in bilateral oligopoly is not a trivial issue. In the next section, we present a version of the model that admits no possibility for distinguishing among the solution concepts on the basis of empirical observation. Following that discussion, a slightly different model is presented in which all three equilibrium concepts are identified so that data on quantity, prices, and exogenous factors do provide a basis for discriminating among them.

An Example in Which the Solution Concept Is Not Identified

Let the retail-demand curve be described (in inverse form) by

$$(1) \qquad P_r/S = a_0 + a_1Q + a_2Z_2 + a_3QZ_3 + \in$$

where P_r is retail price in nominal terms; S is an appropriate price index so that P_r / S is the real price of the retail product; Q is quantity; Z_2 is an exogenous demand shifter, perhaps population or a time trend; Z_3 is an exogenous variable affecting the slope (elasticity) of the demand curve, perhaps the price of a substitute product; $\in$ is an additive random error expressed in real terms; and a_0, a_1, a_2, and a_3 are parameters. The interaction term in Q and Z_3 is included to insure identification of the degree of retail market power.[6] Retailers' and manufacturers' marginal costs, MC_r and MC_m, are given by

$$(2) \qquad MC_r = b_0 + b_1Q + b_2W_2 + S\eta \text{ and}$$

$$(3) \qquad MC_m = c_0 + c_1Q + c_2V_2 + S\mu$$

6 This is Bresnahan's (1982) point. Actually, because we have implicitly modeled demand quantity as a function of real price, changes in the price index would alter the slope of demand plotted as a function of nominal price and identify the degree of oligopoly power even without the interaction term in Q and Z_3.

where W_2 and V_2 are exogenous factor prices; η and μ are random errors, again expressed in real terms; and b_0, c_0, b_1, c_1, b_2, and c_2 are parameters.[7]

Total revenue for the retail industry is P_rQ. Monopoly marginal revenue is $P_r + QdP_r/dQ$ or, using (1), $P_r + (a_1 + a_3Z_3)\,SQ$. In the customary way, we can accommodate a range of oligopoly conduct on the part of retailers by assuming that their perceived marginal revenue, PMR_r, is given by

$$(4) \qquad PMR_r = P_r + \lambda\,(a_1 + a_3Z_3)\,SQ.$$

With $\lambda = 0$, the implication is that firms behave as price-takers in the retail market; $\lambda = 1$ corresponds to pure monopoly conduct. Retailers' perceived marginal revenue net of their marginal cost is given by

$$PNMR_r = PMR_r - MC_r$$

or, using (2) and (4), by

$$(5) \qquad PNMR_r = P_r + \lambda\,(a_1 + a_3Z_3)\,SQ - b_0 - b_1Q - b_2W_2 - S\eta.$$

In the bilateral price-taking (BPT) solution, equilibrium quantity is determined by the condition $PNMR_r = MC_m$. Equating (3) and (5) and rearranging terms yields

$$(6) \qquad P_r/S + [\lambda\,(a_1 + a_3Z_3) - (b_1 + c_1)/S]\,Q - (b_0 + c_0)/S - b_2\,W_2/S - c_2\,V_2/S = (\eta + \mu).$$

Retail price is the demand price evaluated at the quantity consistent with (6) and wholesale price is equal to the common value of $PNMR_r$ and MC_m:

$$(7) \qquad P_r/S - [a_1 + a_3Z_3]Q - a_0 - a_2Z_2 = \epsilon \text{ and}$$

7 For industry-level marginal cost functions to be meaningful, all firms must operate at a common value of marginal cost and the firm-level marginal cost functions must be strictly increasing in firm output.

(8) $P_w/S - c_1/S\,Q - c_0/S - c_2\,V_2/S = \mu.$

Equations (6), (7), and (8), which can be viewed as determining endogenous variables P_r/S, Q, and P_w/S, describe the BPT solution. It can be shown that the parameters are identified by the *a priori* information embodied in the form of exclusion restrictions.[8]

Next consider the manufacturer price-taking (MPT) solution. In this scenario, the manufacturers' response is described by the supply curve $P_w = MC_m$ or, using (3), by

(9) $P_w = c_0 + c_1 Q + c_2 V_2 + S\mu.$

Retailers' total expenditure on the wholesale product is $P_w Q$ and the monopsony marginal expenditure is $P_w + Q dP_w/dQ$ or, using (9), $P_w + c_1 Q$. We can accommodate a range of oligopsony conduct in the wholesale market with the assumption that retailers maximize profit subject to a perceived marginal expenditure, PME_r, given by

$$PME_r = P_w + \delta c_1 Q$$

or, using (9), by

(10) $PME_r = c_0 + (1 + \delta)\,c_1 Q + c_2 V_2 + S\mu.$

As with λ, the meaningful range of values for δ is [0, 1]: $\delta = 1$ identifies the pure monopsony case and $\delta = 0$ corresponds to price-taking behavior by retail firms in the wholesale market. Under MPT, equilibrium quantity is determined by $PNMR_r = PME_r$. Using (5) and (10), we have an alternative to the BPT solution's Equation (6):

(6') $P_r/S + [\lambda(a_1 + a_3 Z_3) - (b_1 + (1 + \delta)\,c_1)/S]\,Q - (b_0 + c_0)/S - b_2\,W_2/S - c_2\,V_2/S = (\eta + \mu).$

8 The only assumption about the joint distribution of the three random error terms that is required for identification is that they are distributed independently of the model's exogenous variables.

Equations (7) and (8) from the BPT solution combine with Equation (6') to determine the MPT equilibrium. Obviously, the parametric restriction $\delta = 0$ reduces the MPT case to the BPT case.

It can be shown that the parameters of the MPT solution are identified with the exception of δ and b_1. Thus, the BPT case ($\delta = 0$) is not observationally distinct from the MPT case with pure monopsony ($\delta = 1$) nor from any MPT case involving an intermediate degree of retailer oligopsony power ($0 < \delta < 1$).

The intuition behind this result is illustrated in Figure 8.2a and 8.2b. Imagine that we have two triples of observations on quantity, wholesale, and retail price, (Q^*, P_w^*, P_r^*) and (Q', P_w', P_r'), with the switch from one to the other caused by a demand shift from D_r^* to D_r' and the attendant shift in retailers' perceived net marginal revenue from $PNMR_r^*$ to $PNMR_r'$. For simplicity, assume that $\lambda = 0$ so $PNMR_r$ is just the vertical difference between D_r and MC_r. Figure 8.2a shows that the two triples of quantity/price observations can be reconciled with bilateral price-taking behavior given the retailer marginal cost curve (MC_r) that appears in the figure.

Figure 8.2b shows exactly the same quantity/price observations, demand curves, and manufacturer marginal cost curve as in Figure 8.2a.

Figure 8.2a. Price and Quantity Observations Reconciled with the BPT Solution

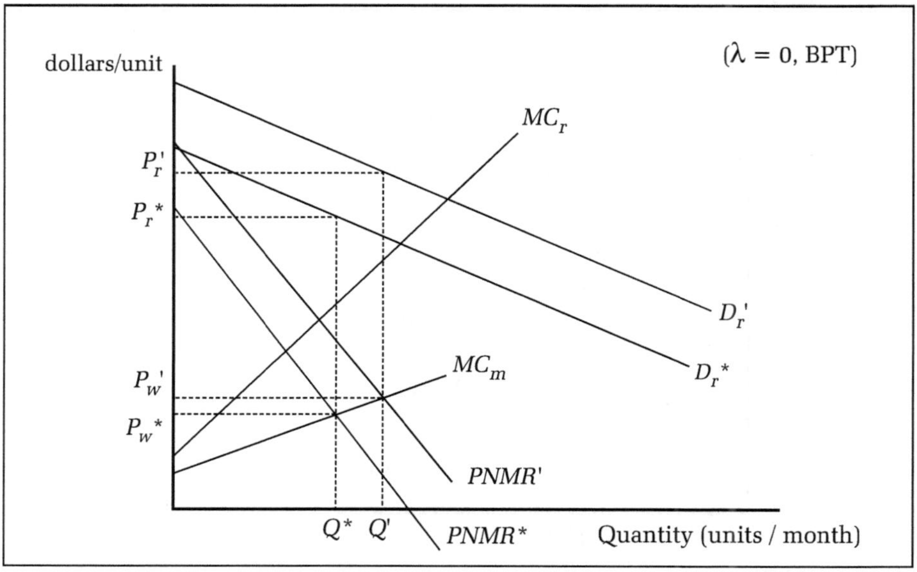

The retailer marginal cost curve, MC_r, differs from that in Figure 8.2a, however, so, via the usual construction, the perceived net marginal revenue curves pertinent to D_r^* and D_r' are correspondingly different from those in Figure 8.2a. Moreover, the retailers' perceived marginal expenditure curve for the pure monopsony case ($\delta = 1$) has been introduced. Note that the change in MC_r results from a change in b_1, the other parameter, along with δ, that is not identified in this formulation. (Q^*, P_w^*, P_r^*) and (Q', P_w', P_r') are equilibria in Figure 8.2b as they were in Figure 8.2a, but this time they are reconciled with the MPT equilibrium with $\delta = 1$. Thus, in the formulation of the model described by Equations (1), (2), and (3), the comparative statics of Q, P_w, and P_r with respect to demand shifts are inadequate to distinguish the BPT ($\delta = 0$) case from the more general MPT case. Similar graphs could be drawn to show that the comparative statics of the model's endogenous variables with respect to rotations of the demand curve, or with respect to shifts of the manufacturers' or retailers' marginal cost curves, are also inadequate to distinguish the two solution concepts.

Figure 8.2b. Price and Quantity Observations Reconciled with the MPT Solution with Pure Monopsony

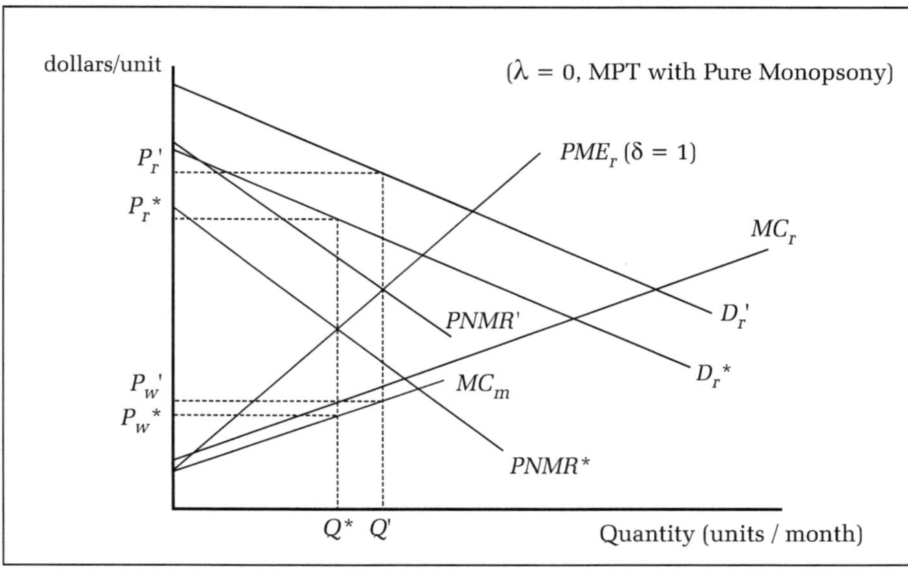

An Example in Which the Solution Concept Is Identified

The model of the previous section must be generalized if it is to support observationally distinct BPT, MPT, and RPT solutions. The simplest generalization that will suffice permits exogenous variables to rotate, as well as to shift, the manufacturers' marginal cost curve. The argument can be made graphically much as in Bresnahan's (1982) discussion of a related identification problem. Consider the point with coordinates (Q^*, P_w^*) in Figure 8.2a and imagine MC_m rotating counterclockwise about this point. Under the BPT solution, and assuming final demand is given by D_r^*, $PNMR_r^*$ is the retailers' derived demand and MC_m is the manufacturers' supply. Since the demand/supply intersection does not change as MC_m rotates, (Q^*, P_w^*, P_r^*) remains the BPT equilibrium.

Now suppose that the same rotation of MC_m were to occur about the (Q^*, P_w^*) point in Figure 8.2b depicting the MPT solution. As MC_m rotates counterclockwise, the position of PME_r also changes. In particular, retailers' perceived marginal expenditure increases in the neighborhood of Q^*. The MPT solutions for Q and P_w would decrease while the MPT solution for P_r would increase. So an exogenous rotation of MC_m does have different implications about equilibrium depending on whether the solution concept is BPT or MPT. In other words, in this particular generalization of the model, the solution concepts are observationally distinct.

We retain Equations (1) and (2) but replace (3) with

(11) $$MC_m = c_0 + c_1Q + c_2V_2 + c_3QV_3 + S\mu$$

where V_3 is another exogenous determinant of manufacturers' marginal cost, perhaps another factor price. Retailers' perceived net marginal revenue is still given by (5). The equilibrium condition for the BPT model is now obtained by equating (5) and (11). Equation (1) and the equation obtained by substituting $P_w = MC_m$ in (11) complete the set of equations describing the BPT solution:

(12) $$P_r/S + [\lambda(a_1 + a_3Z_3) - (b_1 + c_1)/S - c_3V_3/S]\,Q - (b_0 + c_0)/S - b_2W_2/S - c_2V_2/S = (\eta + \mu),$$

(13) $$P_r/S - [a_1 + a_3Z_3]\,Q - a_0 - a_2Z_2 = \in, \text{ and}$$

$$(14)\quad P_w/S - [c_1/S + c_3\,V_3/S]\,Q - c_0/S - c_2\,V_2/S = \mu.$$

In the derivation for the MPT case, retailers' perceived marginal expenditure is a modified version of Equation (10):

$$(15)\quad PME_r = c_0 + (1 + \delta)(c_1 + c_3V_3)\,Q + c_2V_2 + S\mu.$$

Setting $PNMR_r$ (Equation (5)) equal to PME_r (Equation (15)) yields one of the equations of the MPT equilibrium:

$$(12')\quad P_r/S + [\lambda\,(a_1 + a_3Z_3) - (b_1 + (1 + \delta)\,c_1)/S - (1 + \delta)\,c_3\,V_3/S]\,Q \\ - (b_0 + c_0)/S - b_2\,W_2/S - c_2\,V_2/S = (\eta + \mu).$$

Equations (13) and (14) complete the set of equations describing the MPT solution.

In the RPT case, manufacturers face a retailers' derived demand curve given by $PNMR_r$. Marginal revenue for the manufacturing industry is $PNMR_r + Q\,dPNMR_r/dQ$ or, using (1) and (5),

$$PNMR_r + ((1 + \lambda)(a_1 + a_3Z_3)\,S - b_1)\,Q.$$

We allow for a range of oligopoly conduct on the part of manufacturers by representing their perceived marginal revenue as

$$(16)\quad PMR_m = PNMR_r + \gamma\,((1 + \lambda)\,(a_1 + a_3Z_3)\,S - b_1)\,Q$$

where $\gamma \in [0, 1]$ indexes conduct ranging from price-taking ($\gamma = 0$) to pure monopoly in the wholesale market ($\gamma = 1$).

The RPT solution is characterized by $PMR_m = MC_m$. Using (16), (11), and (5), this condition implies that

$$(12'')\quad P_r/S + [(\gamma + \lambda\,(1 + \gamma))\,(a_1 + a_3Z_3) - (b_1\,(1 + \gamma) + c_1)/S - \\ c_3\,V_3/S]\,Q - (b_0 + c_0)/S - b_2\,W_2/S - c_2\,V_2/S = (\eta + \mu).$$

In the RPT model, manufacturers charge a wholesale price equal to the derived demand price at the equilibrium quantity: $P_w = PNMR_r$. Using Equation (5) and substituting for P_r from (12'') yields

$$(14') \quad P_w/S + [\gamma(1+\lambda)(a_1 + a_3Z_3) - (\gamma b_1 + c_1)/S - c_3\, V_3/S]\,Q - c_0/S - c_2\, V_2/S = \mu.$$

Equations (12"), (13), and (14') describe the RPT solution.

The BPT, MPT, and RPT equilibria all have economically meaningful interpretations in terms of firms' conduct in the wholesale market. We will also consider an artificial "nesting" solution (NST) that nests all three. Consider the following equation:

$$(12''') \quad P_r/S + [(\gamma + \lambda(1+\gamma))(a_1 + a_3Z_3) - (b_1(1+\gamma) + (1+\delta)\,c_1)/S - c_3(1+\delta)\,V_3/S]\,Q - (b_0 + c_0)/S - b_2\, W_2/S - c_2\, V_2/S = (\eta + \mu).$$

The NST solution consists of (12'''), (13), and (14'). With $\delta = 0$, (12''') reduces to (12") and the NST solution specializes to the RPT solution. With $\gamma = 0$, (12''') reduces to (12'), (14') reduces to (14), and the NST solution becomes the MPT solution. With $\delta = \gamma = 0$, (12''') reduces to (12) (and (14') reduces to (14)) and the result is the BPT solution.

In each of the four solution concepts, all parameters are identified. The fact that δ is identified in the MPT solution means that bilateral price-taking conduct can be empirically distinguished from retailers' exercise of monopsony power in the wholesale market. Likewise, the fact that γ is identified in the RPT solution means that the data can discriminate between BPT and the exercise of monopoly power by manufacturers. The MPT and RPT solutions cannot be obtained from one another by parametric restriction. But these two solutions also have different implications about comparative statics and so are empirically distinguishable. In the next section, we discuss methods for estimating the model's structural parameters and testing the equilibrium concepts one against the other.

Estimation and Testing Issues

For our model of bilateral oligopoly consisting of structural Equations (1), (2), and (11), we have identified three behaviorally meaningful equilibrium concepts: BPT (characterized by equations (12), (13), and (14)), MPT ((12'), (13), and (14)), and RPT ((12"), (13), and (14')). We have also introduced one artificial hybrid "equilibrium" concept: NST ((12'''), (13), and (14')). The sets of equations for each of these equilibria constitute nonlinear simultaneous-equation models in implicit form. Each can be written as

$$(17) \qquad q\,(y_t, x_t, \theta) = e_t$$

where t subscripts index observations in a time-series of length T, y is a three-by-one vector of endogenous variables:

$$y = [P_r / S, Q, P_w / S]',$$

x is a column vector of the model's exogenous variables, θ is a column vector of the model's unknown parameters, e is a three-by-one vector of structural form error terms:

$$e = [e_1, e_2, e_3]' = [\eta + \mu, \in, \mu]',$$

and $q(\cdot)$ is a three-by-one vector of nonlinear functions of its arguments. The e_ts are assumed to be serially independent and independent of the model's exogenous variables and also are assumed to follow a stationary, zero-mean, multivariate normal distribution with a covariance matrix denoted Ω. Consistent and asymptotically normal estimates of the model's parameters can be obtained by the full-information, maximum likelihood procedure.[9]

Making a judgment about which equilibrium concept is most consistent with observation involves hypothesis testing. It is easy to see that a

9 The log-likelihood function is given by

$$L\,(y_1, y_2, \ldots, y_T; x_1, x_2, \ldots, x_T; \theta, \Omega) = -nT/2 \ln(2\pi) - T/2 \ln(\det \Omega)$$
$$+ \sum_{t=1}^{T} \ln |\det [\partial q/\partial y'\,(y_t, x_t, \theta)]| - 1/2 \sum_{t=1}^{T} q\,(y_t, x_t, \theta)'\,\Omega^{-1}\,q\,(y_t, x_t, \theta)$$

where $n = 3$ is the number of endogenous variables (equations) and $\partial q / \partial y'$ is a three-by-three Jacobian matrix. The concentrated log-likelihood function is

$$\tilde{L}\,(y_1, y_2, \ldots, y_T; x_1, x_2, \ldots, x_T; \theta) = -nT/2\,(\ln(2\pi) + 1)$$
$$+ \sum_{t=1}^{T} \ln |\det [\partial q/\partial y'\,(y_t, x_t, \theta)]| - T/2 \ln(\det [1/T \sum_{t=1}^{T} q\,(y_t, x_t, \theta)\,q\,(y_t, x_t, \theta)']).$$

Estimates of θ are obtained by maximizing this expression by numerical methods. See Amemiya (1977) for consistency and asymptotic normality results.

simple parametric restriction of $\delta = 0$ converts the MPT solution to the BPT solution. So one could test a null hypothesis of bilateral price-taking versus an alternative of manufacturer price-taking coupled with a nonzero degree of retailer oligopsony power by testing $\delta = 0$ versus $\delta > 0$ using an asymptotically valid t-test procedure based on the results of maximum likelihood estimation of the equations of the MPT equilibrium. Likewise, testing $\gamma = 0$ versus $\gamma > 0$ in the equations of the RPT equilibrium would be a means of assessing the viability of the BPT solution compared to the RPT solution. A direct comparison of the MPT and RPT equilibrium concepts is more difficult, however, because the two are not nested; that is, neither can be obtained as a parametric restriction of the other.

One ad hoc approach to testing of nonnested models involves nesting both in a larger artificial model and then testing each nonnested model versus the nesting model, perhaps using the marginal significance levels from these tests to rank the plausibility of the original two nonnested models. The NST model can serve as the nesting model in this approach.[10] Given estimates of the parameters of the NST model, a test of $\delta = 0$ versus $\delta > 0$ is a test of the RPT model versus the more general NST model. A test of $\gamma = 0$ versus $\gamma > 0$ is a test of the MPT model versus the NST model. Of course, one drawback of this approach is that the NST model, unlike the BPT, MPT, and RPT models, has no clear economic behavioral interpretation. So it would be difficult to know what to make of a test result that had us rejecting the MPT model, say, in favor of the NST model.

A more direct approach to comparing the MPT and RPT models makes use of the Cox procedure for testing nonnested hypotheses (1961, 1962). The Cox test's application to the specific case of nonnested, nonlinear

---

10 Up to this point, we have reserved the term "model" for reference to the fundamental structural relations describing consumer demand and manufacturer and retailer marginal cost curves (for example, Equations (1), (2), and (11)). The alternative sets of relations derived from these (for example, Equations (12), (13), and (14)) have been said to be characteristic of the BPT, MPT, or RPT "equilibrium concepts" or "solution concepts." Now that we turn to a discussion of nonnested hypotheses tests, this terminological convention becomes burdensome so, for simplicity, we hereafter refer to the BPT, MPT, RPT, and NST "models."

simultaneous-equation models in implicit form is discussed in Aguirre-Torres and Gallant (1983). To introduce the test procedure, consider the null and alternative hypotheses corresponding to models that are each of the form of Equation (17):

$$\text{(18)} \quad H_0\colon q_0(y_t, x_t, \theta_0) = e_{0t} \text{ for } t = 1, 2, \ldots, T \text{ and some } \theta_0 \in \Theta_0$$
$$H_1\colon q_1(y_t, x_t, \theta_1) = e_{1t} \text{ for } t = 1, 2, \ldots, T \text{ and some } \theta_1 \in \Theta_1$$

where Θ_0 and Θ_1 are the sets of values for θ_0 and θ_1, respectively, and, in our application, we might take H_0 to be the RPT model and H_1 to be the MPT model, for example. Let $f_0(y \mid x, \Omega_0, \theta_0)$ be the conditional density of random vector y associated with the null model, $q_0(y, x, \theta_0) = e_0$.[11] Likewise, let $f_1(y \mid x, \Omega_1, \theta_1)$ be the density of y corresponding to the alternative model, $q_1(y, x, \theta_1) = e_1$. The log-likelihood ratio for the two models is given by

$$1/T \sum_{t=1}^{T} [\ln f_1(y_t \mid x_t, \hat{\Omega}_1, \hat{\theta}_1) - \ln f_0(y_t \mid x_t, \hat{\Omega}_0, \hat{\theta}_0)]$$

where $\hat{\Omega}_0$, $\hat{\Omega}_1$, $\hat{\theta}_0$, and $\hat{\theta}_1$ are maximum likelihood estimators:

$$\hat{\theta}_0 \text{ and } \hat{\Omega}_0 \text{ maximize } 1/T \sum_{t=1}^{T} \ln f_0(y_t \mid x_t, \Omega_0, \theta_0), \text{ and}$$

$$\hat{\theta}_1 \text{ and } \hat{\Omega}_1 \text{ maximize } 1/T \sum_{t=1}^{T} \ln f_1(y_t \mid x_t, \Omega_1, \theta_1).$$

The Cox test rejects H_0 in favor of H_1 if the realization of the log-likelihood ratio is improbably large given its distribution under the null hypothesis—specifically, if

---

11 The density of y corresponding to the null model is derived using the change of variable formula. For example,

$$f_0(y \mid x, \Omega_0, \theta_0) = |\det \partial q_0/\partial y'(y, x, \theta_0)| \bullet \phi_0(q_0(y, x, \theta_0); \Omega_0)$$

where $\phi_0(\bullet\,; \Omega_0)$ is the (multivariate normal) density of the error term, e_0.

$$(19) \quad 1/T \sum_{t=1}^{T} [\ln f_1 (y_t \mid x_t, \hat{\Omega}_1 \hat{\theta}_1) - \ln f_0 (y_t \mid x_t, \hat{\Omega}_0, \hat{\theta}_0)] -$$
$$1/T \sum_{t=1}^{T} E_{f(\cdot \mid t,\, 0)} [\ln f_1 (y \mid x_t, \hat{\Omega}_1, \hat{\theta}_1) - \ln f_0 (y \mid x_t, \hat{\Omega}_0, \hat{\theta}_0)]$$

exceeds an appropriate critical value. $E_{f(\cdot \mid t,\, 0)}$ denotes the expectation with respect to the conditional distribution of y given the null model with period t values for the exogenous variables $\theta_0 = \hat{\theta}_0$ and $\Omega_0 = \hat{\Omega}_0$. The difficulty in application of the test is in the evaluation of these expectations. Aguirre-Torres and Gallant (1983) derive the asymptotic distribution of a test statistic obtained from (19) by replacing these expectations with bootstrap estimates.

To motivate the formula for the Aguirre-Torres and Gallant (1983) test statistic, suppose that one had a random sample of size T from the distribution $f_0 (y \mid x_t, \hat{\Omega}_0, \hat{\theta}_0) : \tilde{y}_1, \tilde{y}_2, \ldots, \tilde{y}_T$. Then a natural Monte Carlo estimate of $E_{f(\cdot \mid t,\, 0)} [\ln f_1 (y \mid x_t, \hat{\Omega}_1, \hat{\theta}_1) - \ln f_0 (y \mid x_t, \hat{\Omega}_0, \hat{\theta}_0)]$ would be given by

$$1/T \sum_{j=1}^{T} [\ln f_1 (\tilde{y}_j \mid x_t, \hat{\Omega}_1, \hat{\theta}_1) - \ln f_0 (\tilde{y}_j \mid x_t, \hat{\Omega}_0, \hat{\theta}_0)].$$

The basic idea is to use this Monte Carlo approach with a pseudo-random sample derived from the sample distribution of the null model's residuals.

More formally, let $y = Y_0 (e, x, \theta_0)$ be the solution for the three-by-one vector y from the system of equations $q_0(y, x, \theta_0) = e$. First, compute the residuals from the null model:

$$(20) \quad \hat{e}_t = q_0(y_t, x_t, \hat{\theta}_0) \qquad t = 1, 2, \ldots, T,$$

a pseudo-random sample from the distribution of e_0. Next, pairing each period's residual vector with each period's vector of exogenous variables, calculate the associated y vectors:

$$\hat{y}_{tj} = Y_0 (\hat{e}_j, x_t, \hat{\theta}_0) \qquad t = 1, 2, \ldots, T, \quad j = 1, 2, \ldots, T.$$

Then, taking $\hat{y}_{tj}$ for a given t and for $j = 1, 2, \ldots, T$ as a pseudo-random sample from $f_0(\cdot \mid x_t, \hat{\Omega}_0, \hat{\theta}_0)$, implement the Monte Carlo estimate of the expectations in (19) to obtain the statistic:

$$(21) \quad \hat{S} = 1/T \sum_{t=1}^{T} [\ln f_1 (y_t \mid x_t, \hat{\Omega}_1, \hat{\theta}_1) - \ln f_0 (y_t \mid x_t, \hat{\Omega}_0, \hat{\theta}_0)] - 1/T \sum_{t=1}^{T} 1/T \sum_{j=1}^{T} [\ln f_1 (\hat{y}_{tj} \mid x_t, \hat{\Omega}_1, \hat{\theta}_1) - \ln f_0 (\hat{y}_{tj} \mid x_t, \hat{\Omega}_0, \hat{\theta}_0)].$$

Aguirre-Torres and Gallant (1983) show that, on the null hypothesis of (18), the statistic is asymptotically normal. Some additional details regarding the calculation of $\hat{S}$ and an expression for $\hat{V}^2$, a consistent estimator of the asymptotic variance of $\sqrt{T}\,\hat{S}$, are given in Appendix A. The test of (18)'s H_0 versus H_1 can be carried out by comparing the sample value of $\sqrt{T}\,\hat{S}/\hat{V}$ to an appropriate critical value of the $N(0, 1)$ distribution.

The Wholesale Market for Beef

In the past few years, the balance of power between manufacturers and retailers has received considerable attention in the business press with many observers suggesting that there has been a recent shift in power toward retailers. Much of this attention has focused specifically on wholesale food markets.[12] Our stylized bilateral oligopoly model with associated estimation and testing procedures provides a means of formally addressing the balance-of-power issue. Our specific application is the wholesale market for beef in the U.S. In this market, manufacturers (beef packers) sell fresh boxed beef to national supermarket chains that have integrated retail with wholesale distribution by purchasing directly from food processors or to independent wholesalers who in turn sell to

12 Ailawadi, Borin, and Farris (1995) discuss the power shift in wholesale markets broadly and Messinger and Narasimhan (1995) focus on the grocery channel in particular. Both cite numerous references in trade journals and popular magazines documenting the widespread perception that power has shifted toward retailers. Messinger and Narasimhan argue that, in spite of recent changes in industry structure and conduct that would appear to enhance the bargaining position of grocery retailers vis-à-vis food manufacturers, there actually is little evidence in either accounting or stock market data that would indicate a shift in channel profitability from manufacturers to retailers.

nonintegrated local or regional chain grocery stores.[13] Latest statistics (USDA 1996) indicate that concentration in boxed-beef production in 1994 was on the order of 85.7 percent for the top four firms, 92.5 percent for the top eight firms, and 97.1 percent for the top twenty firms. In 1990, the beginning of our study period, these figures were 79.3 percent, 88.7 percent, and 95.8 percent, respectively. During the same period, the Herfindahl Index for boxed-beef production increased from 1988 to 2340. Comparable statistics for boxed-beef purchases are not available in such detail. However, a recent study (Johnson et al. 1989) provides some concentration figures for the late 1980s. The top four national supermarket chains purchased close to 60 percent of all fresh beef sold to integrated retailers/wholesalers. Close to 40 percent of the boxed beef sold to independent wholesalers was purchased by the four largest buyers in that market. With such concentration levels on both sides of the market, the wholesale beef market is fairly described as a bilateral oligopoly.

Concerns about the possible behavioral consequences of high concentration have spawned a number of empirical studies of market power in both the beef-packing and the retail grocery industries. In the case of beef packing, most of the attention has focused on the issue of monopsony power in cattle markets but some studies have also addressed monopoly power in the wholesale beef market.[14] In grocery retailing, the main emphasis has been on monopoly power in the retail market.[15] But Hall, Schmitz, and Cothern (1979) examine wholesale-retail marketing margins for beef and obtain results from which they draw some tentative conclusions about the nature of competition in the wholesale

---

13 The state-of-the-art technology in beef packing integrates slaughter with an initial processing stage so that about 90 percent of fresh beef now leaves packing plants as "boxed beef" (primal and subprimal carcass sections that are individually vacuum wrapped and packaged several sections to a box) rather than in the hanging-carcass form that was the industry standard twenty years ago. Almost all boxed beef comes from fed steers and heifers (cattle bred specifically for meat and fed a grain ration at feedlots for approximately four months prior to slaughter). Fed cattle are the predominant source of beef table cuts. Ground beef sometimes comes from other kinds of cattle, primarily culled dairy cows.

14 See Azzam and Anderson (1995) for a comprehensive review of empirical studies of market power in beef packing.

15 Cotterill (1986) investigates the relationship between food prices and supermarket concentration levels in local markets.

market. Using pooled cross-section and time-series data (seven annual observations for each of nineteen standard metropolitan statistical areas), they find a positive correlation between the beef marketing margin and the four-firm concentration ratio for grocery chains.[16] They attribute this, in part, to the buying practices of large food chains that may enable them to exert some monopsony power over beef packers. The positive correlation between the wholesale-retail margin and food chain concentration is a stylized fact of the market that may be suggestive of retailer monopsony power. But it also could be the result of retailer monopoly power over food consumers. To draw a confident conclusion about the nature of competition in the wholesale market for beef, it is necessary to estimate the bilateral oligopoly model and carry out testing to determine which of the candidate equilibrium concepts is most consistent with the data. That is the task to which we now turn.

The equations underlying each of the equilibrium concepts are retail demand (Equation (1)), grocery retailers' marginal cost (2), and meat packers' marginal cost (11). We implement the models using monthly data for the period January 1990 through December 1994 (sixty observations) with the following definitions of variables.[17]

$P_r =$ Retail price of beef (cents per retail pound).

$Q =$ Quantity of beef (millions of pounds, retail weight equivalent).

$P_w =$ Wholesale price of beef (cents per retail pound equivalent).

$S =$ CPI for food (1982 = 100).

$Z_3 =$ Real retail price of pork (1982 dollars per retail pound).

$W_2 =$ Average hourly earnings of grocery store workers (dollars per hour).

---

16 Hall, Schmitz, and Cothern's (1979) time series run only through 1973. Both beef packing and grocery retailing have undergone significant increases in concentration since then.

17 More precise definitions and details about variable measurement appear in Appendix B.

$V_3 =$ A proxy for the price of cattle[18,19] (cents per retail pound equivalent).

$V_2 = Z_2 =$ A time trend.

The results of maximum likelihood estimation for the BPT, MPT, RPT, and NST models are presented in Table 8.1. Because our attention focuses on the task of choosing among competing equilibrium concepts, we address that issue first and later (in footnote 22) interpret individual parameter estimates for the most preferred model. A summary of hypothesis test results appears in Table 8.2.

First, we test the BPT model as parametric restrictions of the MPT and RPT models. In the MPT model, the estimate of δ is 0.99 with an asymptotic t-statistic of 1.375. Hence, we reject H_0: $\delta = 0$ (BPT) versus H_1: $\delta > 0$ (MPT) with a marginal significance level of about 8.5 percent (in a one-tailed test).[20] In the RPT model, the estimate of γ is actually negative (–0.2819) and therefore outside of the economically meaningful range. Nonetheless, the asymptotic t-statistic is –0.802 and the marginal significance level for a test of H_0: $\gamma = 0$ (BPT) versus H_1: $\gamma > 0$ (RPT) is 78.9 percent.[21] So the data appear to favor the MPT model over the BPT model but do not judge the RPT model to be an improvement over the BPT model.

---

18 That the price of cattle should influence packers' marginal cost is intuitively clear. That it should enter the formulation in the role of V_3, not V_2 (that is, as a *rotator*, not a *shifter*, of marginal cost), is perhaps not so clear. We note, however, that the estimate of c_3, the coefficient of the interaction term in Q and V_3, turns out to be significantly different from zero at the 1 percent level in the most preferred model.

19 In one respect, our application does not fit our stylized bilateral oligopoly model: the price of one important manufacturing input, slaughter cattle, is unlikely to be exogenous. (Because there is no use for cattle other than as an input into beef production, one would expect that cattle prices would be jointly determined with wholesale and retail beef prices and beef quantity.) Our ad hoc accommodation of this problem is to replace the cattle-price series with its least squares projection on a set of instruments that are more arguably exogenous or at least predetermined. Appendix B contains details.

20 On the basis of MPT results, the value of the χ_1^2 likelihood-ratio test statistic for a test of H_0: $\delta = 0$ versus H_1: $\delta \neq 0$ is 10.42 with a marginal significance level of 0.1 percent.

21 On the basis of RPT results, the value of the χ_1^2 likelihood-ratio test statistic for a test of H_0: $\gamma = 0$ versus H_1: $\gamma \neq 0$ is 0.662 with a marginal significance level of 41.6 percent.

Table 8.1. Maximum Likelihood Estimates of the BPT, MPT, RPT, and NST Models[a]

	BPT	MPT	RPT	NST
a_0	2.9590***	2.9750***	2.9060***	2.9220***
	0.2630	*0.2384*	*0.2366*	*0.2212*
a_1	–0.1691E–02***	–0.1712E–02***	–0.1560E–02***	–0.1602E–02***
	0.5956E–03	*0.4906E–03*	*0.5709E–03*	*0.4690E–03*
a_2	0.1301E–02	0.1347E–02	0.9526E–03	0.1063E–02
	0.1708E–02	*0.1431E–02*	*0.1672E–02*	*0.1403E–02*
a_3	0.5625E–03**	0.5661E–03**	0.5134E–03*	0.5295E–03**
	0.2724E–03	*0.2379E–03*	*0.2715E–03*	*0.2326E–03*
b_0	–173.7***	–330.5*	–177.2***	–368.8*
	61.81	*197.3*	*63.72*	*212.1*
b_1	0.03201	0.09396	0.03124	0.08719
	0.01974	*0.06574*	*0.02129*	*0.06289*
b_2	29.78***	27.93*	30.43***	34.53**
	9.062	*15.80*	*8.983*	*17.54*
c_0	107.8*	122.4*	151.7**	189.2**
	64.16	*67.74*	*76.54*	*88.77*
c_1	–0.1109***	–0.1057***	–0.08182*	–0.06611
	0.02607	*0.02246*	*0.04407*	*0.05171*
c_2	–0.02890	–0.06349	0.1426	0.06959
	0.1415	*0.1278*	*0.2300*	*0.2128*
c_3	0.1121E–02***	0.1010E–02***	0.9274E–03**	0.6743E–03
	0.3782E–03	*0.3875E–03*	*0.4168E–03*	*0.4755E–03*
λ	0.1090	0.4515	0.1113	0.3856
	0.1238	*0.3685*	*0.1448*	*0.3357*
δ		0.9902		1.465
		0.7203		*1.458*
γ			–0.2819	–0.2112
			0.3511	*0.2749*
Log of Likelihood Function				
	–90.8017	–85.5901	–90.4706	–85.3901
R^2 [b]				
P_r / S	0.7878	0.7959	0.7854	0.7951
Q	0.4142	0.4058	0.4257	0.4197
P_w / S	0.8779	0.8806	0.8789	0.8813

[a] Asymptotic standard errors are in italics. * = significantly different from zero (in a two-tailed test) at the 10 percent level; ** = significantly different at the 5 percent level; *** = significantly different at the 1 percent level.

[b] The estimated structural models were solved simultaneously (assuming zero values for the error terms) for the endogenous variables P_r/S, Q, and P_w / S. The reported R^2s are squared simple correlation coefficients between the fitted and actual series.

Table 8.2. Hypothesis Test Results[a]

Nested Hypothesis Tests (Tests of Parametric Restrictions)	
Tests Based on Estimates of the MPT Model	
H_0: $\delta = 0$ (BPT)	H_0: $\delta = 0$ (BPT)
vs. H_1: $\delta > 0$ (MPT)	vs. H_1: $\delta \neq 0$
t = 1.375	$\chi_1^2 = 10.42$
P = 8.5 percent	P= 0.1 percent
Tests Based on Estimates of the RPT Model	
H_0: $\gamma = 0$ (BPT)	H_0: $\gamma = 0$ (BPT)
vs. H_1: $\gamma > 0$ (RPT)	vs. H_1: $\gamma \neq 0$
t = –0.802	$\chi_1^2 = 0.662$
P = 78.9 percent	P = 41.6 percent
Tests Based on Estimates of the NST Model	
H_0: $\gamma = 0$ (MPT)	H_0: $\delta = 0$ (RPT)
vs. H_1: $\gamma \neq 0$ (NST)	vs. H_1: $\delta \neq 0$ (NST)
$\chi_1^2 = 0.400$	$\chi_1^2 = 10.16$
P = 52.7 percent	P = 0.1 percent
Nonnested Hypothesis Tests	
H_0: RPT	H_0: MPT
vs. H_1: MPT	vs. H_1: RPT
$\sqrt{T}\ \hat{S}/\hat{V} = 10.60$	$\sqrt{T}\ \hat{S}/\hat{V} = 0.662$
P = 0.0 percent	P = 25.4 percent

[a] P values are marginal significance levels.

While the preceding results seem to suggest that the MPT model is superior to the RPT model (as well as to the BPT model), it is conceivable that a different conclusion could be reached through a pairwise comparison of MPT with RPT, the task to which we now turn. As mentioned in the preceding section, one indirect way of effecting this comparison utilizes the artificial "nesting" model from which MPT and RPT can be obtained via parametric restriction. The MPT model is the NST model with $\gamma = 0$. The χ_1^2 likelihood-ratio test statistic for a test of this hypothesis (versus a two-sided alternative) is 0.400 with a marginal significance level of 52.7 percent. The RPT model is the NST model with $\delta = 0$. The χ_1^2 likelihood-ratio test statistic for a test of this hypothesis (again, versus a two-sided alternative) is 10.161 with a marginal significance level of 0.1 percent. Thus we have indirect evidence favoring the MPT model over the RPT model in that the RPT model, but not the MPT model, is rejected (using conventional significance levels) when tested against the encompassing NST model.

A more defensible approach to comparing RPT with MPT involves an application of the Aguirre-Torres and Gallant (1983) nonnested hypothesis testing procedure. When that method is applied with the RPT model taken to be the null hypothesis and the MPT model taken to be the alternative hypothesis, the realized value of the $\sqrt{T}\,\hat{S}/\hat{V}$ test statistic is 10.60. Since on H_0 this statistic has an asymptotic $N(0, 1)$ distribution, the marginal significance level is below 0.1 percent, a result strongly supporting rejection of RPT in favor of MPT. Testing MPT (as the null) versus RPT (as the alternative) yields a test statistic value of 0.662 with a marginal significance level of 25.4 percent.

So the data are more easily reconcilable with the MPT model than with either the RPT or the BPT model.[22] The economic interpretation of this result is that noncompetitive pricing distortions exist in this bilateral oligopoly setting and that they appear to be the result of the exploitation of oligopsony power on the part of retailers while manufacturers exhibit

---

22 Here is a summary of the estimation results for the MPT model. The estimated slope of the demand curve, $\hat{a}_1 + \hat{a}_3 Z_3$, is significantly negative at all sample points. The estimated demand elasticity at sample means is –2.18 with a standard error of 0.59. The significantly positive estimate of a_3 confirms that the beef demand price increases at each quantity as the price of pork increases. Beef and pork are substitutes. The positive estimate of a_2 suggests that demand shifted to the right with time, although significance for this finding is lacking.

The estimated slope of beef packers' marginal cost, $\hat{c}_1 + \hat{c}_3 V_3$, is positive at all sample points but insignificant. The significantly positive estimate of c_3 indicates that the beef packers' marginal cost increases for every quantity as the slaughter-cattle price increases. The insignificantly negative estimate of c_2 provides weak evidence that marginal packing costs have fallen over time, presumably due to technological innovation.

The grocery retailers' marginal cost slopes upward ($\hat{b}_1 > 0$) and shifts upward with increases in the grocery labor wage ($\hat{b}_2 > 0$), though neither of these findings is particularly significant. One disconcerting aspect of the results for the MPT model (but not the BPT or RPT models) is that the point estimates of the retailers' marginal cost are negative, though far from significant, at most sample points. As a result, the model's implied breakdowns of the retail/wholesale margin into retail marginal cost and pricing distortions due to retailers' monopoly and monopsony power are not particularly meaningful.

benign, price-taking conduct. Note that if the analysis had focused on the possibility of monopoly only, maintaining the RPT model as the appropriate one, the outcome of testing would have been a failure to reject the hypothesis of competitive conduct on the part of packers as well. (Based on estimates of the RPT model, we fail to reject $\gamma = 0$ versus $\gamma > 0$.) This result would have "supported" the apparently erroneous conclusion that price and quantity determinations in the wholesale beef market are consistent with perfect competition.

Summary and Conclusions

Implicit in more recent approaches to the measurement of market power is the maintained hypothesis of price-taking behavior on one side of the market or the other. When both sides of the market exhibit significant concentration, one would prefer, however, to leave open the question of whether buyers or sellers (or both) behave competitively. In this chapter, we have investigated whether it is possible to distinguish between the exercise of market power on the buyers' and sellers' sides of a bilateral oligopoly. The econometric focus has been on choosing among three competing equilibrium concepts for a wholesale market: bilateral price-taking (BPT), manufacturer price-taking (MPT), and retailer price-taking (RPT). The choice between the MPT and the RPT hypotheses comes down to a test of nonnested, nonlinear, simultaneous-equation models in implicit form. Our application to the U.S. wholesale beef market showed the MPT solution concept to be most consistent with the data for that specific case.

The main caveat of this work is that it considers only three stylized solution concepts that are recommended primarily by their simplicity and by the convenient and intuitive notions of buyer and seller conduct to which they correspond. The actual mechanism responsible for the determination of wholesale price and quantity could be more complex than any of the three. Perhaps the most conspicuously omitted candidate is a negotiation/bargaining solution in which manufacturers and retailers collaborate in fixing output at its joint profit-maximizing level and then bargain over the wholesale price as the instrument for effecting a particular distribution of profit. Maximizing joint profit would require operating where retailers' marginal revenue net of marginal retail cost equals manufacturers' marginal cost—the bilateral price-taking quantity. In every period, the BPT price would then effect one particular distribution of profit; others, of course, are possible too. To provide a basis for discriminating between the BPT solution and a negotiation/bargaining solution, one would have to maintain some kind of assumption about the stability of the bargaining outcome over time. Perhaps, for example, one could assume that each side of the market gets the same proportionate share of joint profit in every period. Because there would be little to recommend any particular assumption of this nature, we chose not to examine the negotiation/bargaining case.

Appendix A: Details of the Calculation of $\hat{S}$ and $\hat{V}^2$

We wish to test

$$H_0\colon q_0\,(y_t, x_t, \theta_0) = e_{0t} \text{ for } t = 1, 2, \ldots, T \text{ and some } \theta_0 \in \Theta_0$$
$$\text{vs. } H_1\colon q_1\,(y_t, x_t, \theta_1) = e_{1t} \text{ for } t = 1, 2, \ldots, T \text{ and some } \theta_1 \in \Theta_1.$$

Taking the null hypothesis to be the RPT model (given by Equations (12"), (13), and (14')), we have

$$q_0\,(y_t, x_t, \theta_0) = M_0\,(x_t, \theta_0)\,y_t - m\,(x_t, \theta_0)$$

where θ_0, x_t, y_t, and $m\,(x_t, \theta_0)$ are column vectors,

$$\theta_0 = [\lambda, a_0, a_1, a_2, a_3, b_0, b_1, b_2, c_0, c_1, c_2, c_3, \gamma]',$$

$$x_t = [S_t, Z_{2t}, Z_{3t}, W_{2t}, V_{2t}, V_{3t}]',$$

$$y_t = [P_{rt}/S_t, Q_t, P_{wt}/S_t]',$$

$$m\,(x_t, \theta_0) = [(b_0+c_0)/S_t + b_2 W_{2t}/S_t + c_2 V_{2t}/S_t, a_0 + a_2 Z_{2t}, c_0/S_t + c_2 V_{2t}/S_t]',$$

and $M_0\,(x_t, \theta_0)$ is given by

$$\begin{bmatrix} 1 & (\gamma+\lambda\,(1+\gamma))\,(a_1+a_3 Z_{3t}) - (b_1(1+\gamma)+c_1)/S_t - c_3 V_{3t}/S_t & 0 \\ 1 & -(a_1+a_3 Z_{3t}) & 0 \\ 0 & \gamma\,(1+\lambda)\,(a_1+a_3 Z_{3t}) - (\gamma b_1+c_1)/S_t - c_3 V_{3t}/S_t & 1 \end{bmatrix}.$$

With the alternative model taken to be the MPT model (given by equations (12'), (13), and (14)), we have

$$q_1(y_t, x_t, \theta_1) = M_1(x_t, \theta_1)\, y_t - m(x_t, \theta_1)$$

where $\theta_1 = \theta_0$ but with δ replacing γ and $M_1(x_t, \theta_1)$ is given by

$$\begin{bmatrix} 1 & \lambda(a_1 + a_3 Z_{3t}) - (b_1 + (1+\delta)c_1)/S_t - c_3(1+\delta)V_{3t}/S_t & 0 \\ 1 & -(a_1 + a_3 Z_{3t}) & 0 \\ 0 & -(c_1/S_t + c_3 V_{3t}/S_t) & 1 \end{bmatrix}.$$

The null model's residuals (Equation (20)) are given by

$$\hat{e}_t = M_0(x_t, \hat{\theta}_0)\, y_t - m(x_t, \hat{\theta}_0) \qquad \text{for } t = 1, 2, \ldots, T.$$

The pseudo-random samples from the null model's conditional distributions of y_t are generated as

$$\hat{y}_{tj} = [M_0(x_t, \hat{\theta}_0)]^{-1}(\hat{e}_j + m(x_t, \hat{\theta}_0)) \quad \text{for } t = 1, 2, \ldots, T, \quad j = 1, 2, \ldots, T.$$

Owing to the assumed normality of the errors, the individual log-likelihood terms appearing in Equation (21) are, for $i = 0$ and 1, given by

$$\ln f_i(y \mid x_t, \hat{\Omega}_i, \hat{\theta}_i) = -n/2 \ln(2\pi) - 1/2 \ln(\det \hat{\Omega}_i)$$
$$+ \ln|\det[\partial q_i/\partial y'(y, x_t, \hat{\theta}_i)]| - 1/2\, q_i(y, x_t, \hat{\theta}_i)'\, \hat{\Omega}_i^{-1}\, q_i(y, x_t, \hat{\theta}_i)$$

where $\partial q_i/\partial y' = M_i(x, \theta)$ and

$$\hat{\Omega}_i = 1/T \sum_{t=1}^{T} q_i(y_t, x_t, \hat{\theta}_i)\, q_i(y_t, x_t, \hat{\theta}_i)]'.$$

Several of the preceding formulas are used in the evaluation of $\hat{S}$ given by Equation (21). Aguirre-Torres and Gallant (1983) derive the following expression for $\hat{V}^2$, a consistent estimator of the asymptotic variance of $\sqrt{T}\,\hat{S}$:

$$\hat{V}^2 = 1/T \sum_{t=1}^{T} \text{var}^*[\hat{d}_t(y_t^*)]$$

$$+ 1/T \sum_{t=1}^{T} \text{var}^*[\Psi \cdot J^{-1} \cdot \partial/\partial\theta_0 \, \hat{s}_t(y_t^*)]$$
$$- 1/T \sum_{t=1}^{T} 1/T \sum_{m=1}^{T} \text{cov}^*[\hat{d}_t(y_t^*), \hat{d}_m(y_m^*)]$$
$$- 2/T \sum_{t=1}^{T} \text{cov}^*[\hat{d}_t(y_t^*), \Psi \cdot J^{-1} \cdot \partial/\partial\theta_0 \, \hat{s}_t(y_t^*)]$$
$$+ 2/T \sum_{t=1}^{T} 1/T \sum_{m=1}^{T} \text{cov}^*[\Psi \cdot J^{-1} \cdot \partial/\partial\theta_0 \, \hat{s}_t(y_t^*), \; \hat{d}_m(y_m^*)]$$

where, for scalar-valued functions $g(\cdot)$ and $h(\cdot)$,

$$\text{var}^*[g(y_t^*)] \equiv \text{cov}^*[g(y_t^*), g(y_t^*)],$$
$$\text{cov}^*[g(y_t^*), h(y_s^*)] \equiv 1/T \sum_{k=1}^{T} g(\hat{y}_{tk}) h(\hat{y}_{sk}) - E^*[g(y_t^*)] \cdot E^*[h(y_s^*)], \text{ and}$$
$$E^*[g(y_t^*)] \equiv 1/T \sum_{k=1}^{T} g(\hat{y}_{tk}).$$

Furthermore,

$$\hat{s}_t(y) \equiv - \ln f_0(y \mid x_t, \hat{\Omega}_0, \hat{\theta}_0),$$
$$\hat{d}_t(y) \equiv \ln f_1(y \mid x_t, \hat{\Omega}_1, \hat{\theta}_1) - \ln f_0(y \mid x_t, \hat{\Omega}_0, \hat{\theta}_0), \text{ and}$$
$$J \equiv 1/T \sum_{t=1}^{T} \partial^2/\partial\theta_0\partial\theta_0' \; \hat{s}_t(y).$$

Finally,

$$\Psi \equiv -1/T^2 \sum_{t=1}^{T} \sum_{j=1}^{T} [\partial/\partial y' \, \hat{d}_t(\hat{y}_{tj}) \cdot \partial/\partial e' \, Y_0(\hat{e}_j, x_t, \hat{\theta}_0)$$
$$\cdot \partial/\partial\theta_0' \, q_0(y_j, x_j, \hat{\theta}_0) + \partial/\partial y' \, \hat{d}_t(\hat{y}_{tj}) \cdot \partial/\partial\theta_0' \, Y_0(\hat{e}_j, x_t, \hat{\theta}_0)].$$

In evaluating the formula for $\hat{V}^2$, we make use of the following expressions for derivatives:

$$\partial/\partial y' \, \hat{d}_t(\hat{y}_{tj}) = -1/2 \, q_1'(\hat{y}_{tj}, x_t, \hat{\theta}_1)\,(\hat{\Omega}_1^{-1} + (\hat{\Omega}_1^{-1})')$$
$$\cdot \partial/\partial y' q_1(\hat{y}_{tj}, x_t, \hat{\theta}_1) + 1/2 \, q_0'(\hat{y}_{tj}, x_t, \hat{\theta}_0)\,(\hat{\Omega}_0^{-1} + (\hat{\Omega}_0^{-1})')$$
$$\cdot \partial/\partial y' q_0(\hat{y}_{tj}, x_t, \hat{\theta}_0) \text{ and}$$

$$\partial/\partial e' \, Y_0(\hat{e}_j, x_t, \hat{\theta}_0) = [M_0(x_t, \hat{\theta}_0)]^{-1}.$$

In evaluation of Ψ, J, and $\hat{V}^2$, the first- and second-order partial derivatives with respect to the elements of θ_0 were approximated numerically using difference quotients.

Appendix B: Data

Data are monthly from January 1990 through December 1994. Except where otherwise noted, they were obtained from various issues of *Livestock and Poultry Situation and Outlook Report* by the U.S. Department of Agriculture's Economic Research Service. The data series for P_r, P_w, and Q were designed to reflect the retail and wholesale prices and the retail quantity of beef deriving from fed-cattle slaughter, the predominant source of commercially marketed beef.

P_r = Estimated weighted average of prices of retail cuts from choice-yield, grade-three beef carcasses (cents per retail pound).

P_w = Value of wholesale quantity of beef equivalent to one pound of retail cuts (cents per retail pound equivalent).

Q = Quantity of beef (millions of pounds, retail weight equivalent). For steers and heifers separately, the numbers of cattle slaughtered in commercial packing plants were multiplied by the average dressed weight of steers or heifers slaughtered under federal inspection. The total steer and heifer carcass weight was divided by 1.513, the carcass-to-retail conversion factor implicit in the USDA's choice-beef price and spread-calculation procedures.

S = Consumer Price Index (CPI) for food (1982 = 100) (Source: *Economic Report to the President*, various issues).

Z_3 = Estimated weighted-average price of retail cuts from pork carcasses deflated by the CPI for food (1982 dollars per retail pound).

W_2 = Nonsupervisory worker average hourly earnings in Standard Industrial Code 541 (retail trade – grocery stores) (dollars per hour) (Source: *Employment and Earnings*, U.S. Bureau of Labor Statistics, various issues).

V_3 = A proxy for the price of cattle (cents per retail pound equivalent). The basis for this variable was USDA's series on "beef net farm value"—the market value to the producer of 2.4 pounds of live animal (the equivalent of one pound of retail cuts) minus the farm byproduct allowance. Because of the likely endogeneity of this variable, it was replaced with its least squares projection on the following set of instruments: a constant, a time trend, Z_3, W_2, S, and:

1. The wholesale price of no. 2 yellow corn, Chicago (dollars per bushel) (Source: *Agricultural Outlook*, USDA, various issues).

2. The price of feeder steers, Oklahoma City, medium, no. 1, 400–500 pounds (cents per pound).

3. The number of cattle on feed, first of month, for seven states (Iowa, Nebraska, Kansas, Texas, Colorado, Arizona, and California) (thousand head).

References

Aguirre-Torres, V., and A.R. Gallant. "The Null and Non-null Asymptotic Distribution of the Cox Test for Multivariate Nonlinear Regression: Alternatives and a New Distribution-Free Cox Test." *Journal of Econometrics* 21 (1983):5–33.

Ailawadi, K., N. Borin, and P.W. Farris. "Market Power and Performance: A Cross-Industry Analysis of Manufacturers and Retailers." *Journal of Retailing* 71 (1995):211–248.

Amemiya, T. "The Maximum Likelihood and the Nonlinear Three-Stage Least Squares Estimator in the General Nonlinear Simultaneous Equation Model." *Econometrica* 45 (1977):955–968.

Appelbaum, E. "The Estimation of the Degree of Oligopoly Power." *Journal of Econometrics* 19 (1982):287–299.

Azzam, A. "Estimating the Degree of Dominance in a Bilateral Oligopoly." *Applied Economic Letters* 3 (1996):209–211.

Azzam, A.M., and D.G. Anderson. "Assessing Competition in Meatpacking: Economic History, Theory, and Evidence." Washington DC: U.S. Department of Agriculture, Grain Inspection, Packers, and Stockyards Administration, report to Packers and Stockyards Programs, 1995.

Azzam, A., and M. Zhang. "Testing Successive Market Power in Vertically Related Markets." Mimeo, Department of Agricultural Economics, University of Nebraska Lincoln, 1992.

Blair, R.D., D.L. Kaserman, and R.R. Romano. "A Pedagogical Treatment of Bilateral Monopoly." *Southern Economic Journal* 55 (1989):55–76.

Bresnahan, T.F. "The Oligopoly Solution Concept is Identified." *Economic Letters* 10 (1982):87–92.

———. "Competition and Collusion in the American Automobile Oligopoly: The 1955 Price War." *Journal of Industrial Economics* 35 (1987):457–482.

———. "Empirical Studies of Industries with Market Power." *Handbook of Industrial Organization*, vol. 2. R. Schmalensee and R.D. Willig, ed., pp. 1011–1157. Amsterdam Netherlands: Elsevier Science Publishers, 1989.

Corts, K.S. "Conduct Parameters and the Measurement of Market Power." *Journal of Econometrics* 88 (1999):227–250.

Cotterill, R.W. "Market Power in the Retail Food Industry: Evidence from Vermont." *Review of Economics and Statistics* 68 (1986):379–386.

Cox, D.R. "Tests of Separate Families of Hypotheses." *Proceedings of the Fourth Berkeley Symposium in Mathematical Statistics and Probability*. Berkeley CA: University of California Press, 1961.

———. "Further Results on Tests of Separate Families of Hypotheses." *Journal of the Royal Statistical Society* B24 (1962):406–424.

Gasmi, F., J.J. Laffont, and Q. Vuong. "Econometric Analysis of Collusive Behavior in a Soft-Drink Market." *Journal of Economics and Management Strategy* 1 (1992):277–311.

Genesove, D., and W.P. Mullin. "Testing Static Oligopoly Models: Conduct and Cost in the Sugar Industry, 1890–1914." *Rand Journal of Economics* 29 (1998):355–377.

Hall, L., A. Schmitz, and J. Cothern. "Beef Wholesale-Retail Marketing Margins and Concentration." *Economica* 46 (1979):295–300.

Johnson, D.G., J. Connor, T. Josling, A. Schmitz, and G.E. Schuh. *Competitive Issues in the Beef Sector: Can Beef Compete in the 1990s?* Minneapolis MN: University of Minnesota, Hubert H. Humphrey Institute of Public Affairs Report 1, 1989.

Just, R., and W. Chern. "Tomatoes, Technology, and Oligopsony." *Bell Journal of Economics* 11 (1980):584–602.

Messinger, R.R., and C. Narasimhan. "Has Power Shifted in the Grocery Channel?" *Marketing Science* 14 (1995):189–223.

Porter, R.H. "A Study of Cartel Stability: The Joint Executive Committee, 1880–1886." *Bell Journal of Economics* 14 (1983):301–314.

Roberts, M. "Testing Oligopolistic Behavior." *International Journal of Industrial Organization* 2 (1984):367–383.

Schroeter, J.R. "Estimating the Degree of Market Power in the Beef Packing Industry." *Review of Economics and Statistics* 70 (1988):158–162.

U.S. Department of Agriculture, Grain Inspection, Packers and Stockyards Administration. *Packers and Stockyards Statistical Report (1991–1994 Reporting Years)*. Washington DC: U.S. Department of Agriculture Report GIPSA SR 96-1, 1996.

Waterson, M. "Price Cost Margins and Successive Market Power." *Quarterly Journal of Economics* 94 (1980):135–150.

Part V

Spatial and Dynamic Extension of Oligopoly Models

9

A Spatial-Equilibrium Model for Imperfectly Competitive Milk Markets

Tsunemasa Kawaguchi, Nobuhiro Suzuki, and Harry M. Kaiser

Spatial-equilibrium models have been used frequently to analyze interregional competition problems in agriculture, including regional competition issues associated with the dairy industries in the United States (e.g., Chavas, Cox, and Jesse 1994; McDowell 1982; Yavuz et al. 1996) as well as in other countries such as Japan (e.g., Sasaki 1969; Kobayashi 1983; Hayashi 1984) (see also Judge and Takayama (1973)). Originally developed by Enke (1951) and Samuelson (1952) and then refined by Takayama and Judge (1964a, 1964b, 1971), spatial price equilibrium models have assumed that markets are either perfectly competitive or monopolistic. However, the structure of dairy markets in most countries is often neither. Therefore, a more plausible model for analyzing interregional milk movements would be an imperfect-competition spatial-equilibrium model.

The purpose of this chapter is to present a generalization of Takayama and Judge's (1964a, 1964b, 1971) spatial-equilibrium model that allows for the incorporation of any degree of market structure from perfect competition to monopoly. The usefulness of the model is demonstrated with an application to interregional milk movements in the Japanese dairy industry with solutions generated and compared for alternative scenarios regarding the degree of market competition.

This chapter originally appeared as an article titled "A Spatial Equilibrium Model for Imperfectly Competitive Milk Markets" by Kawaguchi, Suzuki, and Kaiser in the *American Journal of Agricultural Economics* in 1997. This material is reprinted by permission of the *American Journal of Agricultural Economics*.

The Japanese Dairy Industry

Dairy policy in Japan features a quota system in the manufacturing-milk market to prevent excess milk production from occurring because of higher-than-competitive market prices. As a result, the Japanese dairy industry can be divided into three distinct markets: the fluid market, the manufacturing market within payment quotas, and the manufacturing market over payment quotas. Prices in the manufacturing markets are set by the government based on a deficiency payment program. For manufacturing milk sold within payment quotas, prefectural milk marketing boards (consignment milk sellers for farmers) receive deficiency payments equal to the difference between the guaranteed price and the standard transaction price for manufacturing milk. Both prices are determined by the national government. The guaranteed price is based on milk production costs while the standard transaction price is based on conditions in the dairy product market, and all buyers of manufacturing milk are required to pay the standard transaction price. To discourage excess production, over-payment-quota manufacturing milk receives the lower standard transaction price. Payment quotas for the guaranteed price are given not to individual producers but to each prefectural milk marketing board. Individual producers are paid a prefecturewide uniform pooled price (weighted-average prices for milk sold in the fluid- and manufacturing-milk markets).

Given manufacturing-milk prices determined by the government, discriminated price formation for fluid milk occurs through negotiations between each prefectural milk marketing board and the processors it supplies. Since the fluid-milk market is more price inelastic than the manufacturing-milk market, the fluid market has higher prices. The structure of the Japanese milk market includes an oligopolistic group of consignment milk sellers (prefectural milk marketing boards) that allocate milk to maximize sales revenues and a large number of perfectly competitive producers who receive pooled returns (blend prices). We refer to this situation as a "dual structure" because dairy farmers are perfectly competitive in producing milk while they are oligopolistic in selling it through the milk marketing boards. Previous spatial price equilibrium models have not accounted for this dual structure in the Japanese milk market.

Conceptual Model

Consider n milk producing and consuming regions for which the geographical scope of producing region i is the same as for consuming region i. In each consuming region, there are three administratively different markets: the fluid-milk market (FMM_i), the manufacturing-milk-market within-payment quotas (WPQ_i), and the manufacturing-milk-market over-payment quotas (OPQ_i). Unit transportation costs for shipping raw milk from producing region i to consuming region j (T_{ij}) are assumed to be the same for both fluid and manufacturing milk.[1]

Buyers of fluid milk in each consuming region are assumed to behave as price-takers, which is reasonable considering the many fluid processors in Japan. Within-payment-quota milk is traded at the fixed guaranteed price, $FP1$, and the quantity of milk is limited to the fixed payment quota. Over-payment-quota milk is traded at the lower fixed standard transaction price, $FP2$, and it is assumed that demand for this milk is perfectly elastic. It also is assumed that each region has a linear marginal raw-milk cost function and a linear fluid demand function with all functions known by all agents (or consignment sellers).

Milk producers in region i consign their annual milk supply, FS_i, to agent i. Agent i's role is to allocate farmers' milk among the $3n$ markets to maximize sales revenue net of transportation costs. The following notation is used based on the variables previously described.

D_j:	Quantity of milk demanded in fluid market j ($j = 1, 2, \ldots, n$).
FS_i:	Quantity of raw milk supplied and consigned in region i ($i = 1, 2, \ldots, n$).
PS_i:	Marginal revenue net of transportation costs for each market for region i ($i = 1, 2, \ldots, n$).

1 Unlike cooperatives in the United States, Japanese cooperatives have only a small share of the manufacturing-milk market and function primarily as raw-milk shippers that negotiate a price for their farmer members. Therefore, the focus of our model is on transactions between raw-milk shippers (cooperatives) and manufacturing companies. Consequently, it is realistic to assume the same transportation costs for raw milk being shipped for fluid and manufacturing-product processing.

X_{ij}: Quantity of raw milk shipped from region i to market j ($i = 1, 2, \ldots, n; j = 1, 2, \ldots, 3n$).

$X_{i(n+j)}$: Quantity of raw milk shipped from region i to the manufacturing-milk-market within-payment quotas (WPQ_j) ($i = 1, 2, \ldots, n; j = 1, 2, \ldots, n$).

$X_{i(2n+j)}$: Quantity of raw milk shipped from region i to the manufacturing-milk-market over-payment quotas (OPQ_j) ($i = 1, 2, \ldots, n; j = 1, 2, \ldots, n$).

PD_j: Demand price in fluid market j ($j = 1, 2, \ldots, n$).

PPP_i: Producer's pooled (blend) price in region i ($i = 1, 2, \ldots, n$).

$D_j = \alpha_j - \beta_j PD_j$: Demand function in fluid market j ($j = 1, 2, \ldots, n$).

$FS_i = -\nu_i + \eta_i PPP_i$: Marginal cost function for raw milk in region i ($i = 1, 2, \ldots, n$) where PPP_i means marginal cost.

T_{ij}: Unit transportation cost of shipping raw milk from producing region i to consuming region j ($i = 1, 2, \ldots, n; j = 1, 2, \ldots, 3n$).

Q_i: The limited quantity (payment quota) that receives the difference between the guaranteed price ($FP1$) and the standard transaction price ($FP2$) ($i = 1, 2, \ldots, n$).

SP_j: Shadow price of the right to sell a unit of milk in the manufacturing-milk-market within-payment quotas (WPQ_j) ($i = 1, 2, \ldots, n$).

R_i: Total milk sales revenue net of transportation costs in region i ($i = 1, 2, \ldots, n$).

Using the preceding notation, agent *i*'s milk sales revenue-maximization problem net of transportation costs can be expressed as

$$(1) \qquad \text{Max:}\, R_i = \sum_{j=1}^{n} PD_j X_{ij} + \sum_{j=1}^{n} FP1 \times X_{i(n+j)} + \sum_{j=1}^{n} FP2 \times X_{i(2n+j)} - \sum_{j=1}^{3n} T_{ij} X_{ij}.$$

The total-revenue maximization problem for all *n* agents is expressed as

$$(2) \qquad \text{Max:}\, \sum_{i=1}^{n} R_i.$$

Agent *i*'s fluid sales revenue in market *j* ($PD_j X_{ij}$) can be written as

$$(3) \qquad \begin{aligned} PD_j X_{ij} &= [\alpha_j/\beta_j - (1/\beta_j)\, D_j]\, X_{ij} \\ &= [\alpha_j/\beta_j - (1/\beta_j)(\sum_{i=1}^{n} X_{ij})]\, X_{ij} \\ &= [\alpha_j/\beta_j - (1/\beta_j)(\sum_{m \neq i} X_{mj} + X_{ij})]\, X_{ij} \end{aligned}$$

where *m* ($m \neq i$) indicates all agents other than *i*. When agent *i* believes that a change in its fluid supply will cause changes in all other agents' fluid supply to market *j*, agent *i*'s "perceived" marginal fluid revenue in market *j* is

$$(4) \qquad \begin{aligned} \partial(PD_j X_{ij})/\partial X_{ij} &= [\alpha_j/\beta_j - (1/\beta_j)\, D_j] - (1/\beta_j)(\partial \sum_{m \neq i} X_{mj}/\partial X_{ij} + 1)\, X_{ij} \\ &= PD_j - (1/\beta_j)(r_{ij} + 1)\, X_{ij} \end{aligned}$$

where r_{ij} is agent *i*'s conjectural variation regarding changes in all other agents' fluid supply to market *j* caused by a change in agent *i*'s supply.

Using the relationship in (4), the total-revenue maximization problem for all *n* agents can be respecified as the following net social-payoff-maximization problem adjusted for imperfectly competitive markets (*ANSP*):

$$(5) \qquad \begin{aligned} \text{Max:}\, ANSP = & \sum_{j=1}^{n} \int [\alpha_j/\beta_j - (1/\beta_j)\, D_j]\, dD_j + \sum_{j=1}^{n} \sum_{i=1}^{n} FP1 \\ & \times X_{i(n+j)} + \sum_{j=1}^{n} \sum_{i=1}^{n} FP2 \times X_{i(2n+j)} - \sum_{j=1}^{n} \sum_{i=1}^{n} (1/\beta_j)\, (r_{ij} + 1) \\ & \int X_{ij}\, dX_{ij} - \sum_{j=1}^{3n} \sum_{i=1}^{n} T_{ij} X_{ij} \end{aligned}$$

(6) subject to $D_j \leq \sum_{i=1}^{n} X_{ij}$ for all j,

(7) $\sum_{i=1}^{n} X_{i(n+j)} \leq Q_j$ for all j,

(8) $\sum_{j=1}^{3n} X_{ij} \leq FS_i$ for all i, and

(9) $D_j \geq 0, X_{ij} \geq 0$ for all i and j.

The difference between *ANSP* in Equation (5) and the net social payoff (NSP) in the conventional spatial competitive equilibrium model by Takayama and Judge (1964b) is the term

$$-\sum_{j=1}^{n} \sum_{i=1}^{n} (1/\beta_j)(r_{ij} + 1) \int X_{ij} dX_{ij}.$$

When the market is perfectly competitive ($r_{ij} = -1$), the term is zero and Equation (5) is equal to the original Takayama and Judge model. When Cournot-Nash behavior is assumed ($r_{ij} = 0$), the term is equivalent to

$$-\sum_{j=1}^{n} \sum_{i=1}^{n} (1/\beta_j) \int X_{ij} dX_{ij},$$

which is shown in Hashimoto's (1985) spatial Nash equilibrium model. Cournot-Nash behavior means that agent i believes that the other agents will not change their supply in response to the agent's action.

Using the Lagrange function (L) with the multipliers λ_j, ω_j, and θ_i for constraints (6), (7), and (8), respectively, the Kuhn-Tucker optimality conditions for the maximization problem can be expressed as

(10) $\partial L/\partial D_j = \alpha_j/\beta_j - (1/\beta_j) D_j - \lambda_j \leq 0, D_j(\partial L/\partial D_j) = 0$, for all j.

(11) $\partial L/\partial X_{ij} = -(1/\beta_j)(r_{ij} + 1) X_{ij} - T_{ij} + \lambda_j - \theta_i \leq 0, X_{ij}(\partial L/\partial X_{ij}) = 0$, for all i and j.

(12) $\partial L/\partial X_{i(n+j)} = FP1 - T_{ij} - \omega_j - \theta_i \leq 0, X_{i(n+j)}(\partial L/\partial X_{i(n+j)}) = 0$, for all i and j.

(13) $\partial L/\partial X_{i(2n+j)} = FP2 - T_{ij} - \theta_i \le 0,\ X_{i(2n+j)}(\partial L / \partial X_{i(2n+j)}) = 0,$ for all i and j.

(14) $-\partial L/\partial \lambda_j = D_j - \sum_{i=1}^{n} X_{ij} \le 0,\ \lambda_j(\partial L / \partial \lambda_j) = 0,$ for all j.

(15) $-\partial L/\partial \omega_j = \sum_{i=1}^{n} X_{i(n+j)} - Q_j \le 0,\ \omega_j(\partial L / \partial \omega_j) = 0,$ for all j.

(16) $-\partial L/\partial \theta_i = \sum_{j=1}^{3n} X_{ij} - FS_i \le 0,\ \theta_i(\partial L / \partial \theta_i) = 0,$ for all i.

The Lagrange multipliers (or dual variables), λ_j, ω_j, and θ_i, measure the fluid demand price (PD_j), the shadow price for the right to sell milk in the within-payment-quota manufacturing market (SP_j), and marginal revenue net of transportation costs for each market (PS_i), respectively. The Kuhn-Tucker conditions represented by Equations (11), (12), and (13) indicate that each agent must equalize marginal revenue net of transportation costs across all markets in which it sells milk. The equilibrium values can be calculated by the quadratic programming model solution.

The term $(1 / \beta_j)(r_{ij} + 1)X_{ij}$ in Equation (11) indicates the difference between the fluid demand price and agent i's marginal revenue in market j. The greater the degree of market power by agents, the larger this difference is. For example, in the case of perfect competition, the term becomes zero because $r_{ij} = -1$. On the other hand, the term becomes $(1 / \beta_j)X_{ij}$ when Cournot-Nash behavior ($r_{ij} = 0$) is assumed. In this chapter, Cournot-Nash behavior is assumed to illustrate the imperfect-competition solution and coalition among agents is treated as follows. Consider Cournot-Nash agent 1, whose "perceived" marginal revenue in fluid market j is $PD_j - (1/\beta_j)X_{1j}$. If agent 1 forms a coalition with agent 2, then marginal revenue for the coalition of agents 1 and 2 is $PD_j - (1/\beta_j)(X_{1j} + X_{2j})$. In the case of monopoly where agent 1 forms a coalition with all other agents, marginal revenue for agent 1 is

$$PD_j - (1/\beta_j)(\sum_{i=1}^{n} X_{ij}).$$

Because any agent can sell the consigned milk individually or in coalition with other agents, as a price-taker or according to Cournot-Nash behavior, many combinations of agents' marketing behavior can be simulated.

To complete the model, the individual farmers' milk supplies need to be incorporated. Unlike the case of oligopolistic marketing behavior of agents, milk production by individual farmers is competitively determined. Producers in region *i*, as price-takers, determine their supply given the pooled producer price. That is, their production level is determined by equating marginal cost to the producer pooled price. Thus,

(17) $PPP_i = R_i / FS_i$ for all *i*, and

(18) $FS_i = -\nu_i + \eta_i PPP_i$ for all *i*.

In the comparative-static equilibrium, FS_i in Equation (18) must be equal to FS_i given in the preceding milk sales-maximization problem. To solve the model, the following iterative solution process is used to find equilibrium values for FS_i.

First, the quadratic programming model is used to generate equilibrium fluid-milk prices and equilibrium quantities of milk shipments in the sales-maximization problem expressed by Equations (5) through (16) based on initial values for FS_i and given patterns of behavior of agents in the oligopolistic milk market. Second, pooled producer prices are calculated in Equation (17). Third, new values of FS_i for the next iteration are computed based on the calculated pooled producer prices and marginal cost functions of producing regions and on the assumption that producers behave as price-takers in Equation (18). Finally, the quadratic programming problem is solved again with new parameter values for FS_i to obtain new equilibrium fluid-milk prices and quantities of milk shipments. This iteration process is continued until values for FS_i become stationary.

Application of the Model to the Japanese Milk Market

The model is applied to the Kyushu area of Japan as a case study. Region 1 includes the Fukuoka, Saga, and Nagasaki prefectures; region 2 is the Kumamoto prefecture; region 3 is the Oita prefecture; and region 4 includes the Miyazaki and Kagoshima prefectures.

Based on the long-run price elasticity of the Kyushu milk supply established by Ito (1989) (0.429), the Kyushu fluid demand price elasticity by Suzuki and Kobayashi (1993) (–0.77), and the regional price and quantity observations shown in Table 9.1, linear marginal cost and fluid-milk demand functions for each region are specified as follows:

Table 9.1. Observations in 1989 (unit: thousand tons and yen/kilogram)

Shipments To	Fluid-Milk Market				Manufacturing-Milk Market within Quota				Manufacturing-Milk Market over Quota				Total
	1	2	3	4	1	2	3	4	1	2	3	4	
From													
1	128.4	19.1	12.0	1.7	34.0	0.0	0.0	0.0	10.9	0.0	0.0	0.0	206.1
2	33.1	74.7	1.5	1.6	0.0	32.9	0.0	0.0	0.0	0.0	0.0	0.0	143.8
3	31.4	0.0	34.3	0.0	0.0	0.0	8.1	0.0	0.0	0.0	1.2	0.0	75.1
4	11.3	8.5	2.0	89.0	0.0	0.0	0.0	39.1	0.0	0.0	0.0	8.3	158.0
Total	**204.2**	**102.3**	**49.8**	**92.3**	**34.0**	**32.9**	**8.1**	**39.1**	**10.9**	**0.0**	**1.2**	**8.3**	**583.0**

Region i or j	Fluid-Milk Price – PD_j	Producers' Pooled Price – PPP_i
1	109.35	101.62
2	118.22	103.21
3	118.22	107.75
4	111.20	99.07
Average	**112.74**	**102.11**

Source: Suzuki and Kobayashi (1993).

$$FS_1 = 135.162 + 0.967PPP_1, \quad D_1 = 361.434 - 1.438PD_1,$$

$$FS_2 = 118.078 + 0.832PPP_2, \quad D_2 = 181.071 - 0.666PD_2,$$

$$FS_3 = 43.490 + 0.293PPP_3, \quad D_3 = 88.146 - 0.324PD_3, \text{ and}$$

$$FS_4 = 119.121 + 0.874PPP_4, \quad D_4 = 163.371 - 0.639PD_4$$

where FS_i and D_j are measured in thousand tons and PPP_i and PD_j are measured in yen/kilogram. Unit transportation costs, T_{ij}, are $T_{12} = T_{21} = 4.58$, $T_{13} = T_{31} = 3.95$, $T_{14} = T_{41} = 7.80$, $T_{23} = T_{32} = 4.71$, $T_{24} = T_{42} = 6.11$, and $T_{34} = T_{43} = 6.00$ yen/kilogram. Because little milk is traded between Kyushu and other regions of Japan, this milk is treated as exogenous to simplify the model. Payment quotas (Q_i) for the four regions are $Q_1 = 34.0$, $Q_2 = 32.9$, $Q_3 = 8.1$, and $Q_4 = 39.1$ thousand tons. The fixed guaranteed price for the within-payment quotas is $FP1 = 79.83$ yen/kilogram and the fixed standard transaction price for over-payment quotas is $FP2 = 67.25$ yen/kilogram.

To demonstrate how solutions vary based on the market-structure assumption, the model is solved for perfect-competition, monopoly, and imperfect-competition scenarios. To represent the perfectly competitive solution, the model is solved assuming that the four agents are all price-takers. For the monopoly solution, the model is solved using the assumption that there is a coalition of four agents. To represent imperfect competition, fifteen separate combinations of price-takers and Cournot-Nash players are solved. In the first case, the four agents are all individual Cournot-Nash players (Cournot-Nash equilibrium). In the next cases, one agent is a price-taker and the other three are individual Cournot-Nash players, thereby creating four different market structures. For cases six through eleven, two agents are price-takers and the other two are individual Cournot-Nash players, thereby creating six new market structures. Finally, in the last four cases, three agents are price-takers and the other is a Cournot-Nash player, thereby creating four market structures. Although other combinations are possible using coalitions, they are not analyzed since our purpose here is simply to demonstrate examples of imperfectly competitive solutions.

The dual-structure, perfect-competition, spatial solution is shown in Table 9.2. In this case, virtually all raw milk is allocated to the fluid market;

Table 9.2. Dual-Structure Spatial Perfect-Competition Equilibrium (unit: thousand tons and yen/kilogram)

Shipments To	Fluid-Milk Market				Manufacturing-Milk Market within Quota				Manufacturing-Milk Market over Quota				Total
	1	2	3	4	1	2	3	4	1	2	3	4	
From													
1	193.2	0.0	0.0	0.0	0.0	0.0	0.0	0.0	0.0	0.0	0.0	0.0	193.2
2	4.8	125.8	0.0	0.0	0.0	0.0	0.0	0.0	0.0	0.0	0.0	0.0	130.5
3	7.0	0.0	61.0	0.0	0.0	0.0	0.0	0.0	0.0	0.0	0.0	0.0	68.0
4	30.5	0.0	0.0	112.4	0.0	0.0	0.0	1.4	0.0	0.0	0.0	0.0	144.3
Total	**235.4**	**125.8**	**61.0**	**112.4**	**0.0**	**0.0**	**0.0**	**1.4**	**0.0**	**0.0**	**0.0**	**0.0**	**536.0**

Region *i* or *j*	**Fluid-Milk Price – PD_j**	**Agents' Marginal Revenue[a] – PS_i**	**Producers' Pooled Price[b] – PPP_i**
1	87.63	87.63	88.23
2	83.05	83.05	87.19
3	83.68	83.68	83.68
4	79.83	79.83	83.32
Average	**84.46**	**84.46**	**86.07**

[a] PS_i is agent *i*'s "perceived" marginal revenue (net of transportation costs) equalized in each market (marginal revenue = market price in perfect competition) (same in Tables 9.3, 9.4, and 9.5).

[b] Exogenously given milk shipments from each region to the outside of Kyushu are taken into account in calculating PPP_i (same in Tables 9.3, 9.4, and 9.5).

a trivial amount is shipped to the within-payment-quota manufacturing-milk market in region 4. There also is only a small amount of interregional shipments of fluid milk, mostly to region 1. The amount of milk allocated to the fluid market in the perfect-competition solution is substantially higher than the actual amount allocated (see Table 9.1). This is due to the assumption that agents act as price-takers, which results in equality of price across markets net of transportation costs instead of equality across markets of "perceived" marginal revenue net of transportation costs. Consequently, fluid-milk prices and pooled producer prices in the perfect-competition case are much lower than actual levels.

The dual-structure monopoly spatial solution is shown in Table 9.3. In this case, the allocation of raw milk to the fluid market is about one-half of the amount allocated under perfect competition and also is less than the actual levels (Table 9.1). Instead, the monopoly solution allocates significant amounts of raw milk to the within- and over-payment-quota manufacturing-milk markets. The model predicts no interregional shipment of milk in all three markets. Because the demand for fluid milk is inelastic, restricting allocations to the fluid-milk market results in higher pooled returns to farmers. In fact, pooled producer prices under monopoly are 30 percent higher than prices under the perfect-competition case, as well as 10 percent higher than actual prices. It should be noted that the monopoly distribution of pooled returns to farmers is based on the assumption that the difference in the pooled producer price among regions is the same as the differentials generated in the perfect-competition solution. Alternatively, one national pooled producer price for all regions could have been allocated. It should also be noted that total milk supply is largest in monopoly equilibrium under the dual structure. Unless agents have power to control supply, individual producers increase milk supply as higher blend prices are given. Consequently, real monopoly rents cannot be realized under the dual structure.

The Cournot-Nash equilibrium is shown in Table 9.4. The regional fluid-milk and pooled producer prices in this solution are the closest to actual prices for the four regions (Table 9.1). Not surprisingly, the allocation of raw milk among the three markets in this case is somewhere between the perfect-competition and monopoly cases. Unlike the two previous cases, however, the Cournot-Nash equilibrium solution results in two regions shipping milk to each other; region 2 ships 51,400 tons of fluid milk to region 1 and region 1 ships 29,500 tons of fluid milk to

Table 9.3. Dual-Structure Spatial Monopoly Equilibrium (unit: thousand tons and yen/kilogram)

Shipments	**Fluid-Milk Market**				**Manufacturing-Milk Market within Quota**				**Manufacturing-Milk Market over Quota**				**Total**
To	**1**	**2**	**3**	**4**	**1**	**2**	**3**	**4**	**1**	**2**	**3**	**4**	
From													
1	132.4	0.0	0.0	0.0	34.0	0.0	0.0	0.0	51.8	0.0	0.0	0.0	218.1
2	0.0	68.1	0.0	0.0	0.0	32.9	0.0	0.0	0.0	50.9	0.0	0.0	152.0
3	0.0	0.0	33.2	0.0	0.0	0.0	8.1	0.0	0.0	0.0	34.3	0.0	75.6
4	0.0	0.0	0.0	60.2	0.0	0.0	0.0	39.1	0.0	0.0	0.0	67.5	166.8
Total	**132.4**	**68.1**	**33.2**	**60.2**	**34.0**	**32.9**	**8.1**	**39.1**	**51.8**	**50.9**	**34.3**	**67.5**	**612.5**

Region *i* or *j*	**Fluid-Milk Price – PD_j**	**Agents' Marginal Revenue – PS_i**	**Producers' Pooled Price[a] – PPP_i**
1	159.30	67.25	114.03
2	169.56	67.25	112.99
3	169.65	67.25	109.48
4	161.46	67.25	109.13
Average	**163.29**	**67.25**	**111.87**

[a] Estimated *PPP* differentials in the perfect-competition equilibrium shown in Table 9.2 are used to allocate monopoly pooled returns and to calculate PPP_i for each region.

Table 9.4. Dual-Structure Spatial Cournot-Nash Equilibrium (unit: thousand tons and yen/kilogram)

Shipments To	Fluid-Milk Market				Manufacturing-Milk Market within Quota				Manufacturing-Milk Market over Quota				Total
	1	**2**	**3**	**4**	**1**	**2**	**3**	**4**	**1**	**2**	**3**	**4**	
From													
1	64.6	29.5	14.5	24.7	34.0	6.6	8.1	0.0	20.5	0.0	0.0	0.0	202.6
2	51.4	29.5	12.8	22.9	0.0	26.3	0.0	0.0	0.0	0.0	0.0	0.0	142.9
3	33.5	17.6	10.1	14.6	0.0	0.0	0.0	0.0	0.0	0.0	0.0	0.0	75.7
4	50.6	27.2	13.2	28.5	0.0	0.0	0.0	39.1	0.0	0.0	0.0	0.0	158.6
Total	**200.1**	**103.8**	**50.6**	**90.7**	**34.0**	**32.9**	**8.1**	**39.1**	**20.5**	**0.0**	**0.0**	**0.0**	**579.7**

Region i or j	Fluid-Milk Price – PD_j	Agents' Marginal Revenue – PS_i	Producers' Pooled Price – PPP_i
1	112.18	67.25	97.94
2	116.10	71.83	102.09
3	115.99	84.96	109.91
4	113.76	69.20	99.70
Average	**113.85**	**70.73**	**100.89**

region 2. While these shipping patterns are unintuitive, they do occur in reality, as is clear from Table 9.1. The other two spatial competition models did not predict these interregional milk shipment patterns. This suggests that the complicated interregional milk movements actually observed may be caused by imperfectly competitive behavior.

To explain why the actual situation in Japan conforms more closely to the Cournot-Nash solution than to the monopoly or the perfectly competitive solution, one needs to examine the market power of the prefectural milk marketing boards. Each board controls total milk supplied in the prefecture and therefore has some market power, particularly within the prefecture. However, at the national level, the prefectural milk marketing boards compete with one another, which lessens the market power of each board. This suggests that the prefectural marketing boards are neither price-takers nor pure monopolists but rather behave at some intermediate level between the two market-power extremes. Therefore, it is reasonable that the actual situation conforms more with the Cournot-Nash solution than with other solutions.

Compared with the other imperfect-competition cases in which at least one region is assumed to be a price-taker (an example is given in Table 9.5), fluid and pooled producer prices in the Cournot-Nash equilibrium solution shown in Table 9.4 are closer to actual prices. Price-takers' returns tend to be greater than those of the Cournot-Nash players when both price-takers and Cournot-Nash players exist, as shown in Table 9.5. This is because Cournot-Nash agents try to keep fluid-milk prices higher based on their "perceived" marginal revenues and price-takers obtain benefits by moving their milk to the fluid-milk market. In this case, acting as a price-taker is like "cheating" in a cartel agreement.

Conclusion

The traditional spatial-equilibrium model assumes that market structure is either perfectly competitive or monopolistic. In this chapter, a new generalized, dual-structure spatial imperfect-competition equilibrium model was developed that incorporates any degree of market structure from perfect competition to monopoly. The model, which was applied to the Japanese milk market as a case study, incorporated a dual structure in which there are oligopolistic consignment sellers and many perfectly competitive small-scale producers that receive pooled returns.

Table 9.5. Dual-Structure Spatial Equilibrium in the Case Where Agent 1 Is a Price-Taker and Others Are Individual Cournot-Nash Players (unit: thousand tons and yen/kilogram)

Shipments To	**Fluid-Milk Market**				**Manufacturing-Milk Market within Quota**				**Manufacturing-Milk Market over Quota**				**Total**
	1	**2**	**3**	**4**	**1**	**2**	**3**	**4**	**1**	**2**	**3**	**4**	
From													
1	107.9	47.8	24.1	28.1	0.0	0.0	0.0	0.0	0.0	0.0	0.0	0.0	207.9
2	36.0	22.8	9.4	20.0	16.3	32.9	0.0	0.0	0.0	0.0	0.0	0.0	137.4
3	30.6	16.7	9.5	17.3	0.0	0.0	0.0	0.0	0.0	0.0	0.0	0.0	74.0
4	36.0	20.9	10.0	26.0	13.0	0.0	8.1	39.1	0.0	0.0	0.0	0.0	153.0
Total	**210.6**	**108.2**	**52.9**	**91.4**	**29.2**	**32.9**	**8.1**	**39.1**	**0.0**	**0.0**	**0.0**	**0.0**	**572.3**

Region *i* or *j*	**Fluid-Milk Price – PD_j**	**Agents' Marginal Revenue – PS_i**	**Producers' Pooled Price – PPP_i**
1	104.89	104.89	103.45
2	109.47	75.25	95.45
3	108.84	79.68	104.16
4	112.69	72.03	93.37
Average	**107.95**	**88.60**	**98.43**

Using the model, many spatial-equilibrium solutions in the Japanese milk market were demonstrated based on alternative sets of imperfectly competitive behavior with the dual structure. The results indicate that, under perfect competition, most milk was shipped to fluid markets, there were very few interregional milk movements, and prices generated for milk were much lower than the observed actual prices. Under monopoly, much less milk was shipped to fluid markets, there was no interregional milk movement, and milk prices were much higher than under the perfect-competition solution and actual prices. The Cournot-Nash equilibrium solutions were the most similar to actual observations, and actual interregional milk movements could be explained by assuming imperfectly competitive behavior.

References

Chavas, J.P., T.L. Cox, and E.V. Jesse. "Regional Impacts of Reducing Dairy Price Supports and Removing Milk Marketing Orders in the U.S. Dairy Sector." Madison WI: University of Wisconsin, Department of Agricultural Economics Staff Paper 367, 1994.

Enke, S. "Equilibrium among Spatially Separated Markets: Solution by Electric Analogue." *Econometrica* 19 (1951):40–47.

Hashimoto, H. "A Spatial Nash Equilibrium Model." *Spatial Price Equilibrium: Advances in Theory, Computation and Application*. P.T. Harker, ed., pp. 20–40. New York NY: Springer-Verlag, 1985.

Hayashi, M. "Market Adjustment of Milk in Japan." *Excessive Supply and Market Adjustment of Agricultural Products*. K. Tsuchiya, ed., pp. 145–165. Tokyo Japan: Norin-Tokei-Kyokai, 1984.

Ito, F. *A Spatial Equilibrium Analysis on Regional Milk Trade*. Sapporo Japan: Snow Brand Research Institute Report 54, 1989.

Judge, G.G., and T. Takayama. *Studies in Economic Planning over Space and Time*. Amsterdam Netherlands: North-Holland Publishing, 1973.

Kobayashi, K. *Prices and Market Adjustment of Milk*. Tokyo Japan: Taimei-do, 1983.

McDowell, F.H. "Domestic Dairy Marketing Policy: An Interregional Trade Approach." Ph.D. Thesis, University of Minnesota, 1982.

Samuelson, P.A. "Spatial Price Equilibrium and Linear Programming." *American Economic Review* 42 (1952):283–303.

Sasaki, K. "Spatial Equilibrium in Eastern Japan Milk Market." *Journal of Rural Economics* 41 (1969):106–116.

Suzuki, N., and K. Kobayashi. "Equilibrium in Interregional Milk Transportation." *Journal of Rural Economics* 64 (1993):221–232.

Takayama, T., and G.G. Judge. "Equilibrium among Spatially Separated Markets: A Reformulation." *Econometrica* 32 (1964a):510–524.

———. "Spatial Equilibrium and Quadratic Programming." *Journal of Farm Economics* 46 (1964b):67–93.

———. *Spatial and Temporal Price and Allocation Models*. Amsterdam Netherlands: North-Holland Publishing, 1971.

Yavuz, F., C. Zulauf, G. Schnitkey, and M. Miranda. "A Spatial Equilibrium Analysis of Regional Structural Change in the U.S. Dairy Industry." *Review of Agricultural Economics* 18 (1996):693–703.

10

A Dynamic Model of Oligopoly in the Coffee Export Market

Larry S. Karp and Jeffrey M. Perloff

We estimate the degree of competitiveness and the adjustment paths of the two largest coffee exporters, Brazil and Colombia, using a subgame-perfect dynamic feedback model of oligopoly.[1] There is evidence that major coffee exporters behave noncompetitively and that dynamics play an important role in determining outcomes (Marshall 1983). Since 1959, most exporting and importing countries have participated in a series of international coffee agreements (ICAs) that set export quotas (Fischer 1972; Bates and Lien 1985). It is alleged, however, that many countries regularly violate these quotas. In his survey of various ICAs, Gilbert (1987, p. 602) quotes critics who claim that the agreements "are no more than an internationally sanctioned producer's cartel." He concludes that the agreements have resulted in higher prices rather than simply more stable prices (1987, p. 604). Greenstone (1981) also asserts that the large coffee producers have behaved as a cartel. Based on econometric models, de Vries (1975), Akiyama and Duncan (1982), Palm and Vogelvang (1986), and Herrmann (1986) argue that the ICAs have resulted in higher prices for member importing countries but lower prices for nonmember importing countries.

This chapter originally appeared as an article titled "A Dynamic Model of Oligopoly in the Coffee Export Market" by Karp and Perloff in the *American Journal of Agricultural Economics* in 1993. This material is reprinted by permission of the *American Journal of Agricultural Economics*.

1 A combination of strategies is a subgame-perfect Nash equilibrium if it is a Nash equilibrium for the game and the relevant strategies are Nash equilibriums for every subgame (subperiod within the game).

We view Brazil and Colombia as a dynamic coffee duopoly facing a fringe with exogenous exports. During the sample period of 1961/62 through 1983/84, Brazil and Colombia's share of total world exports averaged 43 percent. Brazil and Colombia act like large "firms" in that each centrally controls exports. The Brazilian Coffee Institute (IBC) controls supply and price; supervises grading, packing, and weighing; and sets quotas within the country. The Colombian Federation of Coffee Growers (FNCC) buys from small farmers; evaluates, blends, grades, and cleans coffee; and manages the market through prices and taxes. In the absence of an ICA in 1974, Brazil and Colombia attempted to form an explicit producers' cartel and were later joined by other (smaller) producers. At various times, particularly during extended ICA negotiations from 1978 into 1980, the Brazilian and Colombian governments appeared to intervene in the market to maintain stability.

The institutional arrangements in the coffee market constitute circumstantial evidence of an intent by large producers to exert market power. The difficulties of negotiating these agreements and the allegations that many exporting countries fail to comply fully with them suggest that producers do not behave as a perfect cartel. Thus, the hypothesis that the market structure lies between monopoly and competition is plausible. The complexity and inconstancy of coffee marketing institutions and a paucity of data make it unreasonable to attempt to estimate the explicit game that producers play. Instead, we estimate a fairly general set of relations that include, as special cases, the equilibrium conditions associated with perfectly competitive, collusive, and Nash-Cournot behavior. Thus, we are able to determine whether the observed outcome is consistent with any of these equilibria.

The standard static assumption is inappropriate where there are large adjustment costs in training, storage, or capital accumulation. There are two reasons to use a dynamic model for coffee. First, changes in production involve nonlinear costs. A lag of two to five years exists between planting and first harvest; a tree produces its maximum output between five and ten years of age and bears for up to thirty years. This pattern suggests that average adjustment costs increase with the size of the adjustment. Second, Brazil and Colombia maintain large stockpiles. In standard inventory models (e.g., Blinder 1982), the costs of inventory adjustment are assumed to be reasonably approximated by a quadratic function. Costs of adjusting exports, therefore, stem from costly adjustment of production or inventories.[2]

The Model

Our dynamic model is flexible enough to allow for the possibility that Brazil (firm 1) and Colombia (firm 2) act like price-takers, collude, or export oligopolistic levels between those two extremes. For simplicity, other countries' exports are ignored in this section but are accounted for in the estimates. In period t, firms face the inverse residual linear demand curve

$$(1) \qquad p_t = a(t) - bQ_t$$

where p_t is the real price in period t, Q_t is the combined export of Brazil (q_{1t}) and Colombia (q_{2t}), $a(t)$ includes effects of exogenous variables such as exports of other countries, and b is a positive coefficient.

Each firm i has constant marginal costs θ_i with respect to contemporaneous exports q_{it} and a quadratic adjustment cost $\theta_i q_{it} + (\gamma_i + \delta/2\ u_{it})\ u_{it}$ where $u_{i,t} \equiv q_{it} - q_{i,t-1}$ is the change in a firm's export level from period t–1 to period t. In contrast, a static model sets γ_i and δ equal to zero.

Static Model

In most empirical static models of oligopoly, aggregate- or firm-level data are used to estimate a parameter, which we call v, that reflects the markup of price over marginal cost. Given demand Equation (1) for a homogeneous product, firm i's effective marginal revenue curve (the marginal revenue given the degree of market power actually exercised) is $MR_i(v) = p + (1 + v)p'q_i = p_t - (1 + v_i)bq_i$.

If all firms have a common v and a common marginal cost $MC = MC_i = \theta$, we can rewrite the equilibrium equations for each firm, $MR_i(v) = MC \equiv \theta$, as

$$(2) \qquad p = \theta + (1 + v)bq_1.$$

---

2 Our technique could be generalized to allow for adjustment costs in production and inventories. However, because reliable inventory data are not available, we assume the estimated adjustment cost is a proxy for both the cost of adjusting output and the cost of inventories.

Estimating (1) and (2), dividing the coefficient on the q_i term in (2) by the estimate of b from (1), and subtracting one gives an estimate of v. The gap between price and marginal cost, $p - MC = (1 + v)bq_i$, depends on v. If $v = -1$, marginal revenue equals price [$MR = p$] and there is no markup. If $v = 1$, marginal revenue is less than price [$MR = p + p'(2q_i) = p + p'Q$] and the monopoly markup is observed. Intermediate solutions, such as the Nash-Cournot where $v = 0$, are also possible (Carlton and Perloff 1990).

Some authors interpret v as a firm's constant conjectural variation about its rival: $v \equiv dq_j/dq_i$. We prefer the neutral interpretation that v is a measure of market power—the gap between marginal cost and price—so that we do not need to make the behavior assumptions of conjectural-variations models. Moreover, the conjectural-variation interpretation cannot be used in the feedback dynamic model we now describe.

The Linear-Quadratic Dynamic Model

The game-theoretic literature abounds with very general dynamic models of oligopoly that do not lend themselves to estimation. To make the estimation problem tractable, we restrict our model to a specific family of equilibria indexed by a parameter v as in the static model. We estimate the family of equilibria under the assumption that firms use subgame-perfect Markov (feedback) strategies. That is, firms choose strategies (rules) that determine their exports as a function of the state variables.

For comparison, we also estimate the same family of equilibria under the assumption that firms use open-loop strategies. That is, firms choose, at the initial time, a path that they intend to follow thereafter—they do not expect to revise their decisions after an unexpected shock (such as bad weather) affects production. This failure to anticipate revision is irrational, so the open-loop equilibrium is not subgame perfect. The open-loop model is the dynamic generalization of the static model as explained hereafter.

The open-loop and Markov adjustment paths are identical if firms collude ($v = 1$ and firms are of equal size) or act as price-takers ($v = -1$), as shown in Karp and Perloff (1991). In oligopolistic models, such as when firms play Nash within a period, the two models imply different adjustment paths and steady-state export levels. Given cost information that is sufficient to test overidentifying restrictions, one can empirically discriminate between the two types of behavior (Karp and Perloff 1989). Lacking that information, we estimate both models to determine the sensitivity of the measure of market structure to maintained hypotheses regarding behavior. The two assumptions lead to similar results for coffee.

We estimate the Markov model using a variant of the linear-quadratic game solution (Starr and Ho 1969). A general open-loop model can be estimated (Roberts and Samuelson 1988), but using a linear-quadratic specification enables us to compare easily the open-loop and Markov equilibria and to estimate a market-structure parameter.

Our feedback and open-loop models include four common models: collusion, price-taking, open-loop Nash-Cournot, and Markov Nash-Cournot. Other export paths lying between collusion and price-taking could be produced by other dynamic oligopolistic games. For example, Brazil and Colombia might imperfectly collude. Another possibility is that export levels are chosen subject to political pressures (one group wants to maximize export revenues and another wants to increase labor demand), which causes a deviation from Nash-Cournot or collusive equilibria.[3] Rather than try to model explicitly each of these games, we use an index v that allows for intermediate paths and steady-state exports. This index is the dynamic analog of the price/marginal-cost wedge used in static models of oligopoly.

In each period of the dynamic model, firm i's revenues, R_i, equal $p_t q_{it}$. Given a discount factor β, firm i's objective is to maximize its discounted stream of profits,

$$(3) \qquad \sum_{t=1}^{\infty} \beta^{t-1} \left[(p_t - \theta_i) q_{it} - (\gamma_i + \delta/2\, u_{it}) u_{it}\right],$$

where $(p_t - \theta_i) q_{it}$ is the contemporaneous profit and the last term is the adjustment cost. With Markov strategies, firm i chooses changes in

3 Further, the game-theoretic folk theorem (Fudenberg and Maskin 1986) justifies any outcome that is less competitive than Nash.

exports, u_{it}, as a function of current information—its own and its rival's lagged export.

If we define $\boldsymbol{J}_i(\boldsymbol{q}_t; v)$ as the present discounted value of firm i's program, given the state vector $\boldsymbol{q}_t \equiv (q_{1t}, q_{2t})$ and an index v of market power, firm i's dynamic programming equation is

$$(4) \qquad \boldsymbol{J}_i(\boldsymbol{q}_{t-1}; v) = \max_{u_{i,t}} (p_t - \theta_i)q_{it} - \left(\gamma_i + \delta/2\, u_{it}\right) u_{it} + \beta \boldsymbol{J}_i(\boldsymbol{q}_t; v).$$

That is, the present discounted value of the stream of future profits as of the last period equals the present discounted value of future profits as of this period plus the profits from this period.

The first-order condition corresponding to (4) is

$$(5) \qquad p_t = \theta + (1 + v)\, bq_{it} + \gamma_i + \delta u_i - \beta \left[\partial \boldsymbol{J}_i(\boldsymbol{q}_t; v)/\partial q_i + \partial \boldsymbol{J}_i(\boldsymbol{q}_t; v)/\partial q_j\, v\right]$$

where $p - (1 + v)\, bp_{it}$ is marginal revenue and the term in brackets is the discounted shadow value of an extra unit of current exports. Terms in this equation are arranged to emphasize its similarity to the static equilibrium condition (2). The gap between marginal cost and price is the same function of v as in the static model, so v can be interpreted as a market-power measure or index of the family of market structures.

Estimation of Dynamic Models

Our objective is to obtain a consistent estimate of the index v of the market structure in the subgame-perfect dynamic model. We also estimate δ, the quadratic adjustment parameter. We can allow parameters γ_i, θ_i, and $a_i(t)$ to be firm-specific and nonstationary to reflect quality differences, transportation costs, or other firm-specific costs. If $\gamma_i = 0$ (adjustment costs are minimized when there is no adjustment), the open-loop steady-state exports in the collusive, noncooperative Nash-Cournot and price-taking equilibria are identical to the corresponding static equilibria. Because the open-loop and static models have the same equilibria for general cost and revenue functions and not simply quadratic ones, the open-loop equilibrium is the dynamic analog of the static model. The static, open-loop, and feedback models are increasingly complex: the open loop is the dynamic version of the static model and the feedback model replaces the naive strategies of the open-loop (and implicitly the static) model with sophisticated Markov strategies.

We now describe the basic estimation approaches (see the Appendix). The value of firm *i*'s program, $\boldsymbol{J}_i(\boldsymbol{q}_{t-1}; v)$, in (4) is a quadratic function. The equilibrium Markov control rules are linear, and the open-loop strategies can be written as linear functions of lagged export:

(6) $$\boldsymbol{q}_t = \mathbf{g}(t) + \boldsymbol{G}q_{t-1}$$

where $\mathbf{g}(t)$ is a column vector and $\boldsymbol{G}$ is a two-by-two matrix with elements G_{ij} $(i, j = 1, 2)$.[4] In estimating adjustment Equation (6), we make no assumptions regarding whether firms have rational expectations about the *exogenous* variables nor do we impose assumptions about whether the inverse demand intercepts and linear costs are constant over time or across firms. Elements of $\boldsymbol{G}$ are used to infer v. A major advantage of our estimation strategy is its simplicity: market structure is logically distinct from the rational-expectations hypothesis. One could test rational expectations by including the exogenous information in the state vector (Chow 1981). If that hypothesis is true, there is a loss of efficiency from ignoring it but our estimate is still consistent.

Results

We estimate both dynamic models of the coffee export market for crop years 1961/62 through 1983/84. During that period, Brazil and Colombia's share of total world exports ranged from 32 to 50 percent and averaged 43 percent. Brazil's share was, on average, twice that of Colombia's share. The rest of South America's share was 4 percent, the rest of Latin America's share was 17 percent, Asia's share was 6 percent, and Africa's share was 29 percent. Shares of the two largest exporters in Africa were 7 percent (Ivory Coast) and 5 percent (Uganda).

We assume that Brazil and Colombia engage in a dynamic game in which they treat the rest of the world as a fringe with exogenous exports. Most Latin American and some African countries produce Arabica coffee

4 Infinite horizon games may have many subgame-perfect equilibria. We obtain a unique equilibrium by taking the equilibrium strategies that result from the game with a finite horizon T and letting $T \rightarrow \infty$.

(70 percent of all coffee produced). Most African and Asian countries produce Robustas, which are used mainly in instant coffees. A linear demand curve treating Arabica and Robustas from various countries as imperfect substitutes produced results that are similar to the following.

Quantity data are from *Coffee: World Coffee Situation* (various years) published by the U.S. Department of Agriculture's Foreign Agricultural Service. Coffee price is an average of prices for all coffee traded in New York, the major coffee market. Price data for coffee (New York) and tea (London), the world commodity wholesale price index, and the world gross domestic product (GDP) at constant prices are from the International Monetary Fund's *International Financial Statistics* (various years). In the following, we use real prices obtained by deflating nominal prices by the world commodity wholesale price index.

Classical Estimates

We estimate inverse demand curve (1) using an instrumental-variables technique:[5]

$$\begin{aligned}
\text{Real Price of Coffee} = {} & \underset{(2.25)}{4.401} - \underset{(-4.11)}{0.000123}\ \text{World Coffee Exports} \\
& + \underset{(0.53)}{0.0169}\ \text{World GDP} + \underset{(1.53)}{0.559}\ \text{London Price of Tea} + \underset{(1.78)}{0.206}\ t \\
& - \underset{(-2.59)}{0.00380}\ t^2
\end{aligned}$$

where t is a time trend (1, 2, . . .) and the terms in parentheses are t-statistics against the null hypothesis that the coefficient equals zero. The correlation between observed and predicted price is 0.69 and the Durbin Watson statistic is 1.96. The demand curve plays only a minor role because b, the coefficient on exports, affects only δ and not v.

We estimate the $\boldsymbol{G}$ matrix in adjustment Equation (6) using Zellner's seemingly unrelated equations method (Table 10.1). Each country's exports are regressed on its own lagged exports, the other country's

5 Instrumental variables are time; time squared; time cubed; the freeze dummy; the London price of tea; the world gross domestic product; and Brazilian and Colombian rainfall, temperatures, gross domestic products, and populations.

lagged exports, a time trend, and a dummy for the 1977/78 freeze in Brazil.[6] Cross-equation symmetry constraints are imposed in Table 10.1: the coefficients on the own lagged exports are equal $[G_{11} = G_{22} \equiv G_1]$ across equations, as are the coefficients on the other country's lagged exports $[G_{12} = G_{21} \equiv G_2]$. Based on unrestricted estimates, the F-statistic on these restrictions is 0.64 with two and thirty-four degrees of freedom, so we cannot reject the restrictions. The hypothesis that the coefficients on the lagged exports are zero, independently or collectively, is rejected by t-tests, F-tests, and likelihood-ratio tests.

Table 10.1. Adjustment Equations: Regression of Exports on Lagged Exports

	Brazil	**Colombia**
Constant	12,986.00	6,967.90
	4.99[a]	*4.28*
Brazilian Freeze 1977/78	–9,980.70	843.70
	–4.66	*0.74*
Time (1, 2, . . .)	22.40	124.80
	0.30	*2.59*
Lagged Brazil Exports	0.302	–0.192
	2.270	*–2.420*
Lagged Colombia Exports	–0.192	0.302
	–2.420	*2.270*
R^2	0.57	0.74
Durbin Watson	2.23	1.57
Durbin's h	–0.72	1.34

[a] Figures in italics are t-statistics against the null hypothesis that the coefficient equals zero.

6 To test the effects of the ICA agreements, we tried the following experiments: (1) dummies for the time of ICA meetings were included; (2) a one-year lag or lead of the meeting dummies was tried; and (3) a continuous variable measuring time from the previous meeting was used. Estimated coefficients were not statistically different from zero at the 0.05 level individually or collectively. We also divided the period in half and estimated separate adjustment equations for the two subperiods (to capture a fundamental structural change, possibly due to changes in the ICA). Based on a Chow (1981) test, we could not reject the hypothesis of no structural change.

Based on these estimates of **G** and assuming $\beta = 0.95$, $v_c^o = \phi_o(G) = -0.84$ and $v_c^f = \phi_f(G) = -0.80$ [where $\phi_o(\mathbf{G})$ is a nonlinear function of **G**, superscripts o or f indicate open-loop or feedback, and subscript c indicates the classical estimate]. Standard errors of v_c^o and v_c^f, based on a Taylor expansion, are 0.27 and 0.31 (Table 10.2). Thus, we cannot reject the price-taking ($v = -1$) hypothesis, but we can reject Nash-Cournot ($v = 0$) and less competitive models at the 0.05 level. Adjustment coefficients are the same under both models:

$$\delta_o^c = \psi_o(\mathbf{G}, b) = 0.000037 = \delta_f^c = \psi_f(\mathbf{G}, b).$$

For the estimated dynamic system to "make sense," it must have three properties:

- The system is *stable*: $-1 < G_1 + G_2 < 1$ and $-1 < G_1 - G_2 < 1$.
- The *market structure* lies between collusion and price-taking: $1 > v^k = \phi_k(\mathbf{G}) > -1, \quad k = o \text{ or } f$.
- The *adjustment* parameter in each of the models is positive: $\delta^k = \psi_k(\mathbf{G}, b) > 0, \qquad k = o \text{ or } f$.

Our classical point estimates of the elements of **G** and our estimates of v^k and δ^k are consistent with these restrictions.

Bayesian Estimates

Rather than estimate the unconstrained system and hope that the point estimates lie in the desired range, we can impose the preceding three sets of restrictions. Although it would be extremely difficult, if not impossible, to impose such inequalities using a classical approach or to test them, Geweke (1986, 1989) and Chalfant, Gray, and White (1991) show how to impose and test inequality restrictions with Bayesian techniques.

Our Bayesian prior is the product of a conventional uninformative distribution and an indicator function that equals one when the inequality constraints are satisfied and zero elsewhere. The posterior distribution is calculated using Monte Carlo numerical integration with importance sampling. Given a quadratic (absolute difference) loss function, estimates of the parameters consistent with the restrictions are obtained by calculating the mean (median) of the coefficient estimates for all replications in which the constraints are satisfied. Indeed, we obtain the full posterior distributions of v_b^o and v_b^f.

Table 10.2 summarizes results of the classical and Bayesian estimates. The v_a^k based on an absolute-difference loss function (medians) are close to the classical point estimates. The v_b^k based on a quadratic loss function (means) are about 0.2 higher than the classical estimates. Standard deviations on v_b^k are slightly greater than the Taylor approximations for the classical estimates.

Table 10.2. Classical and Bayesian Inequality-Constrained Estimates

	v^o	v^f		
Classical Estimates				
v_c^k (Unrestricted)	–0.84	–0.80		
σ (Taylor Approximation)	0.27	0.31		
Bayesian Inequality-Constrained Estimates	**Importance Sampling (T = 5,000)** v^o	v^f	**Bootstrapping (T = 2,000)** v^o	v^f
Quadratic Loss				
v_b^k (mean)	–0.65	–0.62	–0.68	–0.63
σ	0.35	0.36	0.43	0.44
Precision of Mean of v_b^k $(\sigma/\sqrt{T})$	0.0059	0.0060	0.0096	0.0097
Absolute Loss				
v_a^k (median)	–0.76	–0.73	–0.86	–0.81
σ	0.37	0.37	0.46	0.47
Reject Because (%)				
Unstable	0.002	0.002	0.0	0.0
$\delta^k \le 0$	25.2	23.6	18.7	18.7
$v^k < -1$	24.9	25.4	18.7	18.7
$v^k > 1$	1.7	1.2	11.0	11.0
Total Rejections $(1-p)$	26.5	26.5	29.7	29.7
Asymptotic Standard Error of p $(\sqrt{p(1-p)/T})$	0.0073	0.0073	0.0109	0.0109

To estimate the probability that the restrictions hold, we calculate the (importance weighted) proportion of Monte Carlo replications satisfying the restrictions. Stability conditions are virtually always met (Table 10.2). All three sets of conditions hold in approximately three-quarters of the replications, so imposing the restrictions seems reasonable (Table 10.2). Because the restriction that δ is positive holds in three-quarters of the cases (the odds in favor of a positive δ are three to one), the data indicate that there is dynamic adjustment.[7]

The assumption in this Bayesian approach that the original error terms are normal can be relaxed by using a bootstrapping approach as reported in Table 10.2.[8] The bootstrap estimates show slightly higher standard deviations corresponding to the mean v_s^k estimates (0.43 and 0.44), a lower probability of rejecting due to $v_s^k \leq -1$ (19 percent), and a higher probability of rejecting due to $v_s^k < 1$ (11 percent).

Market Structure

Bayesian estimates provide an entire posterior distribution of market parameter v^k. Some of the information from the importance-sampling histograms is summarized in Table 10.3, which shows the probability that v^k lies within certain ranges. The probability that v_b^k lies between minus one (price-taking) and zero (Nash-Cournot) is greater than 90 percent. There is a slightly higher probability (nearly 50 percent) that v_b^k lies between the classical estimate and zero than between minus one and the classical estimate (more than 40 percent). Two-thirds of the distribution lie below v_b^k. In the feedback model, the posterior odds ratio that the market structure lies between price-taking and Nash-Cournot rather than between Nash-Cournot and collusive is 12.9.

Results for the bootstrap model are similar; however, the bootstrap distribution has thicker tails. The bootstrap posterior odds ratio that the market structure is more competitive than Nash-Cournot versus the opposite is 8.3—less than the odds using the importance-sampling

---

7 We also estimated the corresponding static model in which marginal cost is a function of the freeze and a time trend (but not of dynamic adjustments). In the static optimization equation, the real price was regressed on quantity, a dummy for the 1977/78 Brazilian freeze, and a time trend. The estimated $v < -1$ is outside the plausible range.

8 With lagged endogenous variables but no autocorrelation (as indicated by a Durbin's h test), we can bootstrap by choosing rows (left- and right-hand-side variables) of the original data (Freedman and Peters 1984).

Table 10.3. Distribution of v^k Based on Bayesian Estimates

Proportion of Weight Between[a]			Importance Sampling (percent)		Bootstrapping (percent)	
			v^o	v^f	v^o	v^f
–1	and	0	93.5	92.8	90.3	89.2
0	and	1/2	4.2	4.8	5.8	6.3
1/2	and	1	2.3	2.4	3.8	4.4
–1	and	v_c^k	34.0	35.1	53.7	51.7
v_c^k	and	0	59.5	57.7	36.6	37.6
–1	and	v_b^k	67.2	65.5	72.9	71.1
v_b^k	and	0	26.3	27.3	17.5	18.2

[a] The classic estimate is v_c^k and v_b^k is the Bayesian estimate based on a quadratic loss function ($k = o$ or f).

estimates but still very high. Based on either the classical or the Bayesian approaches, the Brazil-Colombia exports are close to price-taking. The probability that they are at least as noncompetitive as Nash-Cournot is no greater than 11 percent according to all our estimating approaches.

Simulations

We can simulate adjustment paths and steady states based on the preceding estimates.[9] For a given v, the open-loop steady-state export and the corresponding static export are equal. For a constant $v \in (-1, 1)$, the feedback games are more competitive than open-loop games in the sense that steady-state exports are greater for a given v (see Reynolds (1987) for the intuition for the Nash-Cournot model). Based on simulations, differences in steady states between the open-loop and feedback models are maximized at values close to those estimated, –0.8. In absolute terms, however, these differences are small.

9 In the simulations, we choose marginal cost parameters for Brazil and Colombia so the steady-state export under the classically estimated feedback model equals the average export for the sample period. Because the constant marginal costs are not identified in our estimation procedure, we set the marginal cost for Brazil to zero and searched for the marginal cost for Colombia and the residual demand intercept producing the average steady-state outputs for the two countries. Subtracting b times the average output of the rest of the world's producers from the estimated demand curve intercept approximately equals the resulting residual demand intercept.

In the feedback model based on the unconstrained classical estimator v_c^f, steady-state exports are 6 percent lower than in the corresponding price-taker model (Table 10.4). In the open-loop and static models based on the classical estimator v_c^o, steady-state exports are 7 percent below those of the price-taker model. Thus, the feedback model is closer to price-taking than is the open-loop model. This quantity difference leads to large price effects because the demand elasticity evaluated at the mean price and quantity is about 0.2.

The dynamic models based on unconstrained classic estimates deviate from the price-taker level by only about half that of those based on the Bayesian constrained estimates. Using the constrained importance-sampling Bayesian quadratic loss estimates (bootstrapping), the feedback steady-state export is 13 (12) percentage points less than the price-taker level and the open-loop steady-state export is 15 (14) percentage points less (Table 10.4).

The steady states vary in the two estimated dynamic models for two reasons. First, for either model, the lower the estimated v, the closer its steady-state export is to price-taking. As Table 10.4 shows, the steady

Table 10.4. Simulated Exports under Various Models

Model	Brazilian and Colombian Exports (Static or Steady State)	Percentage of Price-Taker Output
Static and Dynamic Price-Taking ($v = -1$)	25,789	100
Feedback (v_c^o)	24,665	96
Feedback (v_c^f)	24,308	94
Open-Loop and Static Models (v_c^o)	23,868	93
Open-Loop and Static Models (v_c^f)	23,469	91
Feedback (v_s^f)	22,601	88
Feedback (v_b^f)	22,504	87
Open-Loop and Static Models (v_s^o)	22,232	86
Open-Loop and Static Models (v_b^o)	21,948	85
Feedback Cournot ($v = 0$)	17,557	68
Open-Loop and Static Cournot ($v = 0$)	17,192	67
Static and Dynamic Collusion ($v = 1$)	12,894	50

Note: Exports are in thousands of sixty-kilogram bags. The v_b^ks are the Bayesian quadratic loss estimates and the v_s^ks are the bootstrap estimates ($k = o$ or f).

state of either the feedback or the open-loop model based on v_c^o is 2 percent higher than that based on v_c^f. The open-loop model's $v_c^o = -0.84$ is slightly closer to price-taking than is the feedback model's $v_c^f = -0.80$.

Second, and more than offsetting the first effect, in the open-loop model there is more collusive behavior for any given v. As Table 10.4 shows, the feedback model (v_c^f) predicts a steady state 6 percentage points less than for price-taking whereas the open-loop model predicts a steady state 9 percentage points less. As a result, the less competitive open-loop model must have a lower estimated v to be consistent with the data.

Adjustment paths vary across models. As Figure 10.1 (where exports of both countries are set to zero in year zero) shows for the classical estimates, the combined exports of the two countries in the feedback model reach a higher steady-state level than in the open-loop model; but in both models exports reach their steady-state level after only three years.

To illustrate a more realistic adjustment path, we solve the dynamic game using estimated v and δ and marginal costs obtained by model calibration (assuming steady-state exports equal average exports in the sample period). In each period, we add the residuals plus the coefficient (from the estimated adjustment equation) times the Brazilian freeze dummies to the simulated $\boldsymbol{q}_t$. These simulations exclude demand-side shocks and

Figure 10.1. Exports under Various Models

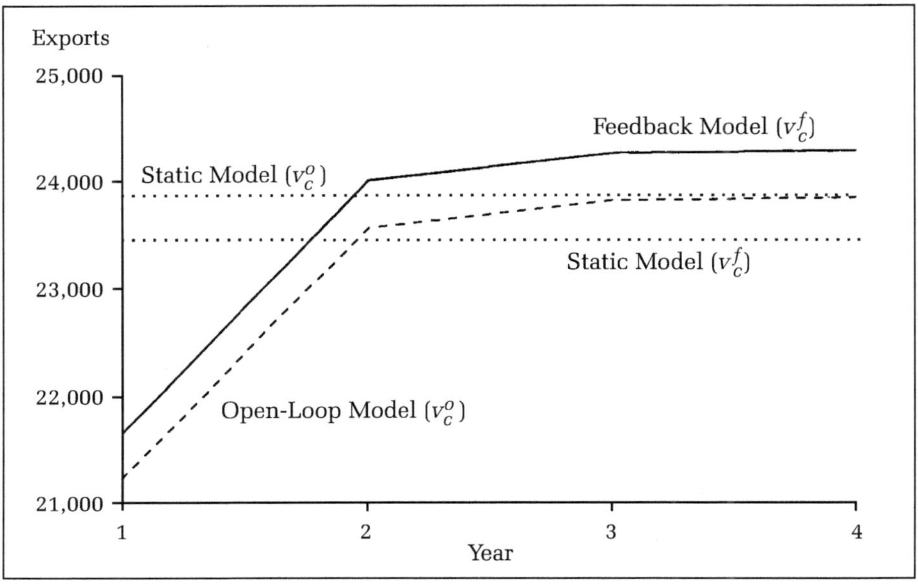

also ignore the upward trend in the demand curve, which may explain why the simulations shown in Figure 10.2 are low toward the end of the period. Even ignoring the demand trend, though, Figure 10.2 shows that the model is capable of emulating reality reasonably well.

Summary and Conclusions

Based on both classical and Bayesian estimates of a dynamic Markov model, Brazil and Colombia compete vigorously with each other in the coffee export market. Steady-state exports based on the classical (Bayesian) estimate are 6 (12 or 13) percent less than if both Brazil and Colombia are price-takers. From the Bayesian analysis, we are reasonably confident that the behavior of Brazil and Colombia is closer to price-taking than to collusion.

Our results from a subgame-perfect dynamic oligopoly model differ only slightly from those based on the much simpler open-loop dynamic model. If these results can be replicated with other data sets, open-loop estimates can be used to approximate Markov strategies without major bias.

Figure 10.2. Actual and Simulated Exports

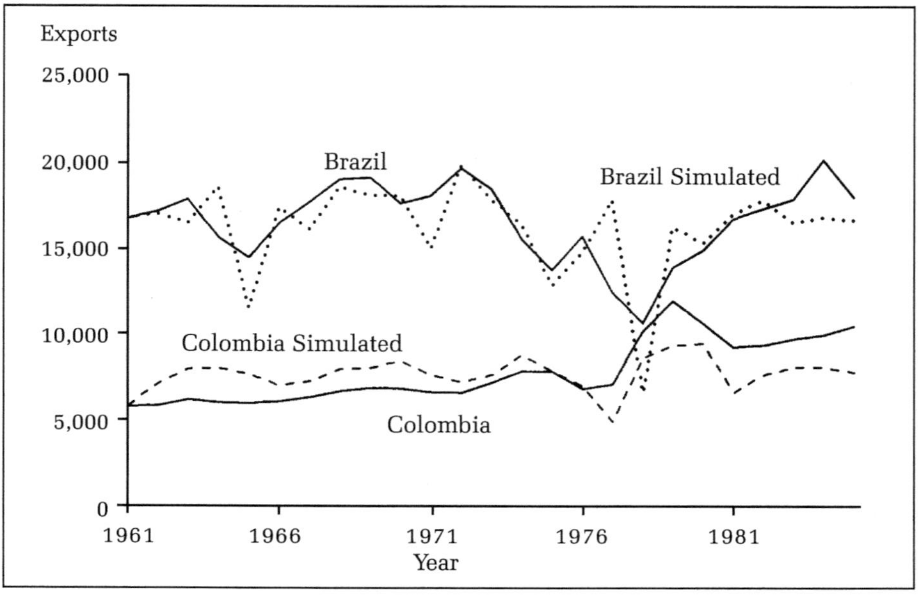

Appendix: Estimating the Dynamic Model

The equilibrium conditions of the dynamic games can be written in matrix notation. Let $\boldsymbol{e}_i$ be the ith unit vector, $\boldsymbol{e}$ equal the column vector of ones, $\boldsymbol{S}_i = \boldsymbol{e}_i\boldsymbol{e}'_i\delta$, and $\boldsymbol{K}_i = b(\boldsymbol{e}\boldsymbol{e}'_i + \boldsymbol{e}_i\boldsymbol{e}')$. The dynamic programming equation for the linear-quadratic model is

$$\boldsymbol{J}_i\,(\boldsymbol{q}_{t-1};\,v) = \max_{\boldsymbol{u}_{it}}\left[\left(a\boldsymbol{e}'_i\boldsymbol{q}_t - 1/2\;\boldsymbol{q}'_t\boldsymbol{K}_i\boldsymbol{q}_t - 1/2\;\boldsymbol{u}'_t\boldsymbol{S}_i\boldsymbol{u}_t\right) + \beta\,\boldsymbol{J}_i(\boldsymbol{q}_t;\,v)\right]$$

where $\boldsymbol{J}_i(\bullet)$ is a quadratic function, $\gamma_i = 0$ for simplicity, and the equilibrium Markov control rules are linear as shown in (6). If the open-loop equilibrium sequence is expressed as a function of the current state, the open-loop equilibrium can be written in "feedback form" as in (6). Firms revise their plans if something unexpected happens, but these revisions are unanticipated. Define $\boldsymbol{v}_i$ as a two-by-one column vector with one in the ith position and v in the other position. Given β, an estimated matrix $\boldsymbol{G}$, and an estimated demand slope b, the open-loop v and δ satisfy

$$\text{(A.1)}\quad \boldsymbol{K}_i\boldsymbol{v}_i = [\boldsymbol{G}^{-1}(\boldsymbol{I}-\boldsymbol{G})(\boldsymbol{I}-\beta\,\boldsymbol{G})]'\,\boldsymbol{e}_i\delta$$

where $\boldsymbol{K}_i$ is of rank two so that the solution to (A.1) is unique. The derivation of (A.1) does not require an assumption of symmetry.

To estimate v and δ in the feedback case, we define the vectors

$$\boldsymbol{x}_i = [\boldsymbol{I}-\beta\,(\boldsymbol{G}'\otimes\boldsymbol{G}')]^{-1}\,[(\boldsymbol{G}'\otimes\boldsymbol{G}')-(\boldsymbol{I}\otimes\boldsymbol{G}')-(\boldsymbol{G}'\otimes\boldsymbol{I})+\boldsymbol{I}]\,[\text{vec}(\boldsymbol{e}_i\boldsymbol{e}'_i)]\text{ and}$$

$$\boldsymbol{w}_i = [\boldsymbol{I}-\beta\,(\boldsymbol{G}'\otimes\boldsymbol{G}')]^{-1}\,[(\boldsymbol{G}'\otimes\boldsymbol{G}')\;(\text{vec}\,(\boldsymbol{K}_i))]$$

where vec($\boldsymbol{Z}$) is a vector of the stacked columns of the matrix $\boldsymbol{Z}$. The inverse vec operation is then used to "rematricize" $\boldsymbol{w}_i$ and $\boldsymbol{x}_i$ to obtain the two-by-two matrices $\boldsymbol{W}_i$ and $\boldsymbol{X}_i$. $\boldsymbol{W}_i$ is linear in firm i's demand coefficients and $\boldsymbol{X}_i$ depends only on β and $\boldsymbol{G}$. If agents use feedback strategies, v and δ must satisfy

$$\text{(A.2)}\quad [\boldsymbol{K}_i + \beta\boldsymbol{W}_i + (\boldsymbol{e}_i\boldsymbol{e}'_i + \beta\boldsymbol{X}_i)\,\delta_i\,]\,\boldsymbol{v}_i = \boldsymbol{G}^{-1}\,\boldsymbol{e}_i\delta_i \equiv \boldsymbol{y}^*_i\delta_i.$$

Given symmetry, the left side of (A.2) is of rank two and the estimate of v is independent of b. Defining matrix $\boldsymbol{A}^i$ such that $b\boldsymbol{A}^i \equiv \boldsymbol{K}_i + \beta\boldsymbol{W}_i$ and matrix $\boldsymbol{B}^i \equiv \boldsymbol{e}_i\boldsymbol{e}_i' + \beta\boldsymbol{X}_i$, $\boldsymbol{A}^i$ and $\boldsymbol{B}^i$ depend only on β and $\boldsymbol{G}$. To recover v and δ, rewrite the ith and the kth ($k \neq i$) equation of (A.2) as

$$\text{(A.3)} \quad b\left(\boldsymbol{A}_{ki} + v \sum_{j\neq i} \boldsymbol{A}_{kj}\right) + \left(\boldsymbol{B}_{ki} + v \sum_{j\neq i} \boldsymbol{B}_{kj}\right)\delta = \boldsymbol{y}^*_{ik}\,\delta \text{ and}$$

$$\text{(A.4)} \quad b\left(\boldsymbol{A}_{ii} + v \sum_{j\neq i} \boldsymbol{A}_{ij}\right) + \left(\boldsymbol{B}_{ii} + v \sum_{j\neq i} \boldsymbol{B}_{ij}\right)\delta = \boldsymbol{y}^*_{ii}\,\delta$$

where subscripts designate the element of $\boldsymbol{A}^i$, $\boldsymbol{B}^i$, and $\boldsymbol{y}$ and the superscript i is suppressed. Equation (A.4) can be solved for δ as a linear function of b and a nonlinear function of v. Substituting this function into (A.3) gives a quadratic in v that is independent of b. Hence a symmetric v can be estimated with knowledge of only β and $\boldsymbol{G}$. Although there are two solutions to (A.3), extensive simulation experiments show that one value is close to the open-loop value and that the other is infeasible ($v < -1$, $v > 1$, or $\delta < 0$). Therefore, in practice, it is easy to choose the correct root.

References

Akiyama, T., and R.C. Duncan. *Analysis of the World Coffee Market.* Washington DC: World Bank Staff Commodity Working Paper 1, 1982.

Bates, R.H., and D.D. Lien. "On the Operations of the International Coffee Agreement." *International Organization* 3 (1985):553–559.

Blinder, A.S. "Inventories and Sticky Prices: More on the Microfoundations of Macroeconomics." *American Economic Review* 72 (1982):334–349.

Carlton, D.W., and J.M. Perloff. *Modern Industrial Organization.* Glenview IL: Scott Foresman, 1990.

Chalfant, J.A., R.S. Gray, and K.J. White. "Evaluating Prior Beliefs in a Demand System." *American Journal of Agricultural Economics* 73 (1991):476–490.

Chow, G.C. *Econometric Analysis by Control Methods.* New York NY: John Wiley and Sons, 1981.

de Vries, J. *Structure and Prospects of the World Coffee Economy.* Washington DC: World Bank Staff Working Paper 208, 1975.

Fischer, B.S. *The International Coffee Agreement: A Study in Coffee Diplomacy.* New York NY: Praeger, 1972.

Freedman, D.A., and S.C. Peters. "Bootstrapping an Econometric Model: Some Empirical Results." *Journal of Business and Economic Statistics* 10 (1984):150–158.

Fudenberg, D., and E. Maskin. "The Folk Theorem in Repeated Games with Discounting with Incomplete Information." *Econometrica* 54 (1986):533–554.

Geweke, J. "Exact Inference in the Inequality Constrained Normal Linear-Regression Model." *Journal of Applied Econometrics* 26 (1986):295–321.

———. "Bayesian Inference in Econometric Models using Monte Carlo Integration." *Econometrica* 57 (1989):1317–1339.

Gilbert, C.L. "International Commodity Agreements: Design and Performance." *World Development* 15 (1987):591–616.

Greenstone, W.D. "The Coffee Cartel: Manipulation in the Public Interest." *Journal of Futures Markets* 1 (1981):1–18.

Herrmann, R. "Free Riders and the Redistributive Effects of International Commodity Agreements: The Case of Coffee." *Journal of Policy Modeling* 8 (1986):597–621.

Karp, L.S., and J.M. Perloff. "Open-Loop and Feedback Models in Dynamic Oligopoly." Working paper, University of California Berkeley, Department of Agricultural and Resource Economics, 1991.

———. "Dynamic Oligopoly in the Rice Export Market." *Review of Economics and Statistics* 71 (1989):462–470.

Marshall, C.F. *The World Coffee Trade*. Cambridge England: Woodhead-Faulkner, 1983.

Palm, F.C., and E. Vogelvang. "A Short-Run Econometric Analysis of the International Coffee Market." *European Review of Agricultural Economics* 13 (1986):451–476.

Reynolds, S.S. "Capacity Investment, Preemption, and Commitment in an Infinite Horizon Model." *International Economics Review* 28 (1987):69–88.

Roberts, M.J., and L. Samuelson. "An Empirical Analysis of Dynamic, Nonprice Competition in an Oligopolistic Industry." *Rand Journal of Economics* 19 (1988):200–220.

Starr, A.W., and Y.C. Ho. "Nonzero Sum Differential Games." *Journal of Optimization Theory and Applications* 3 (1969):184–206.

PART VI

PRODUCT-DIFFERENTIATED OLIGOPOLY MODELS

11

Explaining Pricing Conduct in a Product-Differentiated Oligopolistic Market: An Empirical Application of a Price Conjectural-Variations Model

Junko Kinoshita, Nobuhiro Suzuki,
and Harry M. Kaiser

Most previous empirical studies of imperfectly competitive markets have assumed that all firms within the industry supply a homogeneous product and use quantity as a strategic variable. This framework is basically a generalization of the Cournot model. In this case, there is only one demand function and one common price generated in the market. Examples of such empirical studies include Appelbaum (1982); Azzam and Pagoulatos (1990); Bhuyan and Lopez (1997); Iwata (1974); Karp and Perloff (1989); Lopez (1984); Schroeter (1988); Suzuki, Lens, and Forker (1993); and Liu, Sun, and Kaiser (1995).

However, most imperfectly competitive industries are characterized by multiple differentiated products that compete with each other based on price as the strategic variable rather than quantity. As such, a separate market demand function and unique price exists for each brand, and the brands are incomplete substitutes with each other. In such cases, a product-differentiated oligopoly model with price as the strategic

---

This chapter originally appeared as an article titled "Explaining Pricing Conduct in a Product-Differentiated Oligopolistic Market: An Empirical Application of a Price Conjectural Variations Model" by Kinoshita, Suzuki, and Kaiser in *Agribusiness: An International Journal* in 2002. This material is reprinted with permission of Wiley-Liss Inc., a subsidiary of John Wiley and Sons, Inc. Also, the authors thank the New York State Milk Promotion Advisory Board, which provided partial support for this research.

variable or a generalized Bertrand model should be applied. Several previous studies have developed and applied the Bertrand-type model to various industries; they include Cotterill (1994); Cotterill, Dhar, and Putsis (2000); Cotterill and Putsis (2000); Liang (1987, 1989); Peterson and Cotterill (1998); and Vickner and Davies (1999).

This chapter applies a two-product Bertrand-type linear model to the Japanese dairy sector to gain insights into the pricing conduct for beverage-milk products made from fresh milk and reconstituted milk.[1] The model is similar to one that was originally developed by Liang (1989). The analysis includes some statistical tests to determine whether these Japanese dairy firms have consistent pricing behavior. Japanese milk and reconstituted milk products are chosen for our analysis because many observers have pointed out that increases in Japan in the consumption of reconstituted milk products have resulted in a decrease in the demand for domestically produced raw milk there. While a competitive relationship between milk and reconstituted milk is expected, there currently is little documented empirical evidence of this relationship in Japan.[2]

Theoretical Model

In this section, we derive the conceptual model used in this study and list the assumptions. The model is an n-product generalization of Liang's (1987) original two-product form model. Consider an imperfectly competitive market composed of n firms, each supplying one differentiated product in the market. Since each product is differentiated, the following model is composed of n demand functions for each product i ($i = 1, \ldots, n$) and n price-reaction functions for each firm i ($i = 1, \ldots, n$).

To ensure empirical tractability, the demand function for i is expressed in a linear form[3] as

$$(1) \qquad q_i = a_i + b_i p_i + \sum_{j \neq i} (c_{ij} p_j) + d_i y \qquad \text{for } i \text{ and } j = 1, \ldots, n,\ i \neq j$$

---

1 Unlike the United States, where virtually all beverage-milk products are made exclusively from fresh milk, Japanese beverage-milk products are made from reconstituted milk and from fresh milk.

2 An exception to this is a recent study by Kinoshita et al. (2001).

3 Forms that are more flexible and complicated than the simple linear specification make it impossible to derive an explicit form for the price-reaction function in the cross-equation restraints and identify conjectures. For more discussion on functional forms for demand, see Liang (1987).

where q_i is firm (product) i's quantity supplied/demanded, p_i is product i's price, p_j is the substitute product's price, y is income, and a_i, b_i, c_{ij}, and d_i are parameters. Other variables affecting q_i are omitted here to simplify presentation of the conceptual model but are incorporated into the empirical model that follows.

Firm i's profit-maximization problem using price (p_i) as a strategic variable is

$$\max_{p_i} [p_i q_i - C_i] \tag{2}$$

where C_i is firm i's cost. The first-order condition is expressed as

$$q_i + p_i(dq_i / dp_i) - MC_i(dq_i / dp_i) = 0 \tag{3}$$

where MC_i is firm i's marginal cost, or $dC_i / dq_i = \text{MC}_i$. Firm i's conjectural variation is defined as its perception of how rival firm j will respond to firm i's price change. Mathematically, firm i's conjectural variation, cv_{ji}, is defined as dp_j/dp_i. Using this, dq_i / dp_i in Equation (3) can be written as

$$dq_i / dp_i = b_i + \sum_{j \neq i} c_{ij}\, cv_{ji}. \tag{4}$$

Solving (3) for p_i yields the following price-reaction function, R_i, for firm i:

$$p_i = \frac{-a_i - \sum_{j \neq i} c_{ij} p_j + [b_i + \sum_{j \neq i} c_{ij}\, cv_{ji}] MC_i - d_i y}{2b_i + \sum_{j \neq i} c_{ij}\, cv_{ji}}. \tag{5}$$

The difference in the retail and wholesale (manufacturer's) price should be taken into account in specifying the pricing conduct by manufacturers. Assuming that p_i is the retailer's price and the retailer's margin ratio is m_i, firm i's wholesale price is equal to $(1 + m_i)^{-1} p_i$. Then, firm (manufacturer) i's profit-maximization problem can be rewritten as

$$\max_{p_i} [(1 + m_i)^{-1} p_i q_i - C_i] \text{ or} \tag{6}$$

(7) $$\max_{p_i} \, [p_i q_i - (1 + m_i) C_i]$$

and firm i's price-reaction function, R_i (5), can be rewritten as

(8) $$p_i = \frac{-a_i - \sum_{j \neq i} c_{ij} p_j + (1 + m_i)[b_i + \sum_{j \neq i} c_{ij}\, cv_{ji}]\, MC_i - d_i y}{2b_i + \sum_{j \neq i} c_{ij}\, cv_{ji}}.$$

This price-reaction function represents the manufacturer's behavior.[4]

Using Equation (4), firm i's perceived or conjectured own-price elasticity, e_i, can be defined as

(9) $$\begin{aligned} e_i &= (dq_i / dp_i)\,(p_i / q_i), \\ &= [b_i + \sum_{j \neq i} (c_{ij}\, cv_{ji})]\,(p_i / q_i), \\ &= b_i\,(p_i / q_i) + \sum_{j \neq i} [c_{ij}\,(p_j / q_i)\, cv_{ji}\,(p_i / p_j)], \\ &= e_i^{ow} + \sum_{j \neq i} (e_{ij}^{cr} e_{ji}^{cv}). \end{aligned}$$

Equation (9) shows that e_i is composed of three types of elasticities: (1) e_i^{ow} (partial own-price elasticity), (2) e_{ij}^{cr} (cross-price elasticity), and (3) e_{ji}^{cv} (conjectural-variation elasticity).

If there is no interdependency among firms, e_i is equal to e_i^{ow}. However, oligopolistic firms must recognize e_i as a value in which e_{ij}^{cr} and e_{ji}^{cv} as well as e_i^{ow} are incorporated. The sign of e_i^{ow} is negative, e_{ij}^{cr} is usually positive, and e_{ji}^{cv} can be positive or negative. When $e_{ji}^{cv} < 0$, firm i conjectures that p_j will move in the opposite direction of changes in p_i. Therefore, in absolute value, e_i will be larger than e_i^{ow}. Larger absolute values for e_{ij}^{cr} and e_{ji}^{cv} generate a larger absolute value for e_i when $e_{ji}^{cv} < 0$. On the other hand, when $e_{ji}^{cv} > 0$, firm i conjectures that p_j will move in the same direction of changes in p_i. Therefore, in absolute value, e_i will be smaller than e_i^{ow}. Larger absolute values for e_{ij}^{cr} and e_{ji}^{cv} generate a smaller absolute value for e_i when $e_{ji}^{cv} > 0$.

---

4 Our model does not account for retailer's strategic behavior in determining retail margins. If this type of behavior does occur, the model may derive biased estimates of conduct parameters for the manufacturers.

In addition, when two firms' pricing behaviors are consistent (i.e., firm *i*'s estimated conjecture corresponds to firm *j*'s observed price reaction), firm *i*'s constant conjectural variation, ccv_{ji}, is determined by the condition

(10) $$cv_{ji} = \partial R_j / \partial p_i = -c_{ji} / 2b_j + c_{ji} cv_{ij} \qquad \text{for } i \text{ and } j = 1, 2, i \neq j$$

and is defined by

(11) $$ccv_{ji} = \frac{-b_i b_j + \sqrt{(b_i b_j)(b_i b_j - c_{ij} c_{ji})}}{b_j c_{ij}} \qquad \text{for } i \text{ and } j = 1, 2, i \neq j.$$

(Liang 1989, pp. 37–38).

Data and Empirical Model

This study uses scanner data obtained from Nikkei Quick Information, Inc. The data consist of weekly observations over a two-year period (April 1997 to March 1999) for one of the largest national supermarket chains in Japan. The data include weekly weighted-average prices and quantities sold for each item and number of customers. Since previous studies have found weather to be an important variable, particularly for beverage-milk consumption (e.g., Watanabe, Suzuki, and Kaiser 1997), we also include the following variables: maximum and minimum temperature, average temperature, humidity, precipitation, and number of daylight hours.

Because the data collected were for each store rather than by customer, individual customer-attribute data (e.g., income, household size, age, job, education level) were not available for this study. However, weekly store sales values for milk and related products were used as proxies for customers' expenditure data. Weekly store sales values for milk, reconstituted milk, soda, fruit juice, yogurt, and cheese were included in the milk-related expenditure basket for customers because they were considered to be the most important substitutes of beverage-milk products.

The data set does not include advertising expenditures or any indicators for sales promotion. However, assuming that a store increases its advertising for the item (e.g., the store's advertising booklets inserted in newspapers) during discount sale days for the item, an indicator variable can be constructed as a proxy for advertising intensity. In the model, advertising-indicator variables are equal to one for the week when the item is sold at a discount price and zero otherwise. The advertising-indicator variables for own and rival items were included in the demand function for each item to estimate own and cross effects of advertising.

To focus on the competition between fresh-milk and reconstituted-milk products, the most popular item (based on largest market share) was selected from each of the fresh-milk and reconstituted-milk product categories. For fresh beverage milk, the supermarket's private-label whole milk (3.5 percent fat content, one liter) was used (hereafter referred to as FRESH). For reconstituted beverage milk, a national-label reconstituted low-fat milk product with added calcium (one liter) was used (hereafter referred as RECONST). The market share of each item in each category is 53.3 percent and 28.3 percent, respectively.[5]

One store of the largest national supermarket chain in Japan was selected for our analysis. It has 12,780 square meters of selling area and parking capacity for 900 cars. It is located in Tokyo. We specified the firm's profit-maximization condition for each of the two selected items in the store rather than considering the firm's total profit. This approach was deemed reasonable based on evidence presented at recent confidential hearings that indicated that manufacturers usually look at the profitability of the most popular items in each section of the supermarket for product-mix evaluations.

Average variable costs per product unit were used as a proxy for marginal costs. Average variable costs (or marginal costs) fluctuate in response to changes in major input prices. The cost figures and assumptions were based on evidence presented at recent confidential hearings involving a manufacturer's data (confidential source). The two-year average cost figures obtained from the manufacturer were 147 yen per liter for FRESH and 135 yen per liter for RECONST. Based on these average levels, a weekly marginal-cost data series was constructed by adjusting the two-year averages in proportion to fluctuations in major input prices.

---

5 Inclusion of only two products in the analysis may result in estimation bias. However, it is not possible to include more than two products in our conjectural-variations model. Since the market share of the two products is relatively high, the extent of the potential bias may be minor.

The weekly change in producer raw-milk prices was used as the weekly adjuster for FRESH marginal costs. Weekly import prices (cost, insurance, and freight (CIF)) for nonfat dry milk were used as the weekly adjuster for RECONST marginal costs.

Retailer margins were also obtained from the confidential hearing. The retailer's margin ratio for the private-label product (FRESH) was 12.5 percent. The retailer's margin for the national-label product (RECONST) was set at 17.0 percent.

The empirical model is composed of the following four equations.

FRESH demand equation:

$$q_1 = c_1(1) + c_1(2)p_1 + c_1(3)p_2 + c_1(4)\,expr + c_1(5)\,temp + c_1(6)ad_2.$$

RECONST demand equation:

$$q_2 = c_2(1) + c_2(2)p_2 + c_2(3)p_1 + c_2(4)\,expr + c_2(5)\,temp + c_2(6)ad_1.$$

FRESH price-reaction function R_1:

$$p_1 = \frac{-c_1(1)}{2c_1(2) + c_1(3)c_1(7)} + \frac{-c_1(3)}{2c_1(2) + c_1(3)c_1(7)}\,p_2 + \frac{1.125\,[c_1(2) + c_1(3)c_1(7)]}{2c_1(2) + c_1(3)c_1(7)}\,MC_1 + \frac{-c_1(4)}{2c_1(2) + c_1(3)c_1(7)}\,expr + \frac{-c_1(5)}{2c_1(2) + c_1(3)c_1(7)}\,temp + \frac{-c_1(6)}{2c_1(2) + c_1(3)c_1(7)}\,ad_2.$$

RECONST price-reaction function R_2:

$$p_2 = \frac{-c_2(1)}{2c_2(2) + c_2(3)c_2(7)} + \frac{-c_2(3)}{2c_2(2) + c_2(3)c_2(7)}\,p_1 + \frac{1.170\,[c_2(2) + c_2(3)c_2(7)]}{2c_2(2) + c_2(3)c_2(7)}\,MC_2 + \frac{-c_2(4)}{2c_2(2) + c_2(3)c_2(7)}\,expr + \frac{-c_2(5)}{2c_2(2) + c_2(3)c_2(7)}\,temp + \frac{-c_2(6)}{2c_2(2) + c_2(3)c_2(7)}\,ad_1.$$

where

$_1 =$	Item 1 or FRESH.
$_2 =$	Item 2 or RECONST.
q_1 and q_2 =	Item 1 and 2's weekly quantities (liters) sold per thousand weekly visiting customers.
p_1 and p_2 =	Item 1 and 2's weekly weighted-average prices (yen/liter).
expr =	Customers' weekly expenditures (thousand yen) per thousand weekly visiting customers.
temp =	Weekly average temperature in Tokyo.[6]
ad_1 and ad_2 =	Advertising-indicator variables equal to one for the week when items 1 and 2 were sold at a discount price for more than five days and zero otherwise.[7]
MC_1 and MC_2 =	Item 1 and 2's marginal costs.
$C_{\bullet}(\bullet)$ =	Estimated coefficients (see Table 11.1).

Results

The preceding simultaneous four-equation system was estimated using the three-stage least squares (3SLS) method. Because the Durbin Watson statistics indicated serial correlation of error terms for the two-price reaction functions, a first-order autoregressive model was applied for them, and the estimated serial-correlation coefficients were significant at the 1 percent level. After the adjustment, the Durbin Watson statistics fell in a range between 2.0 and 2.5 for all equations.

---

6 Other weather factors were omitted because they were strongly correlated to each other.

7 Advertising-indicator variables for own items were omitted because they were strongly correlated to own-price variables.

Estimated coefficients are shown in Table 11.1. Critical levels shown by p-values indicate that all of the own- and cross-price coefficients ($c_{\bullet}(2)$ and $c_{\bullet}(3)$) were significant at the 1 percent level. Regarding the other demand coefficients for FRESH, the expenditure coefficient ($c_1(4)$) was significant at the 5 percent level, but the temperature coefficient ($c_1(5)$) and the advertising-indicator coefficient ($c_1(6)$) were not significant. In the demand equation for RECONST, the expenditure coefficient ($c_2(4)$) was not significant while the temperature coefficient ($c_2(5)$) and the advertising-indicator coefficient ($c_2(6)$) were significant at the 5 percent level. The advertising-indicator coefficient suggests that more advertising for FRESH would decrease demand for RECONST. The critical level of the conjectural-variation coefficient for FRESH ($c_1(7)$) was 22.3 percent, which indicates that firm 1's conjecture was not significantly different from zero. On the other hand, the conjectural variation for RECONST ($c_2(7)$) was significant at the 1 percent level. These results suggest that the existence of interdependency among firms is statistically supported.

Table 11.1. Estimated Coefficients

Coefficient	Estimate	p-Value
$c_1(1)$	565.039	0.000
$c_1(2)$	–4.562	0.000
$c_1(3)$	1.344	0.000
$c_1(4)$	0.487	0.015
$c_1(5)$	–	–
$c_1(6)$	–	–
$c_1(7)$	–24.392	0.223
$c_2(1)$	44.362	0.035
$c_2(2)$	–0.747	0.000
$c_2(3)$	0.666	0.000
$c_2(4)$	–	–
$c_2(5)$	0.179	0.014
$c_2(6)$	–8.238	0.009
$c_2(7)$	0.437	0.003

Dash (–): The variable was omitted because the estimated coefficient was insignifficant at the 5 percent level.

That means that the original Bertrand model, which assumes zero conjectural variations for all firms, is inappropriate for our empirical analysis.

Elasticities were calculated based on the sample means shown in Table 11.2. The absolute values of e_i^{ow} (–6.665 and –9.187) may be large compared with U.S. price elasticities for food items. However, they are quite comparable to estimates from a recent study in Japan by Kinoshita et al. (2001). The estimated e_i^{ow} for brand levels in this study are much larger than previous estimates based on aggregate market data because the present results reflect brand-level competition, which is not true in the case of aggregate category data. These estimates support the Japanese retailers' contention that fluid milk is a much more elastic product than previous studies have found.

The signs of e_{ij}^{cr} are both positive and the magnitude of cross-price effects for FRESH is smaller than for RECONST (2.254 and 7.186). This result means that q_i moves in the same direction as the change in p_j; in other words, a 1 percent reduction in the price of RECONST will result in a decrease in demand for FRESH by 2.254 percent and a 1 percent reduction in the price of FRESH will result in a decrease in demand for RECONST by 7.186 percent.

Table 11.2. Sample Means, Elasticities, Price Reactions, and Consistent Conjectures

		FRESH (Item 1)	RECONST (Item 2)
Mean p_i	Item i's average price (yen/liter)	168.0	191.8
Mean q_i	Item i's average quantity sold (liter/thousand customers)	115.2	15.6
e_i	Perceived own-price elasticity	–6.665	–5.608
e_i^{ow}	Partial own-price elasticity	–6.665	–9.187
e_{ij}^{cr}	Cross-price elasticity	2.254	7.186
e_{ji}^{cv}	Conjectural-variation elasticity	0.000	0.498
r_{ij}	Firm i's price reaction	0.147	0.553
cv_{ji}	Firm i's conjecture	0.000	0.437
ccv_{ji}	Consistent conjectural variation	0.479	0.159

The value of e_{ji}^{cv} for FRESH is not significantly different from zero while the value of e_{ji}^{cv} for RECONST is significant and positive (0.498). As shown earlier in Equation (9), a perceived own-price elasticity (e_i) is composed of a partial own-price elasticity (e_i^{ow}), a cross-price elasticity (e_{ij}^{cr}), and a conjectural-variation elasticity (e_{ji}^{cv}). Therefore, the absolute value of e_1 is equal to e_1^{ow} because e_{21}^{cv} is zero, while the absolute value of e_2 is smaller than e_2^{ow} because of the positive e_{12}^{cv}. The positive estimate has the interpretation that firm 2's conjecture of the effects of changes in p_2 on q_2 will be "lessened" by cross-price effects generated through firm 1's price reaction.

Thus, e_1 is larger than e_2 in absolute value while the average price for FRESH (item 1) is much lower than the price for RECONST (item 2) (168.0 and 191.8 yen/liter). This is consistent with the price-discrimination practice of charging higher prices for more price-inelastic markets. If we use the conventional partial own-price elasticity, we cannot reasonably explain this situation because e_1^{ow} is smaller than e_2^{ow} in absolute value.

The estimated conjectural variations were tested to determine whether they represent consistent pricing behavior. Recall that consistent behavior means that firm *i*'s estimated conjecture corresponds to firm *j*'s observed price reaction. Mathematically, consistent pricing behavior can be examined by testing the following two null hypotheses:

$$(12) \qquad c_1(7) = -c_2(4) \,/\, [2c_2(3) + c_2(4)c_2(7)] \text{ and}$$

$$(13) \qquad c_2(7) = -c_1(4) \,/\, [2c_1(3) + c_1(4)c_1(7)].$$

If (12) and (13) both hold, then the conjecture is consistent. The test using chi-square statistics showed that the p-value was 21.2 percent for the first null hypotheses (condition 12), and 4.8 percent for the second null hypotheses (condition 13). While the first null hypothesis could not be rejected, the second null hypothesis was rejected at the 5 percent level. Hence, consistent pricing behavior could not be statistically supported. This result implies that one should be careful concerning the assumption of consistent pricing behavior in building models.

Conclusions

In this chapter, we applied a generalized Bertrand-type model to the Japanese dairy market to gain insights into firms' pricing conduct regarding beverage-milk products made from fresh milk and reconstituted milk. This model provides more detailed insights into pricing conduct in product-differentiated oligopoly markets than previous studies that have assumed that all firms within the industry supply a homogeneous product and use quantity as a strategic variable.

Our empirical results, based on weekly scanner data, reveal that a key assumption of the original Bertrand model was violated. Specifically, the results indicated that the assumption that firms set prices independent of their rivals' reactions was rejected. That is, there is interdependency among firms in strategic price behavior. This result means that the original Bertrand model was inappropriate for our empirical analysis.

In addition, the estimated model resulted in substantially more elastic own-price elasticities than previous estimates for Japan based on aggregate market data. The more elastic results (relative to aggregate data studies) reflect the brand-level competition that is captured in our more micro-level data. These estimates support the Japanese retailers' contention that fluid milk is a more price-elastic product than previous studies had found. Incorporating estimated price conjectures into the model resulted in considerable discrepancies between partial and perceived own-price elasticities. The result shows that a firm's price conjecture can be an important factor in determining the firm's perceived own-price elasticity. Consequently, an approach using perceived (or conjectured) own-price elasticities rather than traditional partial own-price elasticities may improve analyses of firms' conduct in an oligopolistic market.

Finally, the estimated conjectural variations were tested to determine whether they represent consistent pricing behavior. Consistent behavior means that firm i's estimated conjecture corresponds to firm j's observed price reaction. We found that consistent pricing behavior could not be statistically supported. This result implies that one should be careful concerning the assumption of consistent pricing behavior in building models.

References

Appelbaum, E. "The Estimation on the Degree of Oligopoly Power." *Journal of Econometrics* 19 (1982):287–299.

Azzam, A.M., and E. Pagoulatos. "Testing Oligopolistic and Oligopsonistic Behavior: An Application to the U.S. Meat-Packing Industry." *Journal of Agricultural Economics* 41 (1990):362–370.

Bhuyan, S., and R.A. Lopez. "Oligopoly Power in the Food and Tobacco Industries." *American Journal of Agricultural Economics* 79 (1997):1035–1043.

Cotterill, R.W. "Scanner Data: New Opportunities for Demand and Competitive Strategy Analysis." *Agricultural and Resource Economics Review* 23 (1994):125–139.

Cotterill, R.W., R. Dhar, and W.P. Putsis. "Assessing the Competitive Interaction between Private Labels and National Brands." *Journal of Business* 73 (2000):1099–1137.

Cotterill, R.W., and W.P. Putsis, Jr. "Market Share and Price Setting Behavior for Private Labels and National Brands." *Review of Industrial Organization* 17 (2000):17–39.

Iwata, G. "Measurement of Conjectural Variations in Oligopoly." *Econometrica* 42 (1974):947–966.

Karp, L.S., and J.M. Perloff. "Dynamic Oligopoly in the Rice Export Market." *Review of Economics and Statistics* 71 (1989):462–470.

Kinoshita, J., N. Suzuki, T. Kawamura, Y. Watanabe, and H.M. Kaiser. "Estimating Own and Cross Brand Price Elasticities and Price-Cost Margin Ratios using Store-Level Daily Scanner Data." *Agribusiness: An International Journal* 17 (2001):515–525.

Liang, J.N. "An Empirical Conjectural Variation Model of Oligopoly." Washington DC: Federal Trade Commission Working Paper 151, 1987.

———. "Price Reaction Function and Conjectural Variations: An Application to the Breakfast Cereal Industry." *Review of Industrial Organization* 4 (1989):31–58.

Liu, D.J., C.H. Sun, and H.M. Kaiser. "Market Conduct under Government Price Intervention in the U.S. Dairy Industry." *Journal of Agricultural and Resource Economics* 20 (1995):301–315.

Lopez, R.E. "Measuring Oligopoly Power and Production Responses of the Canadian Food Processing Industry." *Journal of Agricultural Economics* 35 (1984):219–230.

Peterson, E.B., and R.W. Cotterill. *Incorporating Flexible Demand Systems in Empirical Models of Market Power.* Storrs CT: University of Connecticut Food Marketing Policy Center Research Report 43, 1998.

Schroeter, J.R. "Estimating the Degree of Market Power in the Beef Packing Industry." *Review of Economics and Statistics* 70 (1988):158–162.

Suzuki, N., J.E. Lenz, and O.D. Forker. "A Conjectural Variations Model of Reduced Japanese Milk Price Supports." *American Journal of Agricultural Economics* 75 (1993):210–218.

Vickner, S.S., and S.P. Davies. "Estimating Market Power and Pricing Conduct in a Product-Differentiated Oligopoly: The Case of the Domestic Spaghetti Sauce Industry." *Journal of Agricultural and Applied Economics* 31 (1999):1–13.

Watanabe, Y., N. Suzuki, and H.M. Kaiser. "Identifying Consumer Characteristics Associated with Japanese Preferences toward Milk Products." *Agribusiness: An International Journal* 13 (1997):357–363.

12

Market Share and Price-Setting Behavior for Private Labels and National Brands

Ronald W. Cotterill and William P. Putsis, Jr.

Competition between "national brand" and "private label" products has been a primary concern of managers in the retail food industry for some time now. More recently, understanding the different factors that influence the competitive interaction between national brands and private labels has taken on greater urgency. Much of this is due to the inroads made by private-label products in many categories. For example, private-label brands in U.S. supermarkets reached an all-time high unit market share of 20.8 percent in the third quarter of 1997 according to Information Resources, Inc. (IRI) (*BrandWeek* 11/24/97). Yet surprisingly little research has been conducted addressing private-label/national-brand interaction.

A number of studies have addressed competitive interaction more generally. Previous empirical research addressing competitive interaction has taken a variety of forms. For example, the "menu" approach requires specifying, *a priori*, the alternative forms of competitive interaction to be considered using nonnested hypothesis tests to ascertain which type of interaction best fits the data (see, for example, Roberts (1984); Gasmi, Laffont, and Vuong (1992); and Kadiyali, Vilcassim, and Chintagunta (1996)). Alternatively, conjectural-variation approaches estimate competitive interaction directly without the need to specify the form of the interaction *a priori* (see, for example, Appelbaum (1979), Putsis and Dhar (1998), and Kadiyali, Vilcassim, and Chintagunta (1998)).

This chapter originally appeared as an article titled "Market Share and Price Setting Behavior for Private Labels and National Brands" by Cotterill and Putsis in *Review of Industrial Organization* in 2000. This material is reprinted with permission of Springer Publishing, Inc.

While previous research addressing the nature of competitive interaction has produced sophisticated models on the supply side, demand specifications used in much of that research, including research on private-label/national-brand interaction, have been rather restrictive in functional form in order to derive unambiguous qualitative results or to simplify estimation. For example, theoretical models developed by Choi (1991) and by Raju, Sethuraman, and Dhar (1995a, 1995b) employ restricted versions of a linear demand model. Further, in recent theoretical work, Lee and Staelin (1997) and Cotterill, Egan, and Buckhold (1999) demonstrated that the type of vertical strategic interaction present depends, in part, on the convexity of the demand curve. Lee and Staelin concluded that a linear demand structure leads to positive vertical (within-channel) reactions, whereas nonlinear demand structures can lead to a positive, a negative, or no reaction within the channel. These concerns are not limited to vertical theoretical models. For example, Genesove and Mullin (1998) observed that new empirical IO (NEIO or new empirical industrial organization) studies "typically impose strong functional-form assumptions on demand, which . . . imply even stronger restrictions on the relationship between price and marginal cost. If these assumptions are erroneous, inferences about market power or marginal cost may be incorrect." Such studies typically assume linear or logarithmic demand with very limited specification of demand-shift factors.

We maintain that developing a complete understanding of the nature of the competitive interaction between national brands and private labels requires (1) a more complete, flexible demand-side specification, and (2) simultaneous examination of demand and strategic pricing behavior. In the next section, we describe the theoretical model that guides the empirical specification and the selection of variables. Using a flexible linear approximate almost ideal demand system (LA/AIDS), we specify corresponding price-reaction equations to analyze observed strategic price conduct in the market. Since market shares (in the LA/AIDS demand system) and prices (in the associated price-reaction equations) are the dependant variables in the empirical analysis, a simultaneous equations system is estimated that allows us to examine (1) consumer's response to firm's pricing and promotion decisions (demand elasticities), (2) the interdependence of national-brand and private-label pricing behavior, and (3) traditional IO supply-side effects of market structure, including market share. Using data for 143 food product categories and

59 geographic markets, we develop a model that captures the variation in private-label/national-brand share and pricing across categories and markets. The results of the empirical analyses are then discussed in detail, followed by a discussion of the managerial implications.

Theoretical Framework

We analyze a category-level model of a duopoly consisting of two firms, one producing a national "branded" product and the other producing a "private label" product. Both products compete in a specific geographic area with price as the sole strategic variable.[1]

Define the following set of variables:

P^1_{ij} = The price per unit volume of the national brand in category i and city j.

P^2_{ij} = The price per unit volume of the private label in category i and city j.

Q^1_{ij} = The quantity of the national brand sold in category i and city j.

Q^2_{ij} = The quantity of the private label sold in category i and city j.

D_{ij} = A vector of demand shift variables for category i and city j.

W^1_{ij} = A vector of supply-side cost-shift variables for the national brand in category i and city j.

W^2_{ij} = A vector of supply-side cost-shift variables for the private label in category i and city j.

---

1 We have not ignored channel issues but we do not explicitly focus on them in this chapter. We have made some simplifying assumptions regarding the strategic interaction between manufacturer and retailer. In the framework presented, imagine that the retailer controls the private-label retail price and that the manufacturer controls the national-brand net retail price through a variety of measures such as trade promotions and on-pack prices. Thus, the price-reaction system developed is derived from retailer and manufacturer profit maximization in a Bertrand (price) differentiated oligopoly. An equivalent specification would have brand and private-label manufacturers setting price with the retailer applying a fixed markup rule (Kadiyali, Vilcassim, and Chintagunta 1998).

Define the demand functions for national-brand and private-label products as

(1) $Q^1_{ij} = Q^1(P^1_{ij}, P^2_{ij}; D_{ij})$.

(2) $Q^2_{ij} = Q^2(P^1_{ij}, P^2_{ij}; D_{ij})$.

The quantities of the national-brand and private-label products demanded are a function of their prices and, following Hoch et al. (1995), a set of demand-shift variables that include per capita expenditures in category *i* and city *j*, the family income level in city *j* (reflecting the overall affluence of the geographic area), the percent of the population that is Hispanic in city *j*, and the median family age in city *j*. Following standard demand theory (Deaton and Muellbauer 1980a), we hypothesize that the own (cross) price is negatively (positively) related to quantity. Similarly, define the cost functions for national-brand and private-label products as

(3) $C^1(Q^1_{ij}, W^1_{ij})$.

(4) $C^2(Q^2_{ij}, W^2_{ij})$.

The total cost of producing Q^1_{ij} is a function of Q^1_{ij} and a vector of supply-side cost-shift variables. We use a vector of cost-shift variables (as in Gasmi, Laffont, and Vuong (1992)) that includes a measure of package size to capture the hypothesis that smaller package sizes have higher costs per pound.

In a Bertrand game where price is the strategic variable, the profit-maximizing problems for the two firms are:

(5) $\text{MAX}\ \Pi_1 = [P^1_{ij}\, Q^1(P^1_{ij}, P^2_{ij}; D_{ij}) - C^1(Q^1(P^1_{ij}, P^2_{ij}; D_{ij}), W^1_{ij})]$.

(6) $\text{MAX}\ \Pi_2 = [P^2_{ij}\, Q^2(P^1_{ij}, P^2_{ij}; D_{ij}) - C^2(Q^2(P^1_{ij}, P^2_{ij}; D_{ij}), W^2_{ij})]$.

The first-order conditions for these maximization problems result in the following two equations:

(7) $\partial\Pi_1 / \partial P^1_{ij} = g(P^1_{ij}, P^2_{ij}, D_{ij}, W^1_{ij}) = 0$ and

(8) $\partial \Pi_2 / \partial P_{ij}^2 = h(P_{ij}^1, P_{ij}^2, D_{ij}, W_{ij}^2) = 0.$

From the first-order condition for national brands, we can solve for the national-brand price P_{ij}^1 as a function of the other variables:

(9) $P_{ij}^1 = R_1(P_{ij}^2, D_{ij}, W_{ij}^1).$

Similarly, using the first-order condition for the private-label price, P_{ij}^2, one obtains

(10) $P_{ij}^2 = R_2(P_{ij}^1, D_{ij}, W_{ij}^2).$

Note that the oligopoly profit-maximization problem can generally be stated as a set of price-reaction curves (Martin 1993) and that this result is independent of the exact functional forms specified. Each price-reaction equation gives a firm's profit-maximizing price as a function of the other firm's price, exogenous demand, and its own cost-shift variables. We hypothesize that the price-reaction curves have positive slopes with regard to the other firm's price (Deneckere and Davidson 1985) because differentiated products are usually strategic complements in price (Tirole 1989).

This model of national-brand/private-label interaction has four equations (two demand equations and two price-reaction equations) and four endogenous variables (two quantities and two prices). Since we also have a set of exogenous variables and this system is, in general, identified, we can use simultaneous-equation estimation techniques to estimate parameters and test hypotheses.

Choice of Functional Form

Demand analysis and functional-form specification have been well developed in the extant literature. Numerous forms have been proposed that are theoretically superior to the linear or logarithmic specification, including the linear approximate almost ideal demand system or LA/AIDS (Deaton and Muellbauer 1980b). The LA/AIDS system is preferable for a number of reasons—it is derived from the underlying choice axioms in utility theory, individual behavior can be aggregated to consistently estimate demand parameters from market-level data, it gives a first-order approximation to any "true" demand-system functional form, and it is sufficiently flexible so as not to unduly constrain channel behavior and empirical estimation of market power.[2]

$$(11) \qquad S^1_{ij} = \alpha_{10} + \alpha_{11} \ln P^1_{ij} + \alpha_{12} \ln P^2_{ij} + \alpha_{13} \ln (E_{ij} / \bar{P}_{ij}) + \alpha_{14} D_{ij}$$

where: S^1_{ij} = the dollar market share of the national brand in category i and city j,

E_{ij} = total per capita expenditure on category i in city j, and

$\bar{P}_{ij}$ = Stone's price index, which is equal to $S^1_{ij} \ln P^1_{ij} + S^2_{ij} \ln P^2_{ij}$.

The ratio of per capita expenditure and Stone's price index is a deflated (real) measure of per capita expenditures. Use of Stone's price index to purge expenditures of price effects gives the "linear approximate" AIDS model and allows linear estimation of the demand system (Deaton and Muellbauer 1980a). We also assume that deflated real expenditures are exogenous and, thus, the coefficient α_{13} gives an estimate of the impact of changes in real expenditures on demand.

Given LA/AIDS demands in (11) and each firm's profit maximization (5 and 6), it is not possible to derive the exact closed-form solution for the price-reaction curves (see Choi (1991) for a discussion of related issues). However, we know that reaction functions exist. Choi (1991) used simulation to explore the theoretical properties of channel models with nonlinear demand forms that did not have closed-form solutions. Here, we examine the empirical properties of one such model by using a logarithmic first-order (Taylor series) approximationo to the reaction functions to produce estimable supply-side relations. While the use of

2 For this last point, see Cotterill, Egan, and Buckhold (1999). The LA/AIDS aggregation properties are especially important for our purposes. First, we note that the LA/AIDS is PIGLOG in form, which does not require the assumption of parallel linear Engel curves. This implies that we are able to consistently estimate expenditure effects using linearly aggregated data. In addition, under the assumption that prices change proportionately from period to period across retailers, taking the first difference of an LA/AIDS demand equation eliminates any linear response aggregation bias (Christen et al. 1997). This point will be discussed further in the discussion section.

such approximations is not uncommon in econometric analysis (e.g., the nonnested P-E test proposed by Davidson and MacKinnon (1981) relies heavily on the same Taylor-series approximation), this does mean that we cannot impose cross-equation constraints between the demand and reaction equation parameters. The implication is that we cannot estimate the conjectural variations that are embedded in the reaction coefficients, but we have no interest in them per se. We are interested in observed price conduct, and for that, estimates of the price-reaction elasticities are sufficient. They enable us to discern how each price reacts to changes in other prices, a focal point of our analysis of strategic pricing behavior. Moreover, we can solve our system for estimates of price-cost margins (Lerner Indices) and thus produce the same measures of market power that a conjectural-variation approach produces.[3] Thus, the loss in analytical detail is minimal when we estimate a more flexible demand system.

Using a Taylor-series expansion to obtain a linear approximate retail-reaction function (analogous to and consistent with the LA/AIDS specification) produces the derived price-reaction equations specified in Equation (12):

$$\text{(12)} \qquad \ln P^1_{ij} = \beta_{10} + \beta_{11} \ln P^2_{ij} + \beta_{12} D_{ij} + \beta_{13} \ln W^1_{ij}.$$

Taking the antilog of Equation 12 gives the price-reaction function for national brands:

$$\text{(13)} \qquad P^1_{ij} = e^{\beta_{10}} \bullet (P^2_{ij})^{\beta_{11}} \bullet e^{\beta_{12} D_{ij}} \bullet (W^1_{ij})^{\beta_{13}}.$$

Note that the slope of the price-reaction function depends on the values of the exogenous variables in the system and their parameter estimates. Since all of these variables are positive, the slope of the reaction is positive if and only if β_{11} is positive. Finally, as explained hereafter, we will also generalize this model to include the impact of market structure and promotion effects on profit-maximizing prices.

---

3 Since we obtain an estimate of each product's own-price elasticity of demand and since a profit-maximizing manager equates margins to the inverse of that elasticity, one can recover the Lerner Index for each product.

Empirical Framework and Model Selection

We develop three empirical models, specifying the parsimonious model derived before as a nested special case of two broader models. The objective is to build on the preceding model in a way that (1) is consistent with theory and (2) allows us to use nested hypothesis tests to select the "best" overall model given the available data. We begin by presenting the most parsimonious model (represented by the preceding derivation), referring to this as the "price model." We then expand this "base" specification by including national-brand market share, a brand-level Herfindahl, and a measure of local retail grocery concentration in the price-reaction functions, thereby creating a fully simultaneous system. We refer to this as the "market structure model." Finally, we expand this to allow for the possibility that the promotion variables impact both the demand and the supply sides. We refer to this third model as the "full power and promotion model."

Empirical Specification I – Base Specification: The Price Model

The parsimonious four-equation price model implied directly by the preceding derivation is as follows (variable definitions for all three specifications can be found in Table 12.1):

$$
\begin{aligned}
\mathit{BRSHARE} = {} & \alpha_{10} + \alpha_{11}\,\mathit{BRPRICE} + \alpha_{12}\,\mathit{PLPRICE} \\
& + \alpha_{13}\,\mathit{EXPENDITURE} + \alpha_{16}\,\mathit{PLDISTN} \\
& + \alpha_{19}\,\mathit{INCOME} + \alpha_{110}\,\mathit{HISPANIC} \\
& + \alpha_{111}\,\mathit{FAMAGE} + \varepsilon_1 \\
\mathit{PLSHARE} = {} & \alpha_{20} + \alpha_{21}\,\mathit{BRPRICE} + \alpha_{22}\,\mathit{PLPRICE} \\
& + \alpha_{23}\,\mathit{EXPENDITURE} + \alpha_{26}\,\mathit{PLDISTN} \\
& + \alpha_{29}\,\mathit{INCOME} + \alpha_{210}\,\mathit{HISPANIC} \\
& + \alpha_{211}\,\mathit{FAMAGE} + \varepsilon_2 \\
\mathit{BRPRICE} = {} & \beta_{10} + \beta_{11}\,\mathit{PLPRICE} + \beta_{12}\,\mathit{BRPRICEREDN} \\
& + \beta_{13}\,\mathit{BRVOLPUN} + \beta_{17}\,\mathit{EXPENDITURE} \\
& + \beta_{112}\,\mathit{PLDISTN} + \beta_{113}\,\mathit{INCOME} + \beta_{114}\,\mathit{HISPANIC} \\
& + \beta_{115}\,\mathit{FAMAGE} + \omega_1 \\
\mathit{PLPRICE} = {} & \beta_{20} + \beta_{21}\,\mathit{BRPRICE} + \beta_{22}\,\mathit{PLPRICEREDN} \\
& + \beta_{23}\,\mathit{PLVOLPUN} + \beta_{27}\,\mathit{EXPENDITURE} \\
& + \beta_{212}\,\mathit{PLDISTN} + \beta_{213}\,\mathit{INCOME} \\
& + \beta_{214}\,\mathit{HISPANIC} + \beta_{215}\,\mathit{FAMAGE} + \omega_2
\end{aligned}
\tag{14}
$$

Table 12.1. Definitions for Variables Used in the Analysis

BRSHARE	Aggregate share of category expenditure for national brands in the *i*th market, *j*th category.
PLSHARE	Aggregate share of category expenditure for private-label products in the *i*th market, *j*th category.
BRPRICE	Natural log of the price of the national brand in the *i*th market, *j*th category.
PLPRICE	Natural log of the price of the private-label product in the *i*th market, *j*th category.
EXPENDITURE	Natural log of per capita category expenditures deflated by Stone's price index.
BRFEATURE	Percent of national-brand volume with feature advertising in the *i*th market, *j*th category.
BRDISPLAY	Percent of national-brand volume with displays and point-of-sale promotion in the *i*th market, *j*th category.
PLFEATURE	Percent of private-label product volume with feature advertising in the *i*th market, *j*th category.
PLDISPLAY	Percent of private-label product volume with displays and point-of-sale promotion in the *i*th market, *j*th category.
INCOME	Average household income in the local market.
HISPANIC	Percent of population in the local market of Hispanic descent.
AGE	Average age of the local market population.
PLDISTN	Private-label percent average distribution in the *i*th market, *j*th category.
BRPRICEREDN	Weighted percent average price reduction for national brands in the *i*th market, *j*th category.
PLPRICEREDN	Weighted percent average price reduction for private-label products in the *i*th market, *j*th category.
BRVOLPUN	Natural log of average volume (weight) per package unit sold for the national brand.
PLVOLPUN	Natural log of average volume (weight) per package unit sold for the private label.
HERFINDAHL	Herfindahl Index of brand concentration in the *i*th market, *j*th category.
SUPERCR4	Percentage of all supermarket sales by the top four chains in the *i*th market, *j*th category.

Note: The choice of variables was influenced by data availability. For example, no coupon or advertising information was available. Also, average age, income, and percent Hispanic were the only local demographic variables available. Also note that the missing numbers in the preceding coefficient sequences (α_{10}, α_{11}, α_{12}, α_{13}, α_{18}, . . .) reflect the fact that the price model is a restricted version of the more general models presented hereafter.

Empirical Specification II – The Market-Structure Model

The estimated own-price demand elasticities in the price model are a direct measure of a manager's ability to elevate price above marginal cost. However, we can extend this framework to provide additional insights on the sources of market power. Deneckere and Davidson (1985) and Willig (1991) demonstrated that market share is positively related to price even in Bertrand models (i.e., models that assume zero price conjectures), suggesting that larger-share brands are able to unilaterally raise prices. Liang (1987), using a linear demand model, demonstrated the general proposition that the slope of the price-reaction curve is positively related to a firm's price conjecture. Here, such conjectures, following Cowling and Waterson (1976) and Kwoka and Ravenscraft (1986), are hypothesized to be functions of observable (but endogenous) market structure. Consequently, we specify national-brand market share and the brand-level Herfindahl Index in each logarithmic price-reaction equation to capture the traditional market-structure effect of changes in oligopolistic interdependence on profit-maximizing prices. National-brand market share is hypothesized to have a positive impact on national-brand price—as national-brand share increases, the market power of national brands also increases due to reduced private-label presence in the category.

Private-label price is also hypothesized to increase when brand share increases. In differentiated product models where price is the strategic variable, an increase in a product's price generally leads to increases in the prices of competing products (Deneckere and Davidson 1985; Levy and Reitzes 1993). Specifying national-brand share in the price-reaction equation makes the model fully simultaneous as opposed to recursive in nature. Now, national-brand price can have a negative (demand relationship) in the brand demand equation. However, a reverse relationship (brand share positively related to prices) is hypothesized to occur in the price-reaction curves, capturing the impact of more dominant national brands.

The brand-level Herfindahl Index is defined as the sum of the square of all individual brand market shares.[4] As such, when introduced jointly

---

4 Note that the brand Herfindahl Index is not the company Herfindahl Index. The brand-level Herfindahl in breakfast cereal, for example, is very low but the company-level Herfindahl is very high because each of the top three companies sells many brands. Company Herfindahls were not available for inclusion in this study.

with national-brand market share, it measures the size dispersion of brands. There are two possible impacts on price. First, a fragmented brand structure may have less pricing power than one with large dominant brands. This suggests that the brand-level Herfindahl would be negatively related to the prices of both national brands and private labels (Putsis 1997). Alternatively, Schmalensee's (1978) analysis of brand proliferation as a barrier to entry suggests that the impact of the brand Herfindahl on private-label prices may be positive. As incumbent firms in these markets build portfolios of brands with small shares, it becomes harder for potential entrants, including private-label entrants, to enter with a "me too" brand. For example, many successful children's breakfast cereal brands capture only 0.006 (0.6 percent) of the cereal market. A private label can at best capture one-third of this. The resulting volume is not sufficient to sustain production and distribution. Previous research has shown that segmentation and multiple-brand strategies in a category can elevate the prices of all brands (Willig 1991; Levy and Reitzes 1993; Werden and Rozanski 1994).

In our empirical analysis, the brand-level Herfindahl is specified as endogenous. We instrument for it following the methodology used by Nevo (1998). Specifically, an instrument is created by forming a weighted average of previous-period Herfindahl Indices (the weights created by time-series regressions). This instrument is correlated with the current-period Herfindahl at more than 90 percent but it is predetermined in the current period.

Finally, we also specify the retail-supermarket four-firm concentration ratio in the price-reaction curves to capture the increased oligopolistic interdependence in cities where a few supermarket chains account for most of the sales. Both national-brand and private-label prices are hypothesized to be higher in these markets. Since supermarket concentration is for sales of all food and nonfood products, it is exogenous when analyzing data at the product-category level (e.g., cookies).

The second extension of the price model addressed in the market-structure model is inclusion of trade-promotion variables in the price-reaction functions. These promotion variables are short-term percent price reduction, percent of volume sold on display, and percent of volume sold with a local newspaper feature ad. While one could model these as additional strategic variables to create a multidimensional game, this would generate six more reaction equations and prevent estimation of the system due to insufficient exogenous cost-shift variables in those equations to identify them. We specify these variables as endogenous strategic factors that each duopolist uses to determine price levels.[5] Since IRI-reported prices are net of promotional price reductions, the level of price reduction is clearly one determinant of reported price. Thus, the percent price reductions for national brands (private labels) are specified in the national-brand (private-label) price-reaction equation. Similarly, display and feature programs are strongly tied to shelf price-reduction strategies and may affect demand primarily via communicating changes in retail prices. This formulation implies specifying the national-brand feature and display variables only in the national-brand price-reaction equation and the private-label feature and display variables only in the private-label price-reaction equation.[6]

Empirical Specification III – The Full Power and Promotion Model

A third model, full power and promotion, specifies the four promotion variables (national-brand feature, national-brand display, private-label feature, and private-label display) in each demand equation as well as in the price-reaction equations. This allows the promotion variables to have a direct share-expanding effect as well as an indirect effect via communicating price changes for specific products in a crowded supermarket. This implies the following most general specification:

---

5 We address the endogeneity of the trade-promotion variables through the use of instrumental variables. The principle is similar to the approach taken by Berry, Levinsohn, and Pakes (1995). Specifically, each promotional vehicle for market i, category j, is expressed as a function of the promotional activity in each of the other j ($j \neq i$) markets using the fitted value as the instrument. Note that for this approach to eliminate the endogeneity bias, the equation errors for each promotion instrument have to be independent. This implies that display and feature decisions, for example, are made on a market-by-market (including chain-by-chain in a market) basis.

6 In the interest of brevity, we do not specify the exact equation for the market-structure model here. The market-structure model restricts α_{14}, α_{15}, α_{17}, α_{18}, α_{24}, α_{25}, α_{27}, α_{28}, β_{110}, β_{111}, β_{28}, and β_{29} to zero in the full power and promotion model specified in Equation (15) (note that this is the restriction tested in the nested hypothesis test employed).

Table 12.2. A Summary of Maintained Hypotheses

Hypothesis	Rationale
H1: $\alpha_{11}, \alpha_{22} < 0; \alpha_{12}, \alpha_{21} > 0$	Standard economic theory predicts negative own-price elasticities and positive cross-price elasticities for substitute goods. Further, effects should be asymmetric (Blattberg and Wisniewski 1989).
H2: $\alpha_{14}, \alpha_{15}, \alpha_{27}, \alpha_{28} > 0;$ $\alpha_{17}, \alpha_{18}, \alpha_{24}, \alpha_{25} < 0$	Increased own promotions have a positive impact on own sales and a negative impact on the rival's sales.
H3: $\alpha_{19} > 0; \alpha_{29} < 0$	As per capita income in a market increases, we expect that the national-brand share increases and the private-label share decreases.
H4: $\alpha_{16} < 0; \alpha_{26} > 0;$ $\beta_{112} < 0; \beta_{212} > 0$	As more supermarkets in a local market carry private labels (increased private-label distribution), the share and price of national brands decrease (due to increased competition) and the share and price of private labels increase.
H5: $\beta_{11}, \beta_{21} > 0$	The slopes of the price-reaction curves are positive (Deneckere and Davidson 1985).
H6: $\beta_{13}, \beta_{23} < 0$	Increasing average package size lowers cost, thereby lowering market price.
H7: $\beta_{14} > 0; \beta_{24} > 0$	Increasing national-brand share increases the market power of branded products, increasing the ability of national-brand manufacturers to raise prices. Private-label prices also increase (Deneckere and Davidson 1985; Levy and Reitzes 1993; Werden and Rozanski 1994; Haller and Cotterill 1996).
H8: $\beta_{16}, \beta_{26} > 0$	Increases in grocery-firm local market concentration increase prices due to increased market power at the retail level (Marion 1979; Weiss 1989; Cotterill 1999a).

$$
\begin{aligned}
BRSHARE = {} & \alpha_{10} + \alpha_{11}\,BRPRICE + \alpha_{12}\,PLPRICE + \\
& \alpha_{13}\,EXPENDITURE + \alpha_{14}\,BRFEATURE + \\
& \alpha_{15}\,BRDISPLAY + \alpha_{16}\,PLDISTN + \\
& \alpha_{17}\,PLFEATURE + \alpha_{18}\,PLDISPLAY + \\
& \alpha_{19}\,INCOME + \alpha_{110}\,HISPANIC + \\
& \alpha_{111}\,FAMAGE + \varepsilon_1 \\
PLSHARE = {} & \alpha_{20} + \alpha_{21}\,BRPRICE + \alpha_{22}\,PLPRICE + \\
& \alpha_{23}\,EXPENDITURE + \alpha_{24}\,BRFEATURE + \\
& \alpha_{25}\,BRDISPLAY + \alpha_{26}\,PLDISTN + \\
& \alpha_{27}\,PLFEATURE + \alpha_{28}\,PLDISPLAY + \\
& \alpha_{29}\,INCOME + \alpha_{210}\,HISPANIC + \\
& \alpha_{211}\,FAMAGE + \varepsilon_2
\end{aligned}
$$

(15)

$$
\begin{aligned}
BRPRICE = {} & \beta_{10} + \beta_{11}\,PLPRICE + \beta_{12}\,BRPRICEREDN + \beta_{13} \\
& BRVOLPUN + \beta_{14}\,BRSHARE + \beta_{15}\,HERFINDAHL + \\
& \beta_{16}\,SUPERCR4 + \beta_{17}\,EXPENDITURE + \\
& \beta_{18}\,BRFEATURE + \beta_{19}\,BRDISPLAY + \\
& \beta_{110}\,PLFEATURE + \beta_{111}\,PLDISPLAY + \\
& \beta_{112}\,PLDISTN + \beta_{113}\,INCOME + \\
& \beta_{114}\,HISPANIC + \beta_{115}\,FAMAGE + \omega_1 \\
PLPRICE = {} & \beta_{20} + \beta_{21}\,BRPRICE + \beta_{22}\,PLPRICEREDN + \\
& \beta_{23}\,PLVOLPUN + \beta_{24}\,BRSHARE + \\
& \beta_{25}\,HERFINDAHL + \beta_{26}\,SUPERCR4 + \\
& \beta_{27}\,EXPENDITURE + \beta_{28}\,BRFEATURE + \\
& \beta_{29}\,BRDISPLAY + \beta_{210}\,PLFEATURE + \\
& \beta_{211}\,PLDISPLAY + \beta_{212}\,PLDISTN + \\
& \beta_{213}\,INCOME + \beta_{214}\,HISPANIC + \\
& \beta_{215}\,FAMAGE + \omega_2
\end{aligned}
$$

Note that our nine hypotheses are not all-inclusive. A few of the coefficient signs are not predicted by our theory. Examples include the effects of age, deflated per capita category expenditures, and the percent Hispanic on own-market shares.

Empirical Estimation

The data used in this study are annual IRI market-level data on food products across 59 geographic markets and 211 categories for 1991 and 1992. Categories were excluded from the analysis if they lacked data or if they were categories for which private labels have not been introduced. This left 143 categories in the sample and 7,197 observations for an average coverage of 50 out of 59 possible cities for a typical category. National-brand share averaged 0.783 while the private-label share was 0.217 in 1992.

These data are merged with independent data from Progressive Grocer on the demographic characteristics of the IRI geographic markets. Thus, we have two principal dimensions on which the data vary—across categories and across geographic markets. Consistent with previous work on private labels (e.g., Sethuraman and Mittelstaedt 1992; Hoch and Banerji 1993; Slade 1995), aggregate national-brand and private-label variables were created for the 143 product categories and 59 markets. Brand-price, feature, display, and price-reduction variables are volume as opposed to dollar market-share weighted averages. IRI reports corresponding aggregate private-label variables for all categories and local markets.

In estimating cross-category price equations, it is important to note that cross-category analysis precludes the use of price levels. Following Kelton and Weiss (1989), we estimated the first-difference form of our model. The parameters estimated in the first-difference model are identical to those in Equations 18 through 20 and thus can be used to compute elasticities. In the following sections, all reported estimates use the annual difference rather than the level of the variable for 1991 to 1992. For example, *BRSHARE* is 1992 *BRSHARE* minus 1991 *BRSHARE* and *BRPRICE* is 1992 ln *BRPRICE* minus 1991 ln *BRPRICE*. Changes in the natural logarithm of price from 1991 to 1992 are percent changes, which can be analyzed across categories.

Estimating a first-difference model is also attractive because it controls for first-order fixed effects due to excluded local-market and category variables in level regressions.[7] Further, to the extent that private-label quality is constant from 1991 to 1992, estimating a first-difference model eliminates the need for inclusion of a category private-label quality measure—an assumed constant level of quality drops out of the analysis when we difference. This is particularly important since quality measurement is such a difficult task (Hoch and Banerji 1993; Narasimhan and Wilcox 1998).

Although our model has four equations, one of the demand equations is redundant for estimation purposes. Since the market shares of national brands and private labels sum to one, any loss of national-brand share due to changes in any variable (e.g., private-label price) must go to private-label share. This general adding-up property of a demand system means that we can recover the estimated coefficients and standard errors (t-ratios) for the dropped equation. We drop the private-label demand equation and estimate the remaining three equations with three-stage least squares. We do not impose the homogeneity and symmetry restrictions of demand theory because they would restrict the four own-price and cross-price coefficients to a common value in this aggregate two-good demand system.

Results

Results are reported in Tables 12.3 through 12.5. Since traditional R^2 measures are not bounded between zero and one in three-stage least squares, Carter and Nagar's (1977) multiple-squared coefficient of correlation for simultaneous systems, $R_w{}^2$, is used here and reported in Tables 12.3 and 12.4.[8] Since the model structure represented by the three models was nested, zero parameter restrictions were tested via an analog of the likelihood-ratio test (Gallant and Jorgenson 1979; Kiviet 1985; Judge 1985). We

---

7 Hausman and Taylor (1981) argue that excluded local-market variables in panel data of this type can bias estimation results for level regressions. They show that this can be avoided by specifying a set of city binary variables. These drop out of the model when one takes the first difference. This is also true for specifying a set of category binary variables in level regressions to control for excluded variables in individual categories.

8 $R_w{}^2$ has a usual R^2 interpretation. Specifically, it measures the percent of systemwide variation in the endogenous variables explained by all independent variables in the system. It is bounded by zero and one. However, we note that this statistic is frequently very high and should be interpreted with caution (Berndt 1991). Further, collinearity may inflate the estimates of Rw^2 throughout.

tested two sets of zero parameter restrictions—the first set of restrictions was α_{14}, α_{15}, α_{17}, α_{18}, α_{24}, α_{25}, α_{27}, α_{28}, β_{110}, β_{111}, β_{28}, and β_{29} all equal zero (for the market-structure model) and the second set added that β_{14}, β_{15}, β_{16}, β_{18}, β_{19}, β_{24}, β_{25}, β_{26}, β_{28}, and β_{29} all equal zero (for the price model). The test statistic T^o, distributed as a chi square (see Gallant and Jorgenson (1979)), leads us to strongly reject both sets of zero parameter restrictions.

Since the inclusion of all market-power and promotion variables is supported by the nested hypotheses tests, we discuss only the full power and promotion model presented in Table 12.4 (for comparison purposes, we do report the estimation results for the price model in Table 12.3). In the interest of brevity, we do not report the results for the market-structure model (these are available upon request from the authors). Note that parameters and t-ratios are recovered for the private-label demand equations as well.[9]

Earlier, in Table 12.2, nine hypotheses were presented with a total of twenty-six predicted signs. All twenty-six of the estimated coefficients were of the predicted signs with twenty-four of the twenty-six coefficients significant at the 5 percent level or better. We now turn to a discussion of these hypotheses.

Price and Expenditure Effects

The own- and cross-price coefficients in the demand equations (**H1**) have the hypothesized signs and are significant at the 1 percent level. The direction and significance levels of the expenditure effects indicate that national brands are viewed as luxuries and private labels as necessities (in an economic sense). As expenditures on a category increase, more goes to national brands than to private labels. This is consistent with recent work on category expenditure (Putsis 1999; Putsis and Dhar 1998). Increases in household income (**H3**) behave like increases in category per capita expenditures—income gains significantly increase (decrease) the national-brand (private-label) share. Private-label distribution measures

9 The adding-up condition of the demand system requires that all coefficients on a particular variable sum to zero. With only two demand equations in the system, the recovered private-label coefficients and t-ratios are the negative of the corresponding national-brand values. For example, if a unit increase in the national-brand price reduces national-brand market share by 5 percent, the private-label market share must fall by a corresponding percentage.

Table 12.3. Estimation Results for the Price Model

	Demand Equations				Price-Reaction Equations			
	Branded Share		Private-Label Share		Branded Price		Private-Label Price	
Branded Price	–0.0144	(–3.699)**	0.0144	(3.699)**			0.0831	(9.332)**
Private-Label Price	0.0081	(2.497)*	–0.0081	(–2.497)*	0.0793	(15.41)**		
Branded Price Reduction					–0.1964	(–11.65)**		
Private-Label Price Reduction							–0.2815	(–15.65)**
Branded Volume/Unit					–0.8856	(151.1)**		
Private-Label Volume/Unit							–0.9101	(–126.1)**
Expenditure	0.0577	(16.17)**	–0.0577	(–16.17)**	0.2229	(35.60)**	0.1297	(15.02)**
Private-Label Distribution	–0.1898	(–35.60)**	0.1898	(35.60)**	0.0479	(5.264)**	–0.0292	(–2.270)*
Income	0.0223	(3.172)**	–0.0223	(–3.172)**	0.1218	(10.11)**	0.0450	(2.644)**
Hispanic	0.0207	(0.284)	–0.0207	(–0.284)	–0.3974	(–3.203)**	0.0010	(0.006)
Family Age	–0.0006	(–0.036)	0.0006	(0.036)	–0.1932	(–7.395)**	–0.0630	(–1.705)

Notes: ** Significant at the 1 percent level. * Significant at the 5 percent level. Number of Observations = 6,717; $R_w^2 = 0.755$.

Table 12.4. Estimation Results for the Full Power and Promotion Model

	Demand Equations				Price-Reaction Equations			
	Branded Share		Private-Label Share		Branded Price		Private-Label Price	
Branded Price	–0.0083	(–2.163)**	0.0083	(2.163)**			0.1140	(4.997)**
Private-Label Price	0.0087	(2.736)**	–0.0087	(–2.736)**	0.0671	(8.874)**		
Branded Price Reduction					–0.1057	(–6.084)**		
Private-Label Price Reduction							–0.2535	(–6.655)**
Branded Volume/Unit					–0.8768	(–112.7)**		
Private-Label Volume/Unit							–0.8712	(–40.86)**
Branded Share					0.9835	(1.393)	0.4636	(2.052)*
Brand Herfindahl					0.0008	(0.062)	–0.0270	(–1.043)
Grocery CR4					0.0439	(3.185)**	0.0545	(2.653)**
Expenditure	0.0544	(15.64)**	–0.0544	(–15.64)**	0.1796	(4.790)**	–0.1271	(–1.002)
Branded Feature	0.1128	(6.491)**	–0.1128	(–6.491)**	–0.3495	(–3.999)**	–0.5327	(–2.071)*
Branded Display	0.1580	(13.38)**	–0.1580	(–13.38)**	–0.6080	(–5.222)**	–0.7715	(–2.182)*
Private-Label Feature	–0.0241	(–2.277)*	0.0241	(2.277)*	0.0381	(1.504)	–0.0455	(–0.608)
Private-Label Display	–0.0869	(–12.50)**	0.0869	(12.50)**	0.0784	(1.260)	0.1498	(0.723)
Private-Label Distribution	–0.1879	(–36.24)**	0.1879	(36.24)**	0.2317	(1.742)	0.8515	(2.005)*
Income	0.0207	(3.004)**	–0.0207	(–3.004)**	0.1111	(6.126)**	–0.0495	(–0.971)
Hispanic	0.0035	(0.050)	–0.0035	(–0.050)	–0.5160	(–3.933)**	–0.1615	(–0.649)
Family Age	–0.0140	(–0.940)	0.0140	(0.940)	–0.1794	(–6.171)**	–0.0342	(–0.539)

Notes: ** Significant at the 1 percent level. * Significant at the 5 percent level. Number of Observations = 6,717; $R_w^2 = 0.874$.

the proportion of supermarkets that sell private labels in a given market. As expected (**H4**), private-label penetration has a strongly significant negative (positive) relation to national-brand (private-label) share.

Each price-reaction equation has a positive and significant slope with regard to the other price, as hypothesized (**H5**). The estimated price-reaction elasticities, however, are not high. A 1 percent increase in the private-label price elicits only a 0.0671 percent increase in the national-brand price while a 1 percent increase in the national-brand price elicits only a 0.1140 percent increase in the private-label price. Price followship between private labels and national brands is positive, as hypothesized, but it is not strong. The volume-per-unit (**H6**) variables behave as hypothesized and are highly significant.

Market Structure and Promotion Effects

In terms of the market-structure variables in the price-reaction equations, an increase in national-brand share has a positive (but not significant) impact on national-brand prices and a positive significant impact on private-label prices (**H7**). The estimated coefficient for national-brand price is more than twice the coefficient for private-label price. These results provide modest support for the proposition that categories dominated by national brands have higher prices for both national brands and for private labels.

The brand-level Hirschman Herfindahl Index has no significant impact on brand or private-label prices. Perhaps our two competing hypotheses cancel each other in this cross-section data set. Categories that are prone to product proliferation, such as cereal, and categories that are prone to large-share leading brands, such as ketchup, may both have high prices. Supermarket four-firm concentration (**H8**) has the hypothesized market-power effect—increases in the share of the top four supermarkets in a city elevate both national-brand and private-label prices.

The four feature and display variables are highly significant in the demand equations, as expected (**H2**). National-brand displays and features strongly increase the national-brand share and decrease the private-label share. Private-label features and displays have the same expected effect for private labels. The results suggest that the promotion share effects are asymmetric. National-brand promotions have a greater effect on national-brand share than private-label promotions have on private-label share.

In the national-brand price-reaction equation, national-brand features and displays have a strong negative estimated coefficient. Feature advertising and point-of-sale displays occur to advertise price cuts. However, private-label displays and features have an interesting and opposite effect in the national-brand price equation. When private-label displays and feature ads are active, national-brand prices are *higher*. Retailers promote private labels when national-brand prices are high. Recent experience in the breakfast cereal industry is consistent with this strategy (Gejdenson and Schumer 1995, 1996; Angrisani 1996; Cotterill 1999b).

In the private-label price-reaction equation, however, we see a different competitive response. When national-brand promotion is in place, private labels respond by lowering price. However, as we discuss later, an examination of the relevant price elasticities suggests that such a strategy may be ineffective—private labels have an exceedingly difficult time stealing share from national brands on the basis of a price cut (estimated cross-price elasticity of 0.0019).

Demand Elasticities and Convergence with Previous Research

From the LA/AIDS specification, we can recover the demand (quantity) price elasticities from the share equations within the LA/AIDS framework for both private-label and national-brand products.[10] Table 12.5 presents estimated demand elasticities. It was reassuring to note that the estimated elasticities are extremely robust with respect to specification. Although the market-power and promotion variables add to the explanatory power of the model, the estimated demand elasticities are almost identical in the full and restricted models.

The estimated own-price elasticities for national-brand and private-label products cluster around unitary elasticity on the lower end of the elasticities reported by Tellis (1988). Our results suggest that national brands and private labels, as groups, are maximizing the revenue from their sales. If marginal costs are constant, then revenue maximization would be consistent with profit maximization. Since a unitary elasticity implies zero marginal costs and a 100 percent price-cost margin, these low elasticities suggest that national brands and private labels in these

---

10 See Green and Alston (1990) and/or Cotterill, Putsis, and Dhar (1999) for additional detail.

Table 12.5. Estimated Demand Elasticities

	Price Model		Full Power and Promotion Model	
	Branded Quantity	**Private-Label Quantity**	**Branded Quantity**	**Private-Label Quantity**
Branded Price	−1.0762 (−174.18)**	0.2633 (12.338)**	−1.0652 (−173.3)**	0.2250 (10.599)**
Private-Label Price	−0.0062 (−1.425)	−0.9785 (−64.757)**	0.0019 (−0.0046)	−0.9842 (−67.064)**
Expenditure	1.0745 (233.38)**	0.7429 (46.72)**	−1.0702 (238.53)**	0.7576 (48.899)**
Branded Feature			0.0092 (6.491)**	−0.0319 (−6.491)**
Branded Display			0.0226 (13.38)**	−0.0781 (−13.377)**
Private-Label Feature			−0.0018 (−2.276)*	0.0063 (2.276)*
Private-Label Display			−0.0129 (−12.50)**	0.0445 (12.50)**
Private-Label Distribution	−0.1920 (−35.60)**	0.6630 (35.60)**	−0.1900 (−36.24)**	0.6560 (36.24)**
Income	0.751e–06 (3.172)**	−0.259e–05 (−3.172)**	0.695e–06 (3.004)**	−0.240e–05 (−3.004)**
Hispanic	0.0021 (0.284)	−0.0073 (−0.284)	0.0004 (0.050)	−0.0013 (−0.050)
Family Age	−0.00002 (−0.036)	0.00007 (0.036)	−0.0005 (−0.940)	0.0019 (0.940)

Notes: ** Significant at the 1 percent level. * Significant at the 5 percent level.

local food markets have substantial market power if, as assumed in these models, they price jointly in their respective groups.

National-brand price does have a significant positive effect on private-label sales in both models. Private-label price variation, however, has negligible effects on national-brand sales. This is consistent with most of the work on asymmetric price competition and price tiers (Blattberg and Wisniewski 1989; Allenby and Rossi 1991); the estimated cross-price elasticity in the full power and promotion model is 0.225, within one standard deviation of the mean cross-price elasticity reported in Sethuraman (1995).

The expenditure elasticities exceed 1.0 in all three models for national brands and fall below 1.0 in all three models for private labels. Household-income elasticities (mean household income in 1992 was $39,358) shown in Table 12.5 indicate that an increase in household income has a very small but significant positive impact on national-brand volume and a very small but significant negative impact on private-label volume. This suggests that higher income implies a lower level of private-label consumption; in other words, it is an inferior good. The fact that both income elasticities are less than one implies that food is a necessity and, as income increases, a smaller portion of the budget is allocated to it.

Discussion

Analysis of panel data such as the IRI Supermarket Review data studied here, combined with consideration of both demand- and supply-side influences, provides considerably more insight into competitive strategies than single-equation cross-sectional studies. Previous single-equation studies have found a positive price-share relationship (share as a function of price). When market-share and price-reaction equations are estimated simultaneously (including market-power variables), it becomes clear that the share-price relationship is multidimensional. Specifically, there are two relationships—the negative demand-side relationship and possibly positive impacts for the national-brand share in the price-reaction equations. As the national-brand share increases, the national-brand group tends to dominate pricing in a fashion that benefits other products more than its own (Deneckere and Davidson 1985).

To get a more complete view of how national-brand and private-label products compete, and in an attempt to produce general results for the food sector, we have taken a traditional cross-section approach to the analysis of categories and geographic markets. We felt it was important to understand "the big picture" first, especially given the recent renewed focus in marketing on generalizability and given that this is the first study to address the impact of demand-side as well as supply-side interactions on private-label/national-brand pricing. In taking this cross-section approach, we recognize the inherent trade-off. While we gain generalizability, we lose much of the richness and precision of disaggregated category-level analysis. Nonetheless, we obtain significant and important findings that generalize across categories; although the demand parameter estimates are low, possibly due to cross-category analysis, they are consistent with theory. It is our hope that these results will encourage more detailed analysis for individual product categories such as breakfast cereal and milk.

Once one addresses the multidimensional nature of national-brand/private-label price strategies, certain implications for brand managers and retailers become clear:

- Brand managers should expect to face traditional demand relationships regardless of whether they are managing a national brand or a private label—an increase in the price of a national brand (private label) lowers the national-brand (private-label) share.

- As national brands increase their shares of category sales, both national-brand and private-label prices increase.

- An increase in retail concentration increases both national-brand and private-label prices. We also find that concentration has a greater positive impact on private-label prices. This suggests that leading supermarket chains are able to establish at least some brand loyalty for their own brands and can effectively narrow the price differential between national brands and private labels when they elevate all prices to improve profitability.

- Promotions increase share while rival promotions lower share. Further, promotion effects are asymmetric. For national-brand products, a 10 percent increase in point-of-sale display activity, for example, increases share by about 1.6 percent. In contrast, a similar 10 percent increase in point-of-sale display activity for private-label products increases share by only 0.87 percent.

- When national brands promote, private labels lower prices in an attempt to compete. However, the demand price elasticities suggest that this is a meager way to capture volume from national brands. Feature and display promotion appear to be much more effective ways of gaining share in such categories.

- Cross-price elasticities are decidedly asymmetric with the national-brand price having a major impact on private-label sales, whereas the private-label price has very little impact on national-brand sales. This is consistent with the work on asymmetric competition and price tiers (Blattberg and Wisniewski 1989; Allenby and Rossi 1991).

- Managers responsible for private labels operating in markets with higher per capita income or categories with a higher level of expenditure will have a more difficult time penetrating the market. More generally, we would expect private labels to suffer during stronger economic times.

As discussed in the introduction, insights into the effectiveness of competitive strategies for national-brand and private-label grocery products entail an understanding not only of the effectiveness of various strategies on the demand side but also of competitive interaction between national brands and private labels as well. In order to assess the viability of such strategies, it is important to differentiate between the direct demand-side effect and the likely response of rival firms (see, for example, Cotterill, Putsis, and Dhar (1999)). The present research represents an initial attempt to address these issues. We encourage future research in this area that addresses competitive interaction on a category-by-category basis with the use of disaggregate data.

References

Allenby, G.M., and P.E. Rossi. "Quality Perceptions and Asymmetric Switching between Brands." *Marketing Science* 10 (1991):185–204.

Appelbaum, E. "Testing Price Taking Behavior." Journal of Econometrics 19 (1979):283–294.

Angrisani, C. "Will Private Labels Crackle?" *Supermarket News* 46 (1996):39.

Berndt, E.R. *The Practice of Econometrics*. New York NY: Addison-Wesley, 1991.

Berry, S., J. Levinsohn, and A. Pakes. "Automobile Prices in Market Equilibrium." *Econometrica* 63 (1995):841–890.

Blattberg, R.C., and K.J. Wisniewski. "Price Induced Patterns of Competition." *Marketing Science* 8 (1989):291–309.

Carter, R.A.L., and A.L. Nagar. "Coefficients of Correlation for Simultaneous Equation Systems." *Journal of Econometrics* 6 (1977):39–50.

Christen, M., S. Gupta, J.C. Porter, R. Staelin, and D.R. Wittink. "Using Market-Level Data to Understand Non-Linear Promotion Effects." *Journal of Marketing Research* 34 (1997):322–334.

Choi, S.C. "Price Competition in a Channel Structure with a Common Retailer." *Marketing Science* 10 (1991):271–296.

Cotterill, R.W. "Market Power and the Demsetz Quality Critique: An Evaluation for Food Retailing." *Agribusiness: An International Journal* 15 (1999a):101–118.

———. "High Cereal Prices and the Prospects for Relief by Expansion of Private Label and Antitrust Enforcement." *Agribusiness: An International Journal* 15 (1999b):229–245.

Cotterill, R.W., L. Egan, and W. Buckhold. "Beyond Illinois Brick: The Economics of Cost Pass-through in the ADM Case." *Review of Industrial Organization* 18 (1999):45–52.

Cotterill, R.W., W.P. Putsis, Jr., and R. Dhar. "Assessing the Competitive Interaction between Private Labels and National Brands." *Journal of Business* 73 (1999):109–137.

Cowling, K., and M. Waterson. "Price-Cost Margins and Market Structure." *Economica* 43 (1976):267–274.

Davidson, R., and J.G. MacKinnon. "Several Tests for Model Specification in the Presence of Alternative Hypotheses." *Econometrica* 49 (1981):781–793.

Deaton, A., and J. Muellbauer. *Economic and Consumer Behavior*. New York NY: Cambridge University Press, 1980a.

———. "An Almost Ideal Demand System." *American Economic Review* 70 (1980b):312–326.

Deneckere, R., and C. Davidson. "Incentives to Form Coalitions with Bertrand Competition." *Rand Journal of Economics* 16 (1985):473–486.

Gallant, A.R., and D.W. Jorgenson. "Statistical Inference for a System of Simultaneous, Non-Linear, Implicit Equations in the Context of Instrumental Variable Estimation." *Journal of Econometrics* 11 (1979):275–302.

Gasmi, F., J.J. Laffont, and Q. Vuong. "Econometric Analysis of Collusive Behavior in a Soft-Drink Market." *Journal of Economics and Management Strategy* 1 (1992):277–312.

Gejdenson, S., and C. Schumer. *Consumers in a Box: A Report of High Breakfast Cereal Prices*. Washington DC: U.S. Congress, 1995.

———. "Consumers Still in a Box." Washington DC: Congressional Cereal Price Forum, 1996.

Genesove, D., and W. Mullin. "Testing Static Oligopoly Models: Conduct and Cost in the Sugar Industry, 1890–1914." *The Rand Journal of Economics* 29 (1998):355–377.

Green, R., and J.M. Alston. "Elasticities in AIDS Models." *American Journal of Agricultural Economics* 72 (1990):442–445.

Haller, L.E., and R.W. Cotterill. "Evaluating Traditional Share, Price and Residual Demand Measures of Market Power in the Catsup Industry." *Review of Industrial Organization* 11 (1996):293–306.

Hausman, J.A., and W.E. Taylor. "Panel Data and Unobservable Individual Effects." *Econometrica* 49 (1981):1377–1398.

Hoch, S.J., and S. Banerji. "When Do Private Labels Succeed?" *Sloan Management Review* (1993):57–67.

Hoch, S.J., B. Kim, A.L. Montgomery, and P.E. Rossi. "Determinants of Store-Level Price Elasticity." *Journal of Marketing Research* 32 (1995):17–29.

Judge, G., R.C. Hill, and T.C. Lee. *The Theory and Practice of Econometrics*. New York NY: Wiley, 1985.

Kadiyali, V., N.J. Vilcassim, and P. Chintagunta. "Empirical Analysis of Competitive Product Line Pricing Decisions: Lead, Follow, or Move Together." *Journal of Business* 69 (1996):459–487.

———. "Product Line Extensions and Competitive Market Interactions: An Empirical Analysis." *Journal of Econometrics* 89 (1998):339–363.

Kelton, C., and L. Weiss. "Change in Concentration, Change in Cost, Change in Demand, and Change in Price." *Concentration and Price*. Leonard Weiss, ed., pp. 41–66. Cambridge MA: MIT Press, 1989.

Kiviet, J.F. "Model Selection Test Procedures in a Single Linear Equation of a Dynamic Simultaneous System and their Defects in Small Samples." *Journal of Econometrics* 28 (1985):327–362.

Kwoka, J., and D. Ravenscraft. "Cooperation v. Rivalry: Price Cost Margins by Line of Business." *Economica* 53 (1986):351–363.

Levy, D., and J. Reitzes. "Product Differentiation and the Ability to Collude: Where Being Different Can Be an Advantage." *The Antitrust Bulletin* Summer (1993):349–368.

Lee, E., and R. Staelin. "Vertical Strategic Interaction: Implications for Channel Pricing Strategy." *Marketing Science* 16 (1997):185–207.

Liang, N. "An Empirical Conjectural Variation Model of Oligopoly." Washington DC: Bureau of Economics, Federal Trade Commission Working Paper 151, 1987.

Marion, B. *The Food Retailing Industry*. New York NY: Praeger, 1979.

Martin, S. *Advanced Industrial Economics*. Oxford England: Blackwell, 1993.

Narasimhan, C., and R.T. Wilcox. "Private-Labels and the Channel Relationship: A Cross-Category Analysis." *Journal of Business* 7 (1998):573–600.

Nevo, A. "Measuring Market Power in the Ready-to-Eat Cereal Industry." NBER Working Paper 6387, Cambridge MA: National Bureau of Economic Research, 1998.

Putsis, W.P. "Share, Price and Category Expenditure – Geographic Market Effects and Private Labels." Unpublished manuscript, London Business School 1999.

———. "An Empirical Study of the Effect of Brand Proliferation on Private Label – National Brand Pricing Behavior." *Review of Industrial Organization* 12 (1997):355–371.

Putsis, W.P., and R. Dhar. "Category Expenditure, Promotion and Competitive Market Interaction: Can Promotions Really Expand the Pie?" London Business School working paper, 1998.

Raju, J.S., R. Sethuraman, and S.K. Dhar. "National Brand-Store Brand Price Differential and Store Brand Market Share." *Pricing Strategy & Practice: An International Journal* (1995a):17.

———. "The Introduction and Performance of Store Brands." *Management Science* 41 (1995b):957–978.

Roberts, M. "Testing Oligopolistic Behavior." *International Journal of Industrial Organization* 2 (1984):367–383.

Schmalensee, R. "Entry Deterrence in the Ready-to-Eat Breakfast Cereal Industry." *Bell Journal of Economics* 9 (1978):305–327.

Sethuraman, R. "A Meta-Analysis of National Brand and Store Brand Cross Promotional Price Elasticities." *Marketing Letters* 6 (1995):275–286.

Sethuraman, R., and J. Mittelstaedt. "Coupons and Private Labels: A Cross-Category Analysis of Grocery Products." *Psychology & Marketing* 9 (1992):487–500.

Slade, M.E. "Product Rivalry with Multiple Strategic Weapons: An Analysis of Price and Advertising Competition." *Journal of Economics and Management Strategy* 4 (1995):445–476.

Tellis, G.J. "The Price Elasticity of Selective Demand: A Meta Analysis of Econometric Models of Sales." *Journal of Marketing Research* 25 (1988):331–341.

Tirole, J. *The Theory of Industrial Organization*. Cambridge MA: MIT Press, 1989.

Weiss, Leonard W., ed. *Concentration and Price*. Cambridge MA: MIT Press, 1989.

Werden, G.J., and G. Rozanski. "The Market Delineation Dilemma." *Antitrust* Summer (1994):40–43.

Werden, G.J., and L.M. Froeb. "Simulation as an Alternative to Structural Merger Policy in Differentiated Products Industries." *Competition Policy Enforcement: The Economics of the Antitrust Process*. M.B. Coate and A.N. Kleit, eds., pp. 65–88. Boston MA: Kluwer, 1996.

Willig, R.D. "Merger Analysis, Industrial Organization Theory and Merger Guidelines." *Brookings Papers on Economic Activity*. M.N. Bailey and C. Winston, eds., pp. 281–312. Washington DC: Brookings Institution Press, 1991.

13

Measuring Oligopolistic Distortion in the U.K. Frozen-Potato-Product Sector: A Calibration Modeling Approach

Abigail L. Tiffin and Philip J. Dawson

Dutch import penetration of the U.K. frozen-potato-product market has increased substantially since 1980, which is of concern to the U.K. potato industry.[1] U.K. processors (Potato Processors' Association 1991) have argued that such imports result from the competitive disadvantage created by the Potato Marketing Scheme, which restricts British potato production and raises the average potato price. However, the effects of imperfect competition on frozen-potato-product trade have been largely ignored in the literature. We estimate these trade effects assuming that neither country has a comparative advantage in frozen-potato-product production. We also estimate the welfare distortion effects of imperfect competition in the U.K. sector. The frozen-potato-products classification represents all chilled and deep frozen potato French fries, chips (excluding those produced from fresh potatoes on the premises where they are to be consumed), slices, quarters, sautés, rissoles, croquettes, waffles, skins, and jacket potatoes.

The Herfindahl (H) Index of concentration (Clarke 1985) is used to describe the structure of the U.K. and Dutch frozen-potato-product sectors. The H-Index varies between unity for a monopoly and zero for

---

This chapter originally appeared as an article titled "Measuring Oligopolistic Distortion in the U.K. Frozen Potato Product Sector: A Calibration Modelling Approach" by Tiffin and Dawson in the *Journal of Agricultural Economics* in 1997. This material is reprinted by permission of the *Journal of Agricultural Economics*.

1 Dutch imports into the U.K. have increased from 16,300 metric tons (54 percent of U.K. imports) in 1980/81 to 168,700 metric tons (77 percent of U.K. imports) in 1992/93 (Potato Marketing Board 1980–1993a).

perfect competition, hence a high value results from a low number of firms and/or a high market share inequality. The U.K. sector has a high H-Index of concentration of 0.36 where five processors account for around 84 percent of the sector's capacity.[2] The largest U.K. processor (McCain Foods) accounts for around 60 percent of total capacity. The Dutch sector is less concentrated with a Herfindahl Index of 0.12: five processors account for around 65 percent of capacity and the largest Dutch processor (Cebeco Handelsraad) accounts for 25 percent.

The differing structure in each sector is indicative of their contrasting development. The U.K. sector began as an offshoot of the rapidly developing frozen-food sector in the 1960s rather than as a separate industry as in the Netherlands. Consequently, it has traditionally been dominated by large, highly diversified (into nonpotato products) processors, whereas the Dutch sector is more equally shared with little diversification.

The calibration modeling approach affords annual comparisons of each sector's competitiveness and estimates of the trade and welfare distortion effects of imperfect competition in the U.K. frozen-potato-product sector.

Measuring Distortion due to Imperfect Competition

High concentration levels in the U.K. and Dutch frozen-potato-product sectors lead us to assume that processors behave interdependently or are oligopolistic. Oligopolistic behavior is analyzed at Nash's equilibrium. Sector i's oligopolistic behavior is described by its conjectural variation (V_i):

$$(1) \qquad V_i = dP_i \, / \, dQ_i \quad i = 1, 2,$$

which is the perceived effect on price i of a change in own aggregate output.

Oligopolistic behavior can be modeled using Cournot's quantity-setting theory or Bertrand's price-setting theory. The behavioral assumption of Cournot (Bertrand) is that each firm knows the output (price) of its rival and selects its own output (price) to maximize own profit, taking its rivals'

---

2 In contrast, the five largest U.K. fruit and vegetable processors accounted for 51 percent of total output in 1992 (Central Statistical Office 1980–1992a).

output (price) as constant. In measuring the conjectural variations at the Nash equilibrium, we avoid choosing between these theories. Moreover, conjectural variations provide theoretical benchmarks for comparison with the actual conjectural variations obtained from the calibration model. Although, use of conjectural variations within a static model does have shortcomings (Friedman 1983), conjectural variations provide a useful index of competition.

The welfare effects of oligopolistic distortion in the U.K. market are measured through effects on consumers' surplus, U.K. and Dutch processors' profit, and deadweight welfare loss to society. Harberger (1954) pioneered measurement of welfare costs of imperfect competition and defined it to equal the deadweight welfare loss. In contrast, Posner (1975) redefined it to also include all monopoly profits. We use both of these measures.

We apply Dixit's (1988a) calibration modeling approach rather than a statistical analysis as it affords direct estimation of the annual welfare effects of imperfect competition without requiring detailed firm-level sales, cost, and profit data, which are limited due to their commercial sensitivity. As McCorriston (1993) notes, this data problem is encountered in many industrial economics studies.

The Calibration Model

A calibration model reproduces a base year's data as its solution. The model can then be used to compare the observable base year's equilibrium and an equilibrium produced by the model under hypothetical policy scenarios. Comparisons of each year's two equilibria lead to an applied welfare analysis for the groups identified in the model. The following description of the model is divided between the market equation systems and the welfare analysis.

The U.K. Frozen-Potato-Product Market

The U.K.-Dutch rivalry in the U.K. market for frozen potato products is characterized by separate demand- and supply-side equation systems. The demand side consists of five equations that are solved simultaneously to obtain parameter estimates. From these estimates we obtain the theoretical conjectural variations for both Cournot competition and monopoly. Two first-order profit-maximizing conditions compose the supply side and are solved simultaneously to obtain estimates of the actual conjectural variations. Each sector's theoretical and actual conjectural variations are compared.

Following Harris (1985) and Dixit (1988a), the U.K.'s aggregate utility function is assumed to be of the form $U = Q_0 + U(Q_1, Q_2)$ where Q_1 and Q_2 are the quantities of U.K. and Dutch products sold in the U.K.[3] Q_0 is a competitive numeraire good with a unit price that includes all other sectors of the U.K. economy. $U(Q_1, Q_2)$ is a subutility function for U.K. and Dutch frozen potato products defined as

(2) $$U(Q_1, Q_2) = a_1Q_1 + a_2Q_2 - 1/2\,[b_1Q_1^2 + b_2Q_2^2 + 2kQ_1Q_2],$$

which implies positive marginal utilities and concavity. As Levy and Nolan (1992) note, zero income elasticities for U.K. and Dutch frozen potato products are implied by the aggregate utility function. Thus, the Dupuit-Marshallian triangle is an exact measure of consumers' surplus and is equal to

(3) $$CS = U(Q_1, Q_2) - P_1Q_1 - P_2Q_2$$

for U.K. consumers of both products.

The first-order utility-maximizing conditions give the inverse demand functions:

(4) $$P_i = a_i - b_iQ_i - kQ_j \qquad i, j = 1, 2 \text{ and } i \neq j$$

where a_i, b_i, and $k > 0$ and $b_i, b_j - k^2 > 0$, which imposes imperfect substitutability.[4]

Given the definition of sector i's conjectural variation in (1) and using (4), we derive the following theoretical conjectural variations:

(5) Monopoly: $V_i^m = b_i$,

(6) Cournot: $V_i^c = b_i / n_i$, and

---

3 Non-Dutch imports of frozen potato products are assumed to be insignificant; they accounted for 6 percent of total U.K. consumption in 1991/92 (Potato Marketing Board 1993a).

4 If Q_1 and Q_2 are perfect substitutes, then $a_1 = a_2$, $b_1 = b_2$, and $b_1, b_2 - k^2 = 0$.

(7) Bertrand/Perfect Competition: $V_i^b = 0$

where n_i is sector i's H-Index number equivalent of equal-sized firms, which is equal to the reciprocal of sector i's H-Index. In (5), for a profit-maximizing monopolist, price changes brought about by a change in output are equal to the slope of the inverse demand function. Under Cournot's behavioral assumption of nonresponsive rivals, (6) results because firms believe they are profit-maximizing monopolies over their shares of the market. In (7), $n_i = \infty$ and firms are price-takers.[5]

The (linear) demand functions associated with (4) are:

$$(8) \qquad Q_i = A_i - B_i P_i + KP_j \quad i, j = 1, 2 \text{ and } i \neq j$$

where A_i, B_i, and $K > 0$ and $B_i, B_j - K^2 > 0$. By estimating (8), we can obtain the parameters in (4) and hence the Cournot and monopoly conjectural-variation values.

However, (8) contains five unknown parameters in two equations so three further identities are required. These are expressions for the own-price elasticity of demand, the import-substitution elasticity, and a homotheticity condition. Dixit (1988a) shows that the own-price elasticity of demand in expenditure terms is

$$(9) \qquad \eta = -(B_1 P_1^2 + B_2 P_2^2 - 2KP_1 P_2) / (Q_1 P_1 + Q_2 P_2).$$

The elasticity of import substitution, indicating the ease with which U.K.-produced products can be substituted for Dutch imports in the U.K., is

$$(10) \qquad \sigma = \frac{P_1/P_2\,(B_1 B_2 - K^2)}{(B_1\, P_1/P_2\ - K)(B_2 - K\, P_1/P_2)}.$$

Finally, the homotheticity condition is

$$(11) \qquad P_1\,(A_1 K + A_2 B_1) = P_2\,(A_2 K + A_1 B_2).^6$$

5 Assuming constant costs, Bertrand and perfect competition are identical at the Nash equilibrium.

6 The tension between the homotheticity restriction and use of the Dupuit-Marshallian triangle as an exact measure of consumers' surplus, assuming quasi linear preferences, has not yet been addressed in the literature using this model.

The five-equation demand system in (8), (9), (10), and (11) can be solved using price and quantity data and estimates for the own-price demand elasticity and the import-substitution elasticity to derive the parameter estimates in (4).

The calibration model's supply side is composed of two aggregate, first-order, profit-maximizing conditions:

$$(12) \quad P_i - c_i - Q_i V_i^a = 0 \qquad i = 1, 2$$

where c is (constant) marginal cost and V_i^a is sector i's actual conjectural variation. As McCorriston (1993) notes, if V_i^a / V_i^c is less (greater) than unity, actual behavior is more (less) competitive than the Cournot outcome.

Measuring Oligopolistic Distortion

Following McCorriston (1993), oligopolistic distortion may be estimated through the change in welfare from its actual level due to the simulated expansion of processors' output to perfectly competitive levels. We assume that U.K. production is encouraged to competitive levels through provision of a hypothetical, welfare-maximizing production subsidy by the government. Therefore, the U.K. sector's postsubsidy, first-order, profit-maximizing condition becomes

$$(13) \quad P_1 - c_1 + s^* - Q_1 V_1^a = 0$$

where s^* is an optimal production subsidy allocated to U.K. processors to eliminate all oligopolistic distortion in the U.K. sector.

Both sectors' prices and outputs are assumed to be endogenous to the allocation of a production subsidy to U.K. processors. The postsubsidy prices and outputs are calculated using the following reduced-form equilibrium conditions derived from (4), (12), and (13):

$$(14) \quad \begin{bmatrix} P_1^* \\ P_2^* \end{bmatrix} = \begin{bmatrix} a_1 \\ a_2 \end{bmatrix} - 1/\Delta' \begin{bmatrix} \Delta + b_1 V_2^a & kV_1^a \\ kV_2^a & \Delta + b_2 V_1^a \end{bmatrix} \begin{bmatrix} a_1 - c_1 + s^* \\ a_2 - c_2 \end{bmatrix}$$

$$(15) \quad \begin{bmatrix} Q_1^* \\ Q_2^* \end{bmatrix} = 1/\Delta' \begin{bmatrix} \beta_2 & -k \\ -k & \beta_1 \end{bmatrix} \begin{bmatrix} a_1 - c_1 + s^* \\ a_2 - c_2 \end{bmatrix}$$

where $\Delta = b_1 b_2 - k^2$, $\beta_i = b_i + V_i^a$, $i = 1, 2$, and $\Delta' = \beta_1 \beta_2 - k^2 \geq \Delta$. To obtain the price and quantity levels associated with a monopoly and Cournot and perfect competition in the U.K., we substitute the values of the theoretical conjectural variations in (5) through (7) for V_i^a in (14) and (15).

An identity for the optimal production subsidy is derived by Dixit (1988b) through the following expression for total U.K. welfare:

$$(16) \quad W = CS + \pi_1 - s^* Q_1$$

where U.K. consumers' surplus (CS) is defined in (3), π_1 is the profit of U.K. processors, and $s^* Q_1$ is the subsidy cost. Substituting in (3) and (13), (16) becomes

$$(17) \quad W = [U(Q_1, Q_2) - P_1 Q_1 - P_2 Q_2] + [Q_1 (P_1 - c_1 + s^*)] - [s^* Q_1].$$

The welfare effect of a small change in the equilibrium prices and outputs in (17), given (4), is

$$(18) \quad dW = (P_1 - c_1)\, dQ_1 - Q_2 dP_2.$$

The two terms on the right-hand side of (18) are, respectively, the U.K. market distortion and terms of trade effects. As well as directly eliminating the domestic distortion, a domestic production subsidy indirectly allows the capture of foreign profits by reducing import demand. Using (13) and (15) and setting $dW / ds^* = 0$ in (18), the optimal subsidy is

$$(19) \quad s^* = Q_1 V_1^a + k Q_2 V_2^a / \beta_2.$$

However, as output is endogenous, substituting (15) into (19) gives

$$(20) \quad s^* = \frac{(a_1 - c_1)(\beta_2^2 V_1^a - k^2 V_2^a) + (a_2 - c_2) k (\beta_1 V_2^a - \beta_2 V_1^a)}{\beta_2^2 (\beta_1 - V_1^a) + k^2 (V_2^a - \beta_2)}.$$

As subsidization by the U.K. government is unrealistic, it is perhaps more appropriate to obtain the U.K. price that is consistent with the optimal subsidy outcome. The subsidy-equivalent U.K. price (P_1^*) is obtained by substituting (20) into (14):

$$(21) \quad P_1^* = a_1 - (a_1 - c_1)\,\Delta''\left[1+ \frac{\beta_2^2 V_1^a - k^2 V_2^a}{\beta_2^2(\beta_1 - V_1^a) + k^2(V_2^a - \beta_2)}\right]$$

$$-(a_2 - c_2)\,k\left[\Delta''\left(\frac{\beta_1 V_2^a - \beta_2 V_1^a}{\beta_2^2(\beta_1 - V_1^a) + k^2(V_2^a - \beta_2)}\right] + V_1^a/\Delta'\right)$$

where $\Delta'' = (b_1\beta_2 - k^2) / (\beta_1\beta_2 - k^2)$. The value of P_1^* is used to calculate the subsidy-equivalent U.K. output, Q_1^*, together with (4) and (12). We may then derive the associated P_2^* and Q_2^*. The values of P_1^*, Q_1^*, P_2^*, and Q_2^* are then used to calculate the new level of welfare and the oligopolistic trade and welfare distortion effects.

Examining the effects of changing policies, such as allocation of an optimal production subsidy, is problematic. As Farmer (1991) explains, any change in policy regime would change the characteristics of the reduced-form response by altering the expectations of agents whose behavior is summarized by the reduced-form model. Following Dixit (1988a) the reduced-form equations in (14) and (15), which assume that industry behavior remains constant with a policy change, are deemed acceptable in this static model.

Data and Results

Annual data for 1980/81 through 1991/92 are used to calibrate the model; they are given in the Appendix. The U.K. sector's output almost doubles over the period. Dutch imports steadily increase until 1989/90, rising to a level tenfold that in 1980/81, after which the level falls slightly. The nominal U.K. price is greater than the Dutch price in each year and the difference widens after 1986/87. The nominal U.K. unit cost is greater than the Dutch unit cost in each year except 1980/81 and 1988/89. The GDP (gross domestic product) deflator at factor cost is used to obtain real values of results where comparisons through time are useful.

The elasticities required to solve the model's demand system are estimated in Dennis (1996). The own-price demand-elasticity estimate for frozen potato products in the U.K. is –0.639 and the substitution elasticity of Dutch imports for U.K.-produced products in the U.K. market is

1.832. Finally, the H-Index number equivalent of firms is estimated in Dennis (1996) as 2.79 in the U.K. sector and 8.45 in the Dutch sector.

Consider the results from the calibration model. Table 13.1 shows the theoretical conjectural variations of monopolistic and Cournot competition in (5) and (6) from the demand side of the model in real terms. The U.K. Cournot conjectural variation is larger than that of the Dutch sector except in 1980/81 and 1981/82.[7] This in part reflects the low H-Index number equivalent of U.K. processors compared with that of Dutch processors. The Cournot conjectural-variation benchmark values are increasingly competitive, reflecting a gradual decrease in the slopes of the inverse demand functions (b_i, i = 1, 2) over the period. A Dutch monopolist would have greater market power than a U.K. monopolist, although this difference falls over time, with both monopoly benchmarks becoming increasingly competitive as b_i decreases.

Table 13.1. Theoretical Conjectural-Variation Estimates (1991/92 Prices)

	U.K.		Netherlands	
Year	$V_1^c(10^{-4})$	$V_1^m(10^{-4})$	$V_2^c(10^{-4})$	$V_2^m(10^{-4})$
1980/81	17.28	48.21	44.84	378.93
1981/82	17.28	48.22	18.10	152.93
1982/83	15.73	43.90	14.49	122.46
1983/84	16.98	47.37	12.70	107.30
1984/85	12.53	34.96	11.49	97.06
1985/86	10.10	28.17	8.01	67.67
1986/87	10.14	28.29	6.54	55.24
1987/88	10.50	29.28	5.49	46.40
1988/89	8.41	23.46	4.01	33.86
1989/90	9.05	25.25	2.95	24.93
1990/91	9.09	25.35	3.25	27.48
1991/92	7.41	20.66	3.02	25.52

7 The production years of 1981/82, 1983/84, and 1989/90 were associated with very high continental raw-potato prices; these results should therefore be treated with caution.

Table 13.2. Actual Conjectural-Variation Estimates (1991/92 Prices)

	U.K.		Netherlands	
Year	$V_1^a(10^{-4})$	V_1^a / V_1^c	$V_2^a(10^{-4})$	V_2^a / V_2^c
1980/81	6.32	0.37	47.87	1.07
1981/82	1.79	0.10	4.05	0.22
1982/83	8.17	0.52	37.30	2.57
1983/84	1.37	0.08	–6.93	–0.55
1984/85	7.73	0.62	37.25	3.24
1985/86	7.69	0.76	29.67	3.71
1986/87	8.12	0.80	21.48	3.29
1987/88	9.90	0.94	23.77	4.33
1988/89	8.79	1.05	12.82	3.20
1989/90	7.59	0.84	4.26	1.44
1990/91	9.40	1.04	9.66	2.97
1991/92	7.43	1.00	9.19	3.04

Table 13.2 shows the actual conjectural-variation values in real terms, V_i^a (i = 1, 2), from the supply-side of the model where V_2^a tends to be greater than V_1^a. There are two reasons for the difference between sectors. First is the coexistence of one dominant processor and a periphery of price-taking small processors in the U.K. sector. Second, the Dutch sector is more equally shared with processors being more interdependent so a small change in a Dutch processor's output will have a significant influence on the price of Dutch frozen potato products in the U.K.

The values of V_1^a / V_1^c in Table 13.2 show that actual behavior in the U.K. sector is more competitive than the Cournot situation in 1980/81. However, as the U.K. sector becomes increasingly concentrated, particularly after 1982/83, it becomes less competitive than under Cournot. This coincides with the merger of McCain Foods (U.K.) and Potato and Allied Services. Actual behavior in the Dutch sector is generally less competitive than under Cournot throughout the time period.

We compare the actual price and quantity levels with those associated with Bertrand/perfect competition, Cournot behavior, and monopoly in the U.K. sector. These price and quantity levels are calculated in Table 13.3 for 1991/92 using (14) and (15) without s^* by substituting values of V_1^b, V_1^c, and V_1^m for V_1^a, the parameter values from the inverse-demand

Table 13.3. Prices and Quantities with Different Competition in the U.K. Sector (1991/92)

	P_1 (£/metric ton)	P_2 (£/metric ton)	Q_1 (metric tons)	Q_2 (metric tons)
Actual	594	374	413,158	143,800
Bertrand	275 (–54)	351 (–6)	573,868 (+39)	119,383 (–17)
Cournot	584 (–2)	373 (–0.2)	418,002 (+1)	143,066 (–1)
Monopoly	856 (+44)	392 (+5)	281,260 (–32)	163,843 (+14)

Note: Percentage change from actual levels in parentheses.

functions, marginal costs, and V_2^a. Actual 1991/92 prices and quantities are between those of Cournot and monopoly, but in both sectors behavior is close to the Cournot benchmark. As expected, the U.K. price is more sensitive to the level of competition in the U.K. sector than is the Dutch price; it would fall by 54 percent and rise by 44 percent from the 1991/92 level under perfect competition and monopoly respectively. Production in the U.K. sector would rise by 39 percent and fall by 32 percent under perfect competition and monopoly respectively. Dutch import penetration would fall by 17 percent if the U.K. sector was perfectly competitive and would rise by 14 percent if it was a monopoly.

We now examine the trade effects and welfare costs of oligopolistic distortion in the U.K. sector. Table 13.4 shows the optimal production subsidy levels that would, if allocated to U.K. processors, eliminate oligopolistic distortion in the U.K. sector. It also reports each sector's subsidy-equivalent or nondistorting prices and quantities. The real optimal production subsidy that would eliminate oligopolistic distortion in the U.K. sector if allocated to U.K. processors is estimated from (20). Its level increases to £500/metric ton in 1988/89 with the fall in competitive behavior in the U.K. sector, after which it fluctuates at slightly lower levels. In 1991/92 it is £449/metric ton. Subsidy-equivalent price and quantity levels for U.K.-produced products and Dutch imports, also shown in Table 13.4, are compared with their actual levels. Oligopolistic behavior in the U.K. sector causes the U.K. price to rise by an average of £249/metric ton and the Dutch price to rise by an average of £20/metric ton over the period. In 1991/92, the actual U.K. and Dutch prices are £336/metric ton and £24/metric ton greater than their respective subsidy-equivalent prices. The U.K. sector's oligopolistic distortion increases Dutch import penetration by an average of 13,300 metric tons and in

Table 13.4. Optimal Production Subsidy and Price and Quantity Effects of Oligopolistic Distortion in the U.K. (1991/92 Prices)

Year	s^* (£/ metric ton)	P_1^* (£/metric ton)	P_2^* (£/metric ton)	Q_1^* (metric tons)	Q_2^* (metric tons)
1980/81	165	561	678	255,973	14,783
1981/82	42	669	605	226,496	36,394
1982/83	253	498	636	287,451	46,884
1983/84	25	709	635	227,981	57,266
1984/85	286	422	564	346,002	51,932
1985/86	326	330	511	395,988	69,153
1986/87	394	370	539	455,832	88,711
1987/88	495	319	492	470,754	97,905
1988/89	500	262	426	545,461	110,831
1989/90	374	340	387	477,126	135,590
1990/91	497	267	382	506,975	122,830
1991/92	449	258	350	582,171	118,121
Average	317	417	517	398,184	79,200

Table 13.5. Actual Welfare Levels (Million Pounds in 1991/92 Prices)

Year	*CS*		π_1		π_2		DWL
1980/81	133.5	(+27.2)	33.0	(−102.6)	1.3	(−17.8)	3.0
1981/82	138.6	(+6.8)	8.9	(−104.1)	0.6	(−5.3)	0.5
1982/83	159.0	(+36.5)	47.2	(−111.0)	10.2	(−19.3)	8.9
1983/84	156.6	(+3.6)	7.1	(−79.0)	−2.3	(−3.2)	−1.3
1984/85	169.5	(+44.7)	60.9	(−110.5)	12.8	(−21.7)	12.2
1985/86	172.2	(+54.1)	71.7	(−112.0)	18.7	(−24.1)	16.9
1986/87	227.5	(+55.6)	97.8	(−110.9)	22.7	(−25.5)	22.8
1987/88	231.2	(+67.7)	117.0	(−111.9)	31.4	(−27.4)	31.0
1988/89	236.2	(+73.9)	132.7	(−108.5)	23.2	(−32.2)	34.3
1989/90	226.3	(+55.0)	100.1	(−105.9)	11.2	(−29.9)	20.9
1990/91	224.1	(+72.5)	123.6	(−108.4)	21.5	(−32.2)	31.9
1991/92	234.0	(+72.7)	131.9	(−107.3)	19.0	(−32.5)	32.1
Average	192.4	(+47.5)	77.6	(−106.0)	14.2	(−22.6)	17.8

Note: Percentage changes from actual to competitive welfare levels in parentheses.

1991/92 by 25,700 metric tons, whereas production of U.K.-produced frozen potato products is reduced by an average of 93,600 metric tons and in 1991/92 by 169,000 metric tons.

Table 13.5 summarizes the changes in real welfare of oligopolistic distortion in the U.K. sector. U.K. consumers' surplus and U.K. and Dutch processors' profits are calculated using (3) and the profit functions $\pi_i = (P_i - c_i)\,Q_i$ $(i = 1, 2)$ respectively. The deadweight loss (DWL) of the U.K. sector's oligopolistic distortion is calculated as the difference between the consumers' surplus loss and processors' profit gain from the actual situation relative to the competitive outcome.

Elimination of oligopolistic distortion in the U.K. sector would increase U.K. consumers' surplus by 73 percent (£170.1 million) in 1991/92 (and by an annual average of 48 percent), reduce U.K. processors' profits by 107 percent (£131.9 million) in 1991/92 (and by 106 percent on average), and reduce Dutch processors' profits by 33 percent (£6.2 million) in 1991/92 (and by 23 percent on average). Using Harberger's definition of welfare costs (equal to the level of DWL), U.K. society's welfare is reduced by an annual average of £17.8 million. If we employ Posner's welfare loss measure, which includes monopoly profits, then the loss to U.K. society is equal to the loss in consumers' surplus. On average, DWL accounts for 9 percent of total U.K. expenditures on U.K. and Dutch products due to oligopolistic distortion in the U.K. sector.[8]

Summary and Conclusions

This chapter has adopted a calibration modeling approach to compare the U.K. and Dutch sectors' competitive behavior over the period 1980/81 through 1991/92. A comparison of the actual and theoretical conjectural-variation values from the calibration model indicates that both sectors, but especially the Dutch sector, are increasingly monopolistic. In 1991/92, the two sectors' behaviors are close to the Cournot benchmark. The U.K. sector's price and quantity are highly sensitive to its competition level; compared with the actual 1991/92 price and quantity levels, a perfectly competitive U.K. sector would produce 39 percent more for a 54 percent lower price, whereas a monopolistic U.K. sector would produce 32

8 McCorriston (1993) estimates lower welfare losses due to oligopolistic distortion for the U.K. fertilizer and tractor industries for 1973 through 1986; DWL accounts for 1.6 and 2.5 percent, respectively, of the value of shipments.

percent less for a 44 percent higher price. The U.K. sector's competition level also significantly affects Dutch import penetration of the U.K. market for frozen potato products. If the U.K. sector was perfectly competitive, Dutch imports would fall by 17 percent from their 1991/92 level, whereas a monopolistic U.K. sector would raise Dutch imports by 14 percent.

Oligopolistic distortion in the U.K. frozen-potato-product sector increases the penetration of Dutch imports in the U.K. market by 25,700 metric tons in 1991/92 and by an annual average of 13,300 metric tons over the period. It also affects the welfare of U.K. consumers, U.K. processors, Dutch processors, and society: U.K. consumers' surplus falls by £99.4 million annually on average due to the U.K. sector's oligopolistic distortion, U.K. processors' profit increases by £77.6 million, Dutch processors' profit increases by £4 million, and the DWL is £17.8 million.

Calibration modeling has a number of shortcomings. First, Dixit (1988a) cites the need for improvement in the functional forms used in the calibration model, particularly for the linear demand and constant marginal cost. Second, the use of a conjectural-variation analysis has been criticized but a suitable alternative remains elusive. Third, the deterministic nature of the approach implies that calibrated parameters lack statistical validity.[9] Two main study-specific deficiencies also can be identified. First, the level of aggregation over a number of quite different chilled and frozen potato products is a concern. Second, it would have been preferable to have the annual Herfindahl number equivalent of processors, but the limited availability of market-share data made it impossible.

Finally, we conclude that the trade and welfare costs of market power in the U.K. sector are high and generally increasing. Such trade and welfare trends are likely to continue due to the U.K. sector's oligopolistic power despite the demise of the Potato Marketing Scheme after 1996/97 and creation of a uniform European-Union-wide potato regime. Therefore, the view of the Potato Marketing Board that the Potato Marketing Scheme has provided potato producers with a required source of countervailing power against processors appears justified.

---

9 To counter this criticism, Dennis (1996) uses Monte Carlo simulation techniques to obtain confidence-interval estimates for the calibrated parameters.

Appendix: Data

Year[a]	U.K. Price[b] (P_1)	Dutch Price[c] (P_2)	U.K. Quantity[d] (Q_1)	Dutch Quantity[d] (Q_2)	U.K. Cost[e] (c_1)	Dutch Cost[e] (c_2)	GDP Deflator[f]
1980/81	377.31	363.61	224,100	16,300	299.25	322.23	53.0
1981/82	408.92	348.08	217,350	37,400	385.49	339.37	57.5
1982/83	438.12	401.40	235,800	52,200	315.79	282.29	61.2
1983/84	473.97	409.09	222,300	58,200	453.39	435.13	64.5
1984/85	450.35	400.45	274,950	58,700	299.82	251.75	68.0
1985/86	418.16	382.52	299,474	79,400	248.86	215.98	70.7
1986/87	498.66	417.78	341,579	102,800	288.51	255.80	73.4
1987/88	538.73	412.29	336,842	114,900	269.67	200.72	77.4
1988/89	521.59	377.65	381,579	134,600	233.79	234.85	82.8
1989/90	566.13	355.18	353,684	161,900	313.82	293.72	89.1
1990/91	605.09	388.09	354,737	149,200	273.66	250.89	95.1
1991/92	593.90	373.68	413,158	143,800	274.77	241.57	100.0
Average	490.91	385.82	304,629	92,450	304.74	277.02	

[a] Crop years: June to May.

[b] Central Statistical Office (1980–1992a) (nominal £/metric ton).

[c] Potato Marketing Board statistics (nominal £/metric ton).

[d] Potato Marketing Board statistics (metric tons).

[e] For a typical year (1987/88 in the U.K. and 1986/87 in the Netherlands), the potato cost accounts for 54 percent of total costs. U.K. and Dutch potato prices are from the Potato Marketing Board and Eurostat statistics. The costs of Dutch transport to the U.K. are £42 per metric ton (Center for European Agricultural Studies 1983) (nominal £/metric ton).

[f] Gross domestic product at factor cost deflator (Central Statistical Office 1980–1992b).

References

Center for European Agricultural Studies. *New Arrangements for Processors within the Potato Marketing Scheme.* London England: University of London Center for European Agricultural Studies, 1983.

Central Statistical Office (CSO). *Business Monitor: Report on the Census of Production.* London England: Her Majesty's Stationery Office (HMSO), 1980–1992a.

———. *Economic Trends: Annual Supplement.* London England: Her Majesty's Stationery Office (HMSO), 1980–1992b.

Clarke, R. *Industrial Economics.* Oxford England: Basil Blackwell, 1985.

Dennis, A.L. "An Economic Analysis of the Structure of the Frozen Potato Product Industry." Doctoral thesis, Newcastle University, 1996.

Dixit, A. "Optimal Trade and Industrial Policies for the U.S. Automobile Industry." *Empirical Methods for International Trade.* R. Feenstra, ed., pp. 141–165. Boston MA: MIT Press, 1988a.

———. "Anti-Dumping and Countervailing Duties under Oligopoly." *European Economic Review* 32 1988b:55–68.

Eurostat. "Quarterly Statistics: Agricultural Prices." Brussels Belgium: Eurostat, 1982–1993.

Farmer, R.E.A. "The Lucas Critique, Policy Invariance and Multiple Equilibria." *Review of Economic Studies* 58 (1991):321–332.

Friedman, J.W. *Oligopoly Theory*. New York NY: Cambridge University Press, 1983.

Harberger, A.C. "Monopoly and Resource Allocation." *American Economic Review* 44 (1954):73–87.

Harris, R. "Why Voluntary Export Restraints are 'Voluntary'." *Canadian Journal of Economics* 18 (1985):799–809.

Levy, S., and S. Nolan. "Trade and Foreign Investment Policies under Imperfect Competition: Lessons for Developing Countries." *Journal of Development Economics* 37 (1992):31–62.

McCorriston, S. "The Welfare Implications of Oligopoly in Agricultural Input Markets." *European Review of Agricultural Economics* 20 (1993)1–17.

Potato Marketing Board. *Potato Statistics in Great Britain*. Oxford England: Potato Marketing Board, 1980–1993a.

———. *Potato Consumption and Processing in Great Britain*. Oxford England: Potato Marketing Board, 1980–1993b.

Posner, R.A. "The Social Costs of Monopoly and Regulation." *Journal of Political Economy* 83 (1975):807–827.

Potato Processors' Association. "Position Paper of the Potato Processors' Association on the Common Organisation of the EEC Potato Market." Grantham England: Potato Processors' Association, 1991.

Part VII
Summary

14

New Empirical Industrial Organization and the Food System: Summary

Harry M. Kaiser and Nobuhiro Suzuki

This book presents a detailed, systematic overview of the new empirical industrial organization (NEIO) framework and numerous case studies of the application of this approach to the food sector, illustrating the usefulness of the NEIO approach in measuring market power. The primary contribution of the NEIO framework to industrial organization is the generalization and extension of conventional perfect-competition and monopoly models to intermediate imperfect-competition models that can be empirically estimated. Since the majority of markets in the real world are neither perfectly competitive nor monopolistic, the NEIO framework is valuable in empirical applications. Indeed, many researchers have used this approach to measure and analyze the degree of market imperfection in various industries, with the bulk of the studies focusing on the food industry.

Synopsis of Findings

In Chapter 1, we present an extensive overview of NEIO. We provide a brief review of various theories and schools of thought for industrial organization (IO) and how NEIO fits into the field of IO. In the detailed discussion, we divide the NEIO models into two broad categories: quantity-setting homogeneous oligopoly, which is a generalization of the Cournot model, and price-setting product-differentiated oligopoly, which is a generalization of the Bertrand model.

In the quantity-setting homogeneous oligopoly section, we discuss the basic theory of NEIO, examine consistency of conjectures and rationality of Cournot behavior, generalize the Dorfman-Steiner Theorem, examine the two interpretations and two estimation approaches for conjectural variations, present a framework for measuring economic welfare losses due to oligopoly, look at the NEIO framework for measuring oligopsony power on the part of processors, review a framework for investigating the market power of farmer cooperatives, construct a framework for analyzing bilateral oligopoly in the food market channel, and apply the NEIO framework to price transmissions under imperfect competition, spatial-equilibrium analysis, and international trade issues.

Regarding the price-setting product-differentiated oligopoly section, we summarize the basic theory, derive the relationship between quantity and price conjectures, and discuss identification problems in estimating these models.

Within the discussion of both types of NEIO models, we review some of the studies that have applied NEIO to the food sector. This chapter serves as the foundation for the remaining chapters, which are case studies applying NEIO to various markets in the food industry.

Quantity-Setting Homogeneous Oligopoly Models

Chapters 2 through 10 provide case studies of empirical research using quantity-setting homogeneous oligopoly models. The first four chapters analyze only oligopoly power, with Chapters 2 and 3 focusing on farmer-side oligopoly and Chapters 4 and 5 looking at processor oligopoly.

Chapters 6, 7, and 8 analyze both oligopoly and oligopsony power simultaneously. Chapter 6 pertains to processor input and output markets only while Chapter 7 examines successive oligopoly throughout the food system, including farmer and retailer sides, but does not include bilateral oligopoly between sellers and buyers. Chapter 8 examines bilateral oligopoly.

Chapters 9 and 10 present spatial and dynamic extensions of oligopoly models.

Product-Differentiated Oligopoly Models

Chapters 11 through 13 offer applications of price-setting product-differentiated oligopoly models, illustrating three types of such models. Chapters 11 and 12 offer related approaches, with Chapter 11 being an application of a generalized Bertrand model where demand and price-reaction equations are simultaneously estimated. Chapter 12 applies a price-reaction function model where demand and price-reaction equations are independently estimated. Chapter 13 involves an application of a Dixit-type calibration model.

The Chapter Models

Chapters 2 and 3 both involve the dairy industry and farmer oligopoly power. The main sources of oligopoly power in the farm sector, where there are many price-taking farmers, are (1) farmer cooperatives, and (2) federal marketing orders or marketing boards. Although raw milk is essentially a homogeneous input in the production of fluid milk and manufactured dairy products, in many countries the price received for fluid-milk usage is higher than the price received for manufactured-product usage. Such differences indicate that prices are not competitively determined. This certainly is the case in the U.S., where federal and state milk marketing order programs establish minimum Class I price differentials (premiums) for most milk marketed. There are also over-order fluid premium payments resulting from negotiations between cooperatives and fluid processors.

In Chapter 2, Suzuki et al. develop the first imperfect-competition model with an endogenous fluid differential for the U.S. dairy market using degree-of-competition parameter estimates. They find that the estimated parameters indicate some degree of imperfection in the U.S. milk market that has been declining over time. Suzuki et al. argue that competitive pressures in the fluid market have increased over time due to improvements in transportation technology and increasing reserves of milk in areas other than Minnesota and Wisconsin. While dairy farmers have tried to increase their oligopoly power by merging cooperatives and milk marketing orders, processors have simultaneously become larger. Thus, the gradually decreasing degree-of-competition parameters could be the consequence of a balancing of power due to these developments.

The usefulness of the authors' model is demonstrated by showing the relative differences in dynamic simulation results from the imperfect-competition model with results from a more conventionally used exogenous fluid differential model. The exogenous fluid differential model overestimates the negative effects of deregulation compared to the imperfect-competition model when assuming that the current degree of competitiveness in the U.S. milk market is maintained after deregulation. Decreases in the manufacturing-milk price exactly correspond to decreases in the fluid-milk price in the exogenous fluid differential model. However, some market imperfection could cause less direct translation of decreases in the manufacturing price into decreases in the fluid price.

In Chapter 3, Suzuki et al. apply a similar imperfect-competition model to the dairy sector in the context of generic advertising evaluation rather than policy deregulation. Prior to this study, generic advertising evaluation studies were based on perfect-competition models. The authors develop a framework for incorporating the degree of market competition into the evaluation of milk promotion effectiveness in Japan. Specifically, the fluid-milk premium is endogenously explained by the degree of market power. Using Japanese generic milk promotion data, Suzuki et al. demonstrate how this framework improves the estimates compared to a conventional exogenous-premium model.

In Japan, estimated marginal rates of return to generic promotion derived from the more conventional exogenous-premium model are 23 to 27 percent smaller than those from the imperfect-competition model. These results indicate that analyses with an exogenous-premium model may significantly underestimate the true rate of return to milk promotion when imperfect competition is present.

An equation is also derived to show how several parameters affect the magnitude of increase in the fluid-milk price as promotion expenditures increase. The results indicate that the increase in milk price associated with an increase in promotion becomes larger as the market-power parameter and promotion elasticity of fluid demand become larger and as the price elasticity of fluid demand becomes smaller. This provides empirical evidence that more monopolistic industries receive a larger return on generic advertising than do competitive industries.

Liu, Sun, and Kaiser in Chapter 4 develop a framework to estimate the market power of an oligopolistic industry where there is a government price-support program impacting firms' output price. The proposed framework is then used to estimate the degree of market power exercised by manufacturing and fluid processors in the U.S. dairy industry. The authors also examine whether government price intervention in the dairy industry has a procompetitive or anticompetitive influence on market conduct.

The results indicate that the conjectural elasticities of manufacturing processors (0.100) are smaller than those of fluid processors (0.176). The finding that fluid processors behave in a less competitive manner than manufacturing processors is intuitive because markets for fluid milk are less national in scope due to the perishability and relatively high transportation costs of the product. While average industry-conduct parameters for manufacturing and fluid processors are statistically different from zero for all quarters, the magnitudes of these parameters are not alarming as they are still closer to zero (perfect competition) than one (monopoly). Furthermore, neither parameter exhibits a strong pattern of increases over time, a finding that is reassuring given that the industry has become more concentrated over the sample period. The results also indicate that manufacturing and fluid processors tend to behave in a more competitive manner in the market-equilibrium regime than in the government-supported regime. Interestingly, this result suggests that further deregulation of the dairy price-support program will have a procompetitive impact on market conduct.

In Chapter 5, Bhuyan and Lopez use an NEIO-based approach to estimate allocative efficiency losses in thirty-eight food and tobacco manufacturing industries in the United States. This was the most comprehensive NEIO study that had been done on the U.S. food sector. All the necessary parameters, such as elasticities of cost and demand, are first estimated independently for each industry using the same empirical model. Subsequently, allocative efficiency losses are computed and compared across industries, as well as with five alternative estimates available in the literature.

The empirical findings show that welfare losses due to oligopoly power in U.S. food and tobacco manufacturing industries amounted to 5.03 percent of sales, slightly more than $15 billion, in 1987. The estimated losses range from 33.43 percent of 1987 industry sales in the cereal breakfast preparation industry (SIC 2043) to 0.07 percent of 1987 sales in the pet food industry (SIC 2047). In terms of food industries alone, estimated welfare losses amounted to approximately 5.2 percent ($14.95 billion) while such losses in the tobacco industries exceeded 3.6 percent of sales or $759 million in 1987. These results are consistent with the findings of previous studies as well as with the fact that most of the industries in the U.S. with the highest estimated welfare losses relative to sales have been the focus of public scrutiny for their alleged anticompetitive behavior. On average, the NEIO-based allocative efficiency loss estimates are significantly larger than previous estimates of allocative efficiency losses in the food sector.

Azzam and Pagoulatos extend the traditional conjectural approach to simultaneous analysis of the degree of market imperfection in both output and factor markets in Chapter 6. This was one of the first studies to empirically measure oligopoly and oligopsony power simultaneously. Azzam and Pagoulatos use a production-function approach, which is flexible in that it relaxes the assumption of a fixed-proportions technology in favor of variable proportions. A unique feature of the authors' approach is that it allows derivation of separate conjectural elasticities for the output and input markets. The authors were able to conduct a parametric test to determine whether the degree of market imperfection was the same in the output and factor markets.

Azzam and Pagoulatos use aggregate time-series data for the U.S. meat-packing industry as a case study for applying their model. As mentioned in Chapter 1 of this book, the U.S. meat-packing industry has been at the forefront in terms of increased concentration over time. The four largest beef packers increased their share of the national market from 30 percent in 1978 to 83.5 percent in 2003. The authors point out that this industry has often been at the center of controversy over conditions of competition in both its livestock procurement and wholesale meat markets.

The empirical results of the Azzam and Pagoulatos study indicate that there are market imperfections in both the output and the factor markets of this industry. Specifically, the authors find that the average degree of noncompetitiveness is similar in the two markets: 0.178 in the input market and 0.223 in the output market. The authors advise caution in interpreting the empirical results due to the aggregate nature of their time-series data.

In Chapter 7, Sexton and Zhang examine the distributional impacts of market power on economic welfare throughout the channels of a food market. Rather than looking at an actual market, the authors develop a simulation model for a food-market prototype to explore distributional impacts from alternative degrees of retailer oligopoly power, manufacturer oligopsony power, and successive oligopoly or oligopsony power. This simulation technique is a useful tool because previous empirical investigations have examined market power at a single stage within an industry like food manufacturing when in reality market power likely exists throughout the entire channel. Consequently, to derive behavioral and economic welfare impacts under simultaneous market power, simulation techniques are essential.

Interestingly, Sexton and Zhang's results demonstrate that even modest levels of market power at multiple stages of the market channel can interact to cause dramatic shifts in the distribution of economic welfare among farmers, marketers, and consumers. For instance, the authors show that marketers receive no surplus under perfect competition because of their assumption of constant returns to scale technologies. However, in the simulated alternative market-power scenarios, marketers are able to capture at least half of the market surplus. The results indicate that pure deadweight loss effects are rather small for moderate levels of market power but that the effects on distribution can be large. This raises a policy issue—should policy-makers be concerned solely about efficiency or should distribution also play a role?

Econometric models that estimate market power usually assume price-taking behavior on one side of the market or the other. However, for bilateral oligopoly models, there may be market power on the buyer and/or seller sides of the market. Schroeter, Azzam, and Zhang address the problem of measuring market power in a bilateral oligopoly model in Chapter 8. They focus on price and quantity determination in the wholesale market. If both sides of the market are highly concentrated, the market may not be competitive. The authors examine three stylized equilibrium concepts as possible candidates for such a market. The first possibility is a competitive wholesale market, which would be characterized by "bilateral price-taking." A second possibility is that manufacturers have no market power while retailers exercise some degree of oligopsony power, which is referred to as "manufacturer price-taking." The last stylized equilibrium candidate, "retailer price-taking," implies that retailers have no market power and that manufacturers exercise some oligopoly power. To determine which equilibrium concept is the most consistent with data, the authors construct a test of nonnested, nonlinear, simultaneous-equation models.

Schroeter, Azzam, and Zhang apply their model to the U.S. wholesale beef market, which has high firm concentration on both the buyer side (primarily retail grocery chains) and the seller side (beef packers). They find that, of the three candidates, manufacturer price-taking behavior is the market-equilibrium concept that is the most consistent with the data.

The authors reject the hypothesis of price-taking conduct on both sides of the market. This is a conclusion that would not have been reached had the authors considered monopoly conduct by sellers as the only alternative to perfect competition.

Spatial-equilibrium models have been used frequently to analyze interregional competition problems in agriculture and the food system. Originally developed by Enke (1951) and Samuelson (1952) and then refined by Takayama and Judge (1971), spatial price equilibrium models have assumed that markets are either perfectly competitive or monopolistic. However, the structure of most markets is often neither. Therefore, a more plausible model for analyzing interregional competition would be a spatial imperfect-competition equilibrium model. In Chapter 9, Kawaguchi et al. develop a new, generalized "dual structure" spatial-equilibrium model of imperfect competition that incorporates any degree of market structure from perfect competition to monopoly. The model, which is applied to the Japanese milk market as a case study, incorporates a "dual structure" in which there are oligopolistic consignment sellers and many perfectly competitive small-scale producers with pooled returns given.

Using the model, numerous spatial-equilibrium solutions in the Japanese milk market are solved by assuming alternative sets of imperfectly competitive behavior with the dual structure. The results indicate that, under perfect competition, most milk is shipped to fluid markets, there are very few interregional milk movements, and milk prices are much lower than actual prices. Under monopoly, much less milk is shipped to fluid markets, there is no interregional milk movement, and milk prices are much higher than under the perfect-competition solution and actual prices. The Cournot-Nash equilibrium solutions are the most similar to actual observations, and actual interregional milk movements can be explained by assuming imperfectly competitive behavior.

In Chapter 10, Karp and Perloff estimate the degree of competitiveness and the adjustment paths of the world's two largest coffee exporters, Brazil and Colombia, using a subgame-perfect dynamic feedback model of oligopoly.[1] Their study was one of the first to incorporate dynamics into an NEIO model. A motivation for Karp and Perloff's study was

1 A combination of strategies is a subgame perfect Nash equilibrium if it is a Nash equilibrium for the game and the relevant strategies are a Nash equilibrium for every subgame (subperiod within the game).

evidence that these two major coffee exporters behave noncompetitively and that dynamics play an important role in determining the outcome. They view Brazil and Colombia as a dynamic coffee duopoly facing a fringe with exogenous exports. The institutional arrangements in the coffee market constitute circumstantial evidence of an intent by large producers to exert market power. The difficulties of negotiating these agreements and allegations that many exporting countries fail to comply fully with them suggest that producers do not behave as a perfect cartel. Thus, the hypothesis that the market structure lies between monopoly and competition is plausible.

The complexity and inconstancy of coffee-marketing institutions and the paucity of available data made it unreasonable to estimate the explicit game that producers play. Instead, Karp and Perloff estimate a general set of relations that include, as special cases, the equilibrium conditions associated with perfectly competitive, collusive, and Nash-Cournot behavior. Thus, the authors are able to determine whether the observed outcome is consistent with any of these equilibria.

Karp and Perloff find that Brazil and Colombia compete vigorously with each other in the coffee export market. Exports are 6 to 13 percent lower than if both Brazil and Colombia are price-takers based on classical and Bayesian dynamic market models respectively. Based on the Bayesian analysis, the behavior of Brazil and Colombia appears to be closer to price-taking than to collusion. Their results from a subgame-perfect dynamic oligopoly model differ only slightly from those based on the much simpler open-loop dynamic model.

In Chapter 11, Kinoshita, Suzuki, and Kaiser apply a generalized Bertrand-type model to the Japanese dairy market to gain insights into firms' pricing conduct regarding beverage-milk products made from fresh milk and reconstituted milk. This model provides more detailed insights about pricing conduct in product-differentiated oligopoly markets than previous studies that have assumed that all firms within the industry supply a homogeneous product and use quantity as a strategic variable.

The authors' empirical results, based on weekly scanner data, reveal that a key assumption of the original Bertrand model is violated. Specifically, the results indicate that the assumption that firms set prices independent of their rivals' reactions is rejected. That is, there is interdependency among firms in strategic price behavior. This result means that the original Bertrand model is inappropriate for their empirical analysis.

In addition, the estimated model results in substantially more elastic own-price elasticities than previous estimates in Japan based on aggregate market data. The more elastic results (relative to aggregate data studies) reflect the brand-level competition that is captured in the more micro-level data used by the authors. These estimates support contentions by Japanese retailers that fluid milk is a more price-elastic product than has previously been found. Incorporating estimated price conjectures into the model results in considerable discrepancies between partial and perceived own-price elasticities. The result shows that a firm's price conjecture can be an important factor in determining a firm's perceived own-price elasticity. Consequently, an approach using perceived (or conjectured) own-price elasticities rather than traditional partial own-price elasticities may improve analyses of firms' conduct in an oligopolistic market.

Finally, the estimated conjectural variations are tested to determine whether they represent consistent pricing behavior. Consistent behavior means that firm *i*'s estimated conjecture corresponds to firm *j*'s observed price reaction. Kinoshita, Suzuki, and Kaiser find that consistent pricing behavior could not be statistically supported. This result implies that one should be careful concerning the assumption of consistent pricing behavior in building models.

Competition between "national brand" and "private label" products has been a primary concern of managers in the retail food industry for some time now. In Chapter 12, Cotterill and Putsis use a flexible linear approximate almost ideal demand system (LA/AIDS) and specify corresponding price-reaction equations to analyze observed strategic price conduct in the market. Since market shares (in the LA/AIDS demand system) and prices (in the associated price-reaction equations) are the dependant variables in the empirical analysis, a simultaneous-equations system is estimated that allows for examination of (1) consumer's response to firm's pricing and promotion decisions (demand elasticities), (2) the interdependence of national-brand and private-label pricing behavior, and (3) traditional IO supply-side effects of market structure, including market share. Using data for 143 food product categories and 59 geographic markets, the authors develop a model that captures the variation in private-label/national-brand share and pricing across categories and markets.

Cotterill and Putsis' conclusions are:

- Brand managers should expect to face traditional demand relationships regardless of whether they are managing a national brand or a private label—an increase in the price of a national brand (private label) lowers national-brand (private-label) share.

- As a national brand increases its share of category sales, both national-brand and private-label prices increase.

- An increase in retail concentration increases both national-brand and private-label prices. Also, concentration has a greater positive impact on private-label prices.

- Promotions increase share while rival promotions lower share. Further, promotion effects are asymmetric. For national-brand products, a 10 percent increase in point-of-sale display activity, for example, increases share by about 1.6 percent. In contrast, a similar 10 percent increase in point-of-sale display activity for private-label products increases share by only 0.87 percent.

- When national brands promote, private labels lower prices in an attempt to compete. However, the demand price elasticities suggest that this is a meager way to capture volume from national brands. Features and display promotion appear to be much more effective ways of gaining share in such categories.

- Cross-price elasticities are decidedly asymmetric with national-brand price having a major impact on private-label sales whereas private-label price has very little impact on national-brand sales.

In the final chapter, a calibration modeling approach is adopted to compare the U.K. and Dutch sectors' competitive behavior in the frozen-potato-product market over the period 1980 through 1992. Tiffin and Dawson's conjectural-variation values from an NEIO calibration model indicate that both sectors, but especially the Dutch sector, are increasingly monopolistic. In 1991/92, both sectors' behavior was close to the Cournot benchmark. The U.K. sector's price and quantity were highly sensitive to its competition level—compared with actual 1991/92 price and quantity levels, a perfectly competitive U.K. sector would produce 39 percent more for a 54 percent lower price while a monopolistic U.K. sector would produce 32 percent less for a 44 percent higher price. The U.K. sector's competition level also significantly affects Dutch import penetration of the U.K. market for frozen potato products. If the U.K. sector is perfectly competitive, Dutch imports fall by 17 percent from their 1991/92 level, whereas a monopolistic U.K. sector raises Dutch imports by 14 percent.

Oligopolistic distortion in the U.K. frozen-potato-product sector increased the penetration of Dutch imports into the U.K. market by 25,700 metric tons in 1991/92 and by an annual average of 13,300 metric tons over the period. It also affects the welfare of U.K. consumers, U.K. processors, Dutch processors, and society: U.K. consumers' surplus falls by £99.4 million annually on average due to the U.K. sector's oligopolistic distortion, U.K. processors' profit increases by £77.6 million, Dutch processors' profit increases by £4 million, and the deadweight loss is £17.8 million.

Finally, the authors conclude that the trade and welfare cost of market power in the U.K. sector is high and generally increasing. Such trade and welfare trends are likely to continue due to the U.K. sector's oligopolistic power despite the demise of the Potato Marketing Scheme after 1996/97 and creation of a uniform European-Union-wide potato regime. Therefore, the view of the Potato Marketing Board that the Potato Marketing Scheme provided potato producers with a required source of countervailing power against processors appears justified.

Conclusions from the NEIO Food-Industry Studies

There have been numerous studies conducted on the food industry using NEIO. These empirical studies have provided valuable information to policy-makers in charge of crafting and enforcing government competition policies. This section briefly provides some generalizations of the findings from this body of work.

The majority of NEIO studies on the food system have indicated some degree of market imperfection at all channels in the market (an exception to this is Muth and Wohlgenant (1999a, 1999b), who did not find evidence of market power in the U.S. beef industry). For instance, one of the most comprehensive studies on oligopoly power in the U.S. food industry based on quantity conjectural elasticities was Bhuyan and Lopez (Chapter 5). Table 14.1 shows their estimated parameters with some comparable results from other chapters in this book. The conjectural elasticities ranged from a low of 0.024 for chewing gum to a high of 0.55 for breakfast cereal. Note that 0.219 for meat packing is very close to the 0.223 by Azzam and Pagoulatos (Chapter 6), 0.128 for cheese is very close to the 0.1 for manufactured dairy products by Liu, Sun, and Kaiser (Chapter 4), and 0.182 for fluid milk is very close to the 0.176 by Liu, Sun, and Kaiser (Chapter 4).

However, the majority of studies have also found that food markets are significantly closer to perfect competition than they are to monopoly. For example, the majority of conjectural elasticities in Table 14.1 are less than 0.2. Similarly, Sexton and Zhang (Chapter 7) state that the majority of past NEIO food-market studies have market-power point estimates that are less than 0.20, which they indicate is equivalent to that of a five-firm symmetric Cournot equilibrium. Hence, there is empirical evidence of weak market power within the food system.

The market within the food sector that has received the most attention is the U.S. meat industry. This is not surprising since this market has undergone the greatest increase in firm concentration of any food market in the United States. The tremendous increase in concentration has resulted in a large-scale policy debate at federal and state levels with respect to implementing procompetition policies such as laws that restrict corporate farming. With all the visibility of the public discussions on concentration in the meat market, researchers have played a valuable role in providing estimates of market power in this industry. As for the food sector in general, these studies have generally supported the conclusion of limited market power (except for Muth and Wohlgenant's (1999a, 1999b) finding of no market power).

While the U.S. meat market has been studied the most, U.S. food retailing has received the least attention in NEIO studies. This is because retailers buy and sell thousands of different products, which obviously makes the application of NEIO very difficult. Some researchers have approached this problem by aggregating products into broad categories such as dairy, meat, fruits, vegetables, and other products. For instance, Gohin and Guyomard's (2000) study of French retailers estimated market

Table 14.1. Estimated Market-Power Parameters (Quantity Conjectural Elasticities) for the U.S. Food Industry

		Processor's Market Power	
SIC	**Industry**	**Bhuyan and Lopez's Oligopoly**	**Others**
2011	Meat Packing	0.219	0.223 (oligopoly), 0.178 (oligopsony) Azzam and Pagoulatos
2013	Saus. & Prep. Meats	0.146	
2016	Poultry & Egg Proc.	0.289	
2022	Cheese	0.128	0.100 (manufactured products) Liu, Sun, and Kaiser
2023	Cond. & Evap. Milk	0.192	
2024	Ice Cream	0.112	0.100 (manufactured products) Liu, Sun, and Kaiser
2026	Fluid Milk	0.182	0.176 Liu, Sun, and Kaiser. 0.055 (farmer coops) Suzuki et al.
2032	Canned Specialties	0.063	
2033	Canned Fruit & Veg.	0.176	
2034	Dried Fruit & Veg.	0.035	
2035	Pickled, Sauces, etc.	0.348	
2041	Flour & Grain Mill	0.482	
2043	Cereal Breakfast Prep.	0.550	
2044	Rice Milling	0.032	
2047	Pet Food	0.014	
2048	Prepared Feeds	0.173	
2051	Bread & Bakery	0.145	
2061	Refined Sugar	0.146	
2065	Candy & Confectionary	0.078	
2066	Chocolate & Cocoa Pr.	0.110	
2067	Chewing Gum	0.024	
2074	Cottonseed Oil Mills	0.063	
2075	Soybean Oil Mills	0.484	
2076	Vegetable Oil Mills	0.131	
2077	Anim. & Marine Fats & Oils	0.139	
2079	Edible Fats & Oils	0.167	
2082	Malt Beverages	0.252	
2084	Wine & Brandy Sp.	0.065	
2085	Distilled Liquor	0.225	
2086	Soft Drinks	0.327	
2087	Flavor Extr. & Syrups	0.107	

Note. The data come from Chapters 2,4,5, and 6 in this book.

power for dairy products, meat products, and other food items and rejected the hypothesis that French retailers were perfectly competitive. However, in the United States, the market power of food retailers has not been well studied.

Limitations of NEIO and Future Research Directions

In this final section we examine some of the shortcomings and limitations of previous NEIO studies of the food system. We also recommend several directions for future research that may improve the reliability of future NEIO studies.

One of the biggest challenges to NEIO applications is how to develop an improved empirical model of bilateral oligopoly that incorporates both horizontal and vertical competition. Because there are many bilateral oligopolies in the real world, this is an important problem to address even though finding a solution may be difficult. Most previous studies have assumed price-taking behavior on either the output or input side and then estimated market power on the other side. Consequently, models that build on the work of Azzam (1996) will be fruitful. For instance, Kinoshita, Suzuki, and Kaiser (2004) developed a practical model that estimated the degree of balance of vertical power between fluid-milk processors and retailers and between dairy cooperatives and processors with simultaneous estimation of the degree of horizontal competition in each stage of the market. The authors' results were tentative due mainly to data constraints, but their study constituted the first econometric evidence supporting the general perception that retailers, though facing nearly perfect horizontal competition among themselves, have extremely dominant vertical market power over fluid-milk processors. They also offered the first econometric evidence that processors may have some vertical market power over dairy cooperatives.

Another criticism that has been leveled at past NEIO studies is that the majority of them have used ad hoc specifications of functional forms for estimation of demand and supply or technology relationships (Sexton and Zhang, Chapter 7). This implies that, in addition to testing the hypothesis of price-taking, the NEIO model is jointly testing the maintained hypothesis of functional form for demand and supply (or technology). There is conflicting evidence as to how sensitive the market-power estimates are to functional form. For example, Genesove and Mullin (1998) found market power to be insensitive to functional form, but Bettendorf and

Verboven (2000) found the opposite result. One approach for addressing this problem is to use flexible functional forms in the estimation of these relationships. Another approach is for researchers to run sensitivity analyses by estimating alternative functional forms for demand and supply relationships to determine the sensitivity of the market-power estimates to functional form. Either of these approaches would mitigate this criticism in future research.

NEIO models have been criticized because they are static models of games that are inherently dynamic. That is, in reality, firm's reactions to other firms' price and quantity decisions are dynamic in nature and the majority of NEIO studies have used static models to proxy this decision-making environment. Dynamic models, however, generally require firm-level data and are more difficult to solve than static models, which has greatly limited the use of dynamic models in empirical analyses in industrial organization (Sheldon and Sperling 2003).

In a move to deflect this criticism, many recent NEIO studies have abandoned use of the more dynamic sounding term "conjectural variation" in favor of a more static sounding concept such as "degree of market imperfection." There has also been some recent movement toward harmonization of conjectural-variation models and dynamic game-theoretic models (Cabral 1995; Pfaffermayer 1999; Itaya and Shimomura 2001). For example, Cabral (1995) showed that an exact correspondence exists between the conjectural-variations solution and the solution of a quantity-setting repeated game under some conditions. This justifies use of the conjectural-variations solution as the reduced form of the equilibrium of a dynamic game in some cases. A major challenge in future studies will be to incorporate dynamic aspects of the decision environment into NEIO models along the lines of Karp and Perloff (1989, 1993), Deodhar and Sheldon (1996), and Katchova, Sheldon, and Miranda (2005).

Another criticism is that many of the NEIO food system studies have assumed a fixed-proportions production technology, which implies that the elasticity of substitution between farm and nonfarm inputs is zero. For instance, Muth (1996) and Muth and Wohlgenant (1999a, 1999b) have argued that NEIO studies of the U.S. meat sector are flawed because they have assumed a fixed rather than a variable proportions technology. These authors used a model with variable proportions and estimated no market power, and they have argued that the difference between their results and past results is the assumption of no input substitution in previous works. This may be a valid criticism, but as Sexton (2000) points out, it may pale in comparison to the aggregate nature of the data used in these studies. Also, in our view, this criticism does not seem to apply as much to industries like the dairy industry that have very strict standards of identity, resulting in very little substitution possibilities for the farm input.

Sexton and Zhang (Chapter 7) discuss another related criticism of NEIO models of the food sector—that they fail to adequately account for technological change. The authors correctly point out that the majority of NEIO applications use long time series of annual data and that technological change, which likely has occurred, has been inadequately accounted for in the estimation. While this problem is not unique to NEIO, it is nevertheless an important one that deserves attention. More dynamic models, in theory, may help, but such models are significantly more complicated and often have estimation problems.

Finally, a last criticism of past NEIO food studies has to do with data aggregation (again, this criticism also applies to other applied economic analyses). The majority of studies have used data that are highly aggregated over space, product form, and time.[2] For instance, many studies have used data aggregated at the national level and have examined market power that actually occurs at a more regional level. For example, the fluid-milk industry is regional, not national, in scope. Hence, reliance on national data to estimate individual firm market power at the regional level probably biases the estimates. Many studies have relied on highly aggregated data for the commodity(ies) in question. Consider, for example, Azzam and Pagoulatos' study in Chapter 6, which used census data for aggregate meat output. This degree of product aggregation makes it impossible to distinguish market power in the individual industries that make up the meat sector. And, as was previously noted, many past studies have used annual time-series data over fairly long periods in which structural change in the industry has likely taken place. It therefore appears that an important future research direction will be to obtain and use significantly more disaggregated data. Firm-level panel data would be an ideal data set with which to measure market power.

---

2 An exception to this is Kawaguchi, Suzuki, and Kaiser's (1997, Chapter 9, 2001) spatial-equilibrium models with regionally disaggregated data.

Perhaps chief among potential directions for future research is how to develop an improved empirical model of bilateral oligopoly that incorporates both horizontal and vertical competition. Because there are many bilateral oligopolies in the real world, this is an important problem to address despite the difficulties in resolving it. Another important topic that is related to bilateral oligopoly is how, as each stage of the food system becomes more oligopolistic, to evaluate countervailing power because such power can create a near-competitive price even in bilateral monopoly. A subsequent challenge is harmonization of NEIO with game-theoretic IO approaches. Efforts to incorporate dynamic aspects into NEIO models (see, for example, Karp and Perloff (1989, 1993), Deodhar and Sheldon (1996), and Katchova, Sheldon, and Miranda (2005)) should be intensified. Lastly, development of a more theoretically acceptable and empirically feasible model for analyzing price-setting product-differentiated oligopoly would be greatly beneficial to the NEIO framework.

References

Azzam, A.M. "Estimating the Degree of Dominance in a Bilateral Oligopoly." *Applied Economics Letters* 3 (1996):209–211.

Bettendorf, L., and F. Verboven. "Incomplete Transmission of Coffee Bean Prices: Evidence from the Netherlands." *European Review of Agricultural Economics* 27 (2000):1–16.

Cabral, L.M.B. "Conjectural Variations as a Reduced Form." *Economic Letters* 49 (1995):397–402.

Deodhar, S.Y., and I.M. Sheldon. "Estimation of Imperfect Competition in Food Marketing: A Dynamic Analysis of the German Banana Market." *Journal of Food Distribution Research* 27 (1996):1–10.

Enke, S. "Equilibrium among Spatially Separated Markets: Solution by Electric Analogue." *Econometrica* 19 (1951):40–47.

Genesove, D., and W.P. Mullin. "Testing Static Oligopoly Models: Conduct and Cost in the Sugar Industry, 1890–1914." *Rand Journal of Economics* 29 (1998):355–377.

Gohin, A., and H. Guyomard. "Measuring Market Power for Food Retail Activities: French Evidence." *Journal of Agricultural Economics* 51 (2000):181–195.

Itaya, J., and K. Shimomura. "A Dynamic Conjectural Variations Model in the Private Provision of Public Goods: A Differential Game Approach." *Journal of Public Economics* 81 (2001):153–172.

Karp, L.S., and J.M. Perloff. "Dynamic Oligopoly in the Rice Export Market." *Review of Economics and Statistics* 71 (1989):462–470.

———. "A Dynamic Model of Oligopoly in the Coffee Export Market." *American Journal of Agricultural Economics* 75 (1993):448–457.

Katchova, A.L., I.M. Sheldon, and M.J. Miranda. "A Dynamic Model of Oligopoly and Oligopsony in the U.S. Potato-Processing Industry." *Agribusiness: An International Journal* 21 (2005):409–428.

Kawaguchi, T., N. Suzuki, and H.M. Kaiser. "A Spatial Equilibrium Model for Imperfectly Competitive Milk Markets." *American Journal of Agricultural Economics* 79 (1997):851–859.

———. "Evaluating Class I Differentials in the New Federal Milk Marketing Order System." *Agribusiness: An International Journal* 17 (2001):527–538.

Kinoshita, J., N. Suzuki, and H.M. Kaiser. "The Degree of Vertical and Horizontal Competition among Dairy Cooperatives, Processors, and Retailers in Japanese Milk Markets." Unpublished, Cornell University, 2004.

Muth, M. "Identifying the Degree of Oligopoly Power in the Beef Packing Industry." Unpublished, North Carolina State University, 1996.

Muth, M.K., and M.K. Wohlgenant. "Measuring the Degree of Oligopsony Power in the Beef Packing Industry in the Absence of Marketing Input Quantity Data." *Journal of Agricultural and Resource Economics* 24 (1999a):299–312.

———. "A Test for Market Power Using Marginal Input and Output Prices with Application to the U.S. Beef Processing Industry." *American Journal of Agricultural Economics* 81 (1999b):638–643.

Pfaffermayer, M. "Conjectural Variation Models and Supergames with Price-Competition in a Differentiated Product Oligopoly." *Journal of Economics* 70 (1999):209–326.

Samuelson, P.A. "Spatial Price Equilibrium and Linear Programming." *American Economic Review* 42 (1952):283–303.

Sexton, R.J. "Industrialization and Consolidation in the U.S. Food Sector: Implications for Competition and Welfare." *American Journal of Agricultural Economics* 82 (2000):1087–1104.

Sexton, R.J., and M. Zhang. "An Assessment of the Impact of Food Industry Market Power on U.S. Consumers." *Agribusiness: An International Journal* 17 (2001):59–79.

Sheldon, I., and R. Sperling. "Estimating the Extent of Imperfect Competition in the Food Industry: What Have We Learned?" *Journal of Agricultural Economics* 54 (2003):89–109.

Takayama, T., and G.G. Judge. *Spatial and Temporal Price and Allocation Models.* Amsterdam Netherlands: North-Holland Publishing, 1971.

Contributors

Azzedine M. Azzam is a professor in the Department of Agricultural Economics at the University of Nebraska.

Sanjib Bhuyan is an associate professor in the Department of Agricultural, Food and Resource Economics at Rutgers University.

Ronald W. Cotterill is a professor in the Department of Agricultural and Resource Economics at the University of Connecticut.

Philip J. Dawson is a reader in the School of Agriculture, Food and Rural Development at the University of Newcastle.

Olan D. Forker is a professor emeritus in the Department of Applied Economics and Management at Cornell University.

Harry M. Kaiser is a professor in the Department of Applied Economics and Management at Cornell University and director of the Cornell Commodity Promotion Research Program.

Larry S. Karp is a professor in the Department of Agricultural and Resource Economics at the University of California, Berkeley.

Tsunemasa Kawaguchi is a professor emeritus in the Department of Agricultural and Resource Economics at Kyushu University, Japan, and a professor in the Department of Economics at Kyushu Sangyo University, Japan.

Junko Kinoshita is a senior researcher at the Policy Research Institute, Ministry of Agriculture, Forestry, and Fisheries, Japan.

Kohei Kobayashi is a professor emeritus in the Department of Agricultural and Resource Economics at Kyushu University, Japan.

John E. Lenz was a research associate in the Department of Applied Economics and Management at Cornell University.

Donald J. Liu is an associate professor in the Department of Applied Economics at the University of Minnesota.

Rigoberto A. Lopez is a professor in the Department of Agricultural and Resource Economics at the University of Connecticut.

Emilio Pagoulatos is a professor and department head in the Department of Agricultural and Resource Economics at the University of Connecticut.

Jeffrey M. Perloff is a professor in the Department of Agricultural and Resource Economics at the University of California, Berkeley.

William P. Putsis, Jr. is a professor in the Department of Marketing in the Kenan-Flagle Business School at the University of North Carolina.

John R. Schroeter is an associate professor in the Department of Economics at Iowa State University.

Richard J. Sexton is a professor in the Department of Agricultural and Resource Economics at the University of California, Davis.

Chin-Hwa Sun is a professor and director in the Institute of Applied Economics at National Taiwan Ocean University.

Nobuhiro Suzuki is a professor in the Department of Agricultural and Resource Economics at Kyushu University, Japan.

Abigail L. Tiffin is a research fellow in the Department of Agricultural and Food Economics at the University of Reading.

Mingxia Zhang is a lead market monitoring specialist with the California Independent System Operator.

Index